Current Topics in Microbiology and Immunology

223

Editors

Springer
Berlin
Heidelberg
New York
Barcelona
Budapest
Hong Kong
London
Milan
Paris
Santa Clara
Singapore
Tokyo

The Coxsackie B Viruses

Edited by S. Tracy, N.M. Chapman and B.W.J. Mahy

With 37 Figures

Springer

STEVEN TRACY, Ph.D.
Professor

NORA M. CHAPMAN, Ph.D.
Assistant Professor

University of Nebraska Medical Center
Department of Pathology and Microbiology
600 South 42nd Street, Omaha, NE 69198-6495, USA

BRIAN W.J. MAHY, Ph.D., Sc.D.
Director of Viral and Rickettsial Diseases
National Center for Infectious Diseases
Centers for Disease Control and Prevention
Clifton Road, Atlanta, GA 30333, USA

Cover Illustration: Coxsackievirus B3 image is a computer representation derived from x-ray coordinates (Mucklebauer et al. Structure 3:653, 1995). The PDB ENTRY for the virus is 1COV. The GRASP protein molecular surface is radially depth-cued to visually highlight surface features. The canyon is clearly evident. Note that the canyons are not continuous around the 5-fold icosahedral axes. (Courtesy of Dr. Jean-Yves Sgro, University of Wisconsin-Madison; sgro@rhino.bocklabs.wisc.edu).

Photomicrographs of 0.6 micron thick hemotoxylin and eosin stained sections of heart (above) and pancreas (below) taken from a C3H/HeJ male mouse, 10 days post-inoculation with the cardiovirulent wild-type strain of CVB3, CVB3/DO. Note significant and widespread regions of necrosis and calcification in heart section, typical of results induced by infection with a cardiovirulent CVB3. Pancreas has sustained severe damage to acinar cells. Original magnification, x100.

Cover Design: Design & Production GmbH, Heidelberg

ISSN 0070-217X
ISBN-13: 978-3-642-64507-5 e-ISBN-13: 978-3-642-60687-8
DOI: 10.1007/978-3-642-60687-8

Springer-Verlag Berlin Heidelberg 1997
Library of Congress Catalog Card Number 15-12910
Softcover reprint of the hardcover 1st edition 1997

SPIN: 10498132 27/3136 – 5 4 3 2 1 0

Contents

List of Contributors

(Their addresses can be found at the beginning of their respective chapters.)

BABOONIAN, C.
BECK, M.A.
BOOTH, J.C.
CHAPMAN, N.M.
CROWELL, R.L.
CURREY, K.M.
DAVIES, M.J.
DUNN, J.J.
GAUNTT, C.J.
HUBER, S.A.
KNOWLES, R.J.
LANDAU, B.J.
LEVANDER, O.A.
MCCAULEY, J.W.
MCKENNA, W.J.
MODLIN, J.F.
MUCKELBAUER, J.K.
PALLANSCH, M.A.
PRICE, C.
RAMSINGH, A.
ROMERO, J.R.
ROSSMANN, M.G.
ROTBART, H.A.
SCHULTHEISS, H.-P.
SCHWIMMBECK, P.L.
SHAPIRO, B.A.
TRACY, S.

Preface

S. TRACY

Late in the 1940s, a virus was isolated from a young patient with a flaccid paralysis in the sleepy Hudson River town of Coxsackie in the state of New York. Within the next few years, it was apparent that this and other similar viruses were not polioviruses but were indeed a new group of viruses, viruses that by the mid-1950s had been found to be commonly associated with pediatric inflammatory heart disease. Two groups of coxsackieviruses (A and B) were differentiated on the basis of the type of paralysis induced in suckling mice by these viruses. Group B coxsackieviruses, because of their primacy as etiologic agents of human acute viral myocarditis and its relatively common sequela, dilated cardiomyopathy, are the focus of this volume.

As the end of the century approaches, the massive international effort to eradicate polioviruses through vaccination as causes of human disease has been successful in the Western Hemisphere and in many parts of Europe, and it is expected that worldwide eradication may be achieved within the near future. While this is wonderful news, there are sadly no similar efforts being planned to combat the numerous other human enteroviruses that daily incur widespread morbidity and mortality throughout the world. While this is due in part to the lack of specific knowledge about the other human enteroviruses, it is also due to the perceptions of industry that there is insufficient profit to be made by developing these vaccines. Of these non-polio enteroviruses, the coxsackie B viruses (CVBs) are the best studied. The purpose of this volume is both to provide the reader with an up-to-date summary of the interesting and clinically significant biology of the CVBs as well as to help illuminate current research efforts and identify future interests.

The chapter by Crowell and Landau introduces the CVBs with a short history of the development of the field, from the discovery by Dalldorf and Sickles of what would be called the coxsackieviruses through recent efforts to identify the protein(s) that serve as the cellular virus receptor. Pallansch discusses the epidemiology of the CVBs and demonstrates that these viruses are far from minor infectious events in our lives. Baboonian and colleagues review the roles that CVBs are known, or are thought, to play in human heart disease on the basis of molecular and serological investigations, while Modlin and Rotbart continue and expand this discussion into the diseases of children that are caused by or associated with CVBs. Both chapters illuminate the need for efficient anti-viral drugs as well as provide a sound argument for the consideration of vaccine development. Recent findings regarding the potentially serious impact of nutritional deficiencies upon CVB heart disease are discussed in the chapter by Beck and Levander. Using as their starting point the story of heart disease due to dietary selenium deficiency in a region of China and ending with their data on the effects of selenium deficiency in mice, this chapter points to a new field of interest that concerns both nutrition and virology. Romero and colleagues provide a close look at the current state of knowledge of CVB genetics in a series of comparative studies across the viral genome. This is expanded in detail in the chapter by Knowles and McCauley in their discussion of the story of CVB5 and its close genetic similarity to swine vesicular disease virus, SVDV. The chapter by Currey and Shapiro presents recent computer-modeled studies on the 5' nontranslated region (NTR) of the CVB genome using both the well-known MFOLD as well as a newer algorithm, GA, to probe the specific structure of this vital regulatory sequence. In work that provided the first view of any non-poliovirus human enterovirus capsid, Mucklebauer and Rossman discuss the structure of the CVB3 capsid. One of the insights gained from this structure was the observation of prominent planar regions found astride two-fold axes of symmetry, the so-called dimples. Kuhn examines the potential role for both this region and the canyons in virus-receptor interactions while reviewing and comparing recent findings in the hunt for the CVB group receptor. Although CD55, or decay-accelerating (DAF), has been implicated as one component in this game, it does not appear to be the primary player. The next chapter discusses and compares current data on CVB virulence and also compares the concepts of understanding artificially induced changes in virulence phenotypes versus natural determination of viral virulence. Findings from studies of cardiac and pancreatic disease in murine models suggest a strong role for a primary site controlling virulence to be in the capsid proteins, although to date the role of the 5' NTR has not been excluded. The final two chapters examine the roles played in CVB disease by humoral and cell-mediated immune responses. In examining the role of the humoral response, Gauntt discusses the evidence for molecular mimicry playing a key role in CVB-induced heart disease. Schultheiss and colleagues review recent findings in T cell biology in both murine and human inflammatory heart disease, including the fascinating evidence that T cells from humans with myocarditis can induce a pathology in the hearts of SCID mice.

The editors hope that the reader will find that there is yet much of a unique nature to be discovered and exploited in the study of the CVBs, human enteroviruses that remain a significant health problem for children and adults alike worldwide.

Acknowledgements. The editors very much appreciate the excellent organizational aid provided in the preparation of this volume by Ms. Mary Davis. This effort was also supported in part by grants from the Public Health Service (HL40303) and by the National American Heart Association as well as by the Nebraska Affiliate of the American Heart Association.

January 1997

Steven Tracy
Nora Chapman
Omaha, Nebraska

Brian W.J. Mahy
Atlanta, Georgia

A Short History and Introductory Background on the Coxsackieviruses of Group B

R. L. Crowell and B. J. Landau

1 Discovery, Definition and Taxonomy of the Coxsackieviruses

Almost 50 years ago, during a time of epidemics of paralytic poliomyelitis in the United States, a search was carried out by Gilbert Dalldorf and associates for "parapoliomyelitis viruses." This project had been initiated by Dalldorf at the Westchester, NY County Hospital Laboratories and transferred to the State Laboratories following Dalldorf's appointment as Director of the Division of Laboratories and Research of the N.Y. State Department of Health in Albany, NY, (Sexton 1967). This search for viruses similar to poliovirus bore fruit when viruses in stools of two children exhibiting signs resembling poliomyelitis were isolated in newborn mice (Dalldorf and Sickles 1948). These isolates constituted a new group of enteroviruses that were named coxsackie viruses (originally two words, now combined into one following a convention adopted by the International Enterovirus Study Group in 1963) in recognition of the small village located on the Hudson River south of Albany in which the children lived. The coxsackieviruses were recognized as unique because they were not neutralized by antisera against the polioviruses, and polioviruses were not known to grow in mice. Parenthetically, it should be remembered that Enders, Weller, and Robbins had not yet published their discovery of growing polioviruses in cultures of human embryonic tissues (1949), for which they won the Nobel prize in 1954, and there was considerable interest in developing a more rapid and less expensive test for isolation of polio-

Department of Microbiology and Immunology, Allegheny University of the Health Sciences, MCP Hannemann School of Medicine, Philadelphia, PA 19129, USA

viruses than subhuman primates. It was in this context that a decision was made to try to isolate "parapolioviruses" in newborn mice since mice of this age were frequently more susceptible to virus infection (e.g., encephalitis viruses, yellow fever virus) than were older mice. The success of this approach, leading to the unexpected discovery of an inexpensive laboratory animal for the recovery of the coxsackie viruses, was a major development in the annals of virology.

Soon after the initial report on the recovery of coxsackieviruses, Dalldorf recognized that some of the coxsackievirus isolates caused generalized skeletal muscle destruction leading to rapid (within 2 days postinfection) fatal paralysis of the infected newborn (less than 24 h of age) mouse (GIFFORD and DALLDORF 1951). This histopathology was distinct from that observed from other coxsackievirus isolates. Thus, the different viruses were placed into two groups, A and B. The group A viruses caused generalized myositis, whereas the group B viruses caused infection of many tissues and organs of the newborn mouse leading to a slower, spastic paralytic death (5–7 days) with destruction of pancreatic acinar cells, heart muscle cells, hepatocytes, lesions in the central nervous system (CNS), adrenal tissue and focal myositis. Soon thereafter, Dr. Joseph Melnick and his group at Yale University School of Medicine isolated additional serologically distinct strains of coxsackieviruses giving them the names of cities from which the viruses were isolated. Thus, a series of numbers (Dalldorf used groups A and B, depending on the histopathology in newborn mice and sequential numbers) and names of cities from which new viruses were isolated dominated the scene in the early 1950s, leading to considerable confusion. Fortunately, the more logical Dalldorf system was adopted for designating the prototype strains of coxsackieviruses, because these viruses needed to be organized in some logical manner. The name coxsackie was a provisional name, which surprisingly has survived nearly 50 years. An international committee on virus nomenclature has agreed that new enterovirus isolates would be designated by sequential numbering as enteroviruses, without renumbering the previously identified polioviruses, coxsackieviruses, and Echo viruses (MELNICK et al. 1962).

In the early 1950s, there was a major effort to prepare standardized reagents, antisera and complement fixation antigens to simplify identification of new isolates and to determine the properties of the coxsackieviruses. It was soon recognized that the group B viruses were the etiologic agents of severe human diseases referred to as Iceland's pleurodynia, or "devil's grip," and Bornholm's disease, named after the Danish island, Bornholm. The group A viruses were identified with herpangina, severe sore throat, and with some serotypes (e.g., 5, 10, 16) causing hand-foot-and-mouth disease. As more strains of virus were isolated from human fecal and throat specimens and sorted into their taxonomic places based on histopathology produced in newborn mice, new concerns developed. Namely, that a number of laboratory personnel were becoming infected, through careless handling of the preparations. These viruses grew in high titers in both mouse tissues and titers of $>10^6$ infectious doses could be recovered from the human specimens. The severity of the human diseases which developed, i.e., 2–4 weeks in bed, with severe headache, painful chest, fever, generalized weakness, etc., from infection with the group B viruses prompted a number of investigators to avoid working with these viruses. Moreover, these viruses were not easy to study because there were too many serotypes and a jumble of new strains with which to reckon. Furthermore, funds to study these viruses decreased once a vaccine had been developed for the polioviruses (SALK 1953; SABIN 1955). As a result of decreasing funds, many enterovirus investigators turned their attention to the study of cancer viruses, since that was where the new research funding was to be found. Nevertheless, the coxsackieviruses were too widespread around the world and caused a significant amount of severe infections for them to remain unnoticed. It is hoped that the development of a CVB vaccine (e.g., FOHLMAN et al. 1993; TILLES 1994), a project that has not yet caught the attention of commercial vaccine developers, will help to bring these viruses under control.

2 Properties of the Group B Coxsackieviruses

The CVB comprise but six serotypes of approximately 213 serotypes of other picornaviruses (RUECKERT 1990) and they are recognized separately because of their unique pathogenicity for newborn outbred mice. Infection of these animals causes a variety of syndromes, ranging from focal myositis and neurological disease to fatal acute myocarditis and pancreatitis. In addition, many disease syndromes can be produced by infection of young adult inbred mice, with the Balb/C mouse being one of the most susceptible to lethal infection, often dying from liver disease (CROWELL et al. 1988; LANDAU et al. 1990). There appears to be a strong genetic component to the diseases produced in experimental animals, as determined by the strain of mouse infected and the infecting strain of virus (WEBB and MADGE 1980; WOLFGRAM et al. 1986; GAUNTT 1988).

The CVB are human enteroviruses and consequently are resistant to low pH (i.e., pH 3.0). As a result, they pass unscathed through the gastric acidity of the stomach and replicate in the mucosal cells of the small intestine prior to being disseminated to produce systemic infections of humans. The CVB are approximately 28 nm in diameter, exhibit icosahedral symmetry, are resistant to diethyl ether inactivation, have a single "plus" strand RNA genome with VPg-linked to the 5' end of the genome and a poly A tail at the 3' end, typical of picornaviruses. The genome is translated into a large polyprotein, which is cleaved into functional proteins by virus encoded proteases. The capsid is comprised of 60 copies each of four polypeptides, designated VP1-4 in their order of decreasing molecular size. The RNA of most CVB serotypes has been sequenced and cDNA molecules shown to be capable of replication (KANDOLF and HOFSCHNEIDER 1985; IIZUKA et al. 1987, 1991; RAMSINGH et al. 1990, 1992; LINDBERG et al. 1992; TRACY et al. 1992; CAGGANA et al. 1993; ZHANG et al. 1993; CHAPMAN et al. 1994). The structure of the RNA genome is identical to that of poliovirus in its organization and gene functions. The CVB have been crystallized and the structure of the capsid proteins of CVB3 has recently been determined (MUCKELBAUER et al. 1995). Like polioviruses and rhinoviruses, CVB3 has a canyon surrounding each of the 12 pentameric units of the capsid, which is the binding site for specific cellular receptors (ROSSMANN and PALMENBERG 1988). As for other picornaviruses, CVB RNA is infectious for both susceptible and naturally resistant cell types (HOLLAND et al. 1959). Nevertheless, the host range of the CVB, as for polioviruses, is restricted by the presence of specific cellular receptors (CROWELL and TOMKO 1994). In this context, the CVB are more pantropic than polioviruses, which generally are restricted to growth in primate cells (HOLLAND et al. 1959) and in transgenic mice which have received the human gene for the poliovirus receptor (REN et al. 1990; KOIKE 1991). In contrast to polioviruses, the CVB replicate readily in mice causing severe diseases. The study of viral pathogenesis of human diseases in the mouse model has facilitated a better understanding of the morbidity and mortality caused by the CVB (BENDINELLI and FREIDMAN 1988; MELNICK 1990).

An exciting and more recent discovery has shown that CVB RNA genomes can be found latent in cells, especially heart tissue, where no intact virus or viral proteins can be detected. Thus, the development of molecular probes (HAASE et al. 1984) has facilitated a novel insight into the biological mechanisms of human heart disease (BOWLES et al. 1986b; KANDOLF 1988) and perhaps of other diseases of unknown etiology such as diabetes, polymyositis, and congenital anomalies (BROWN 1972; BOWLES et al. 1986a; YOON et al. 1979; FRISK and DIDERHOLM 1992).

3 Importance of the Coxsackieviruses of Group B in Human Diseases

The CVB are known to produce a variety of human diseases which occur worldwide (Chaps. 3, 4). In temperate climates, infections are predominantly spread in epidemic proportions during the late summer-early fall. However, in the tropics, these indigenous infections appear to occur throughout the year, especially in young children. The more severe diseases are readily recognized by combinations of fever, severe unilateral chest pain (associated with infection of the intercostal muscles), pleural inflammation, headache of prominent proportions, and sore throat. In addition, there may be abdominal tenderness associated with pancreatitis and hepatitis. Most persons are able to cope with the severe common cold-like symptoms; however, a significant number of infections are quite debilitating, in which the person will be acutely ill for 2–3 weeks, followed by 1–2 months of low energy and fatigue. It is important that these individuals not exercise excessively until recovery is complete, since heart disease may develop. Another concern amongst males is the development of orchitis about 1–2 weeks after apparent recovery, not unlike that associated with mumps.

The CVB are also associated with a variety of other human diseases, as mentioned above, such as virus-induced heart disease, and undoubtedly contribute to a significant number of patients in need of heart transplants. As a result, there has been a great deal of interest in solving the pathogenetic mechanisms of CVB-induced heart disease. Several studies have supported the notion that immunologic sequelae are the major factors associated with myocardial injury (HUBER et al. 1992; 1994b; LANE et al. 1992; RABAUSCH-STARZ et al. 1994). These studies have suggested that cytotoxic T cells, lymphokines, cytokines, autoimmune antibodies and complement combine to kill or injure myocardial cells. In addition, a number of studies have linked cross-reacting antibodies between antiviral antibodies and heart muscle proteins which develop following CVB infection (BEISEL et al. 1991; CUNNINGHAM et al. 1992; HUBER et al. 1994a; GAUNTT et al. 1995). Furthermore, antimyosin antibodies alone have been shown to mimic myocardial injury in a mouse model (NEU et al. 1987; ROSE et al. 1993). One of the most significant arguments against the role of immunological mechanisms as major factors in CVB-induced myocarditis is based on the results of a large clinical trial, which showed that immunosuppresive therapy did not increase patient survival (MASON et al. 1995). In addition, Kandolf and his colleagues (HEIM et al. 1994) have found some evidence that interferon therapy may be beneficial to these patients, suggesting that direct viral injury to the myocardium is the most important disease-causing factor. Without question, an understanding of the mechanisms in CVB-induced mycocarditis continues to be difficult, since the process of viral pathogenesis is so complex (KLINGEL et al. 1992; RAMSINGH et al. 1992; HYYPIA et al. 1993; JONGEN et al. 1994). There is ample evidence to indicate that the strain of virus used to initiate infection (BECK et al. 1994; CHAPMAN et al. 1994; TU et al. 1995) together with the genetic background of the individual (RAMSINGH et al. 1989) influences the severity of the produced disease (KANDOLF et al. 1993).

Other manifestations of CVB infection of humans include injury to a number of organs and tissues. The role of CVB in viral-induced insulin-dependent diabetes melitus (IDDM) has long been suspected (YOON et al. 1979; D'ALESSIO 1992; FOHLMAN and FRIMAN 1993). Indeed a number of studies in mice have revealed this outcome, provided the strain of mice and the strain of virus are complementary (KANG et al. 1994; TILLES 1995). One of the major problems in finding conclusive evidence for the etiology of IDDM in humans is the absence of infectious virus in tissues at the time of diagnosis, coupled with the fact that this is a low incidence disease (1:3000) with a high incidence of infection. Thus, the routine sampling for serum antibodies during an epidemic of CVB reveals a population with significant levels of antibodies to the given serotype of virus, but only a few diabetics. Some studies have focused on the finding of IgM class antibodies to the virus serotype as evidence of viral infection if the titer remains elevated for about a year, suggesting persistent infection. This is in contrast to individuals with resolved acute infections that reveal high levels of IgG class antibodies with minimal or no IgM antibodies

(Keeling et al. 1994). A similar study follows patients with CVB-induced heart disease; however, substantial evidence of persistent infection of some of them is revealed by the presence of viral RNA in biopsy specimens (Koide et al. 1992) as detected by cDNA or RNA probes.

The finding of CVB-like particles by electron microscopy together with serological evidence of CVB infection prompted the consideration that CVB may cause a number of chronic muscle diseases, such as polymyositis and dermatomyocytis (Bowles et al. 1986a; Beressi et al. 1994). Hepatitis (Gregor et al. 1975) pancreatitis (Ursing 1973), meningitis and encephalitis (Gauntt et al. 1985; Kamei et al. 1990) and chronic fatigue syndrome (Cunningham et al. 1990) are additional syndromes associated with CVB infections. Again, most of these severe manifestations of infection are of relatively low incidence compared to a high incidence of infection in the population, making difficult an absolute diagnosis of CVB-induced disease. The recent advances in the use of molecular viral probes should facilitate a more rapid and definitive diagnosis (Rotbart 1992). Nevertheless, the CVB are major causes of severe human diseases and warrant the development of a protective vaccine.

4 Areas of Current Research and Future Directions

These are exciting times for further studies of the CVB in that there are methodologies available now for the solution to a number of unanswered questions of viral pathogenesis that were posed many years ago. The development of molecular biological and molecular genetic techniques have provided the thrust into this "new" biology.

Early studies had clearly defined the presence of infectious virus in a wide variety of mouse tissues during infection with CVB. Thus, the CVB have been considered to be pantropic viruses. This wide tropism of the CVB has provided suggestive evidence that these viruses can be the causative agents of a large number of different diseases, as mentioned above. The mechanisms for this variation in certain individuals has not, however, been explained. There is circumstantial evidence that the presence of specific virus receptors on cells greatly influences their susceptibility (Kunin and Halmagyi 1961; Crowell and Landau 1983; Crowell and Tomko 1994). There also is evidence that the capacity of receptors to eclipse virus infectivity is an important function of receptors (Crowell and Siak 1978; Kaplan et al. 1990; Zibert et al. 1992). The finding of multiple receptor specificities for virus variants has provided further complexity to the understanding of the roles of receptors in viral pathogenesis (Reagan et al. 1984). Thus, it is possible that genetic differences among individuals in the human population will provide selective environments for amplification of certain virus variants during infection. Further study of this idea is warranted, since very large populations of different virus strains are known to exist in CVB infections, and an explanation for the low incidence of disease in the presence of high incidence infections is needed. It is clear that receptor binding sites are insufficient to explain the above phenomena (FREISTADT et al. 1990); however, the capacity of certain cells to eclipse and uncoat viral infectivity are tightly coupled events in the early stages of infection. It has been shown that receptor binding of virus may not necessarily lead to infection of cells, since differentiation of cells plays a major role in the subsequent phases of infection (SCHULTZ and CROWELL 1983). The events that sequentially result in virus uncoating are in need of clarification.

The use of "knockout" mice (CAPECCHI 1989) to determine the function of cellular receptors to the host is a potentially exciting area warranting application to our further understanding of the receptors for the CVB. This approach, however, must await the cloning of the receptor genes, a search that has been ongoing in our laboratory for several years. The recent suggestion of a nucleolin-related membrane protein as the elusive receptor for the CVB (RAAB et al. 1995; Chap. 10) must await the results of transfection of receptor-negative cells with the corresponding gene before this protein can be accepted as the group B receptor.

The CVB are distributed world-wide and cause a significant amount of human suffering. Introduction of more rapid techniques into diagnostic virology laboratories should improve the recognition of these viruses and the severe diseases that are associated with them. The oldfashioned methods of newborn mouse inoculation are too expensive, time consuming and awkward. The more widespread use of non-radioactive nucleic acid probes in the diagnostic laboratory should facilitate detection and better epidemiological studies of the CVB.

Knowledge of the atomic structure of CVB3 should help facilitate the development of better chemotherapeutic compounds, based on structural specificity (MUCKELBAUER et al. 1995; Chap. 9). However, this approach, although exciting in an academic sense, has not been very useful at the practical level, since the picornaviruses are capable of mutation with the result that variants are selected that are resistant to drugs (HEINZ et al. 1989). Hopefully, the combined use of immunomodulaters, such as cytokines, lymphokines and interferons, together with the potential for use of soluble receptor preparations to serve as virus decoys, will aid in the control of severe diseases produced by the CVB. Nevertheless, it is likely that the development of virus vaccines will continue to be the most effective route to the control of the CVB and their diseases.

5 Summary

The past 50 years have revealed an array of significant developments in our documentation and understanding of viruses and their associated diseases. The CVB, as enteroviruses, were discovered in the search for poliomyelitis-related viruses by the inoculation of newborn mice. Future strategies for the discovery of additional viruses will undoubtedly come through the application of differentiating cell culture systems with increased susceptibility to infection by specific viruses. Developments in regulation of the cell cycle also will contribute to the better definition of events controlling persistent infections caused by the CVB. Methods utilizing molecular biological probes in situ will prove to be major aids in identifying the molecular events in CVB pathogenesis. Virology of the CVB continues to be an exciting area for research and application of preventive measures to lesson human suffering. The chapters in this book which follow will amplify most of the themes briefly presented here.

Acknowledgements. The authors are grateful to the National Institute of Allergy and Infectious Diseases of the U. S. Public Health Service for continuous support over 32 years for the study of coxsackieviruses and their cellular receptors (AI-30771). We also acknowledge our many graduate students, postdoctorates, research associates and colleagues, whose studies helped advance our knowledge of the coxsackieviruses.

References.

Beck MA, Kolbeck PC, Shi Q, Rohr LH, Morris, VC, Levander OA (1994) Increased virulence of a human enterovirus (coxsackievirus B3) in selenium-deficient mice. J Inf Dis 170:351 – 357

Beisel KW, Srinivasappa J, Prabhakar BS (1991) Molecular cloning of a heart antigen that cross-reacts with a neutralizing antibody to coxsackievirus B4. Eur Heart J 12 (Suppl D):60 – 64

Bendinelli M, Friedman H (eds) (1988) Coxsackieviruses: a general update. Plenum, New York

Beressi A, Sunheimer RL, Huish S, Finck C, Pincus MR (1994) Acute severe rhabdomyolysis in an human immunodeficiency virus-seropositive patient associated with rising anti-coxsackie B viral titers. Ann Clin Lab Sci 24:278 – 281

Bowles NE, Dubowitz V, Sewry CA, Archard LC (1986a) Dermatomyositis, polymyositis, and coxsackie-B-virus infections. Lancet 1:1104 – 1107

Bowles NE, Richardson PJ, Olsen EGJ, Archard LC (1986b) Detection of coxsackie-B-virus-specific RNA sequences in myocardial biopsy samples from patients with myocarditis and dilated cardiomyopathy. Lancet 1:1120 – 1123

Brown GC, Karunas RS (1972) Relationship of congenital anomalies and maternal infection with selected enteroviruses. Am J Epidemiol 95:207 – 217

Caggana M, Chan P, Ramsingh A. (1993) Identification of a single amino acid residue in the capsid protein VP1 of coxsackievirus B4 that determines the virulent phenotype. J Virol 67:4797-803

Capecchi MR (1989) Altering the genome by homologous recombination. Science 244:1288 – 1292

Chapman NM, Tu Z, Tracy S, Gauntt CJ (1994) An infectious cDNA copy of the genome of a non-cardiovirulent coxsackievirus B3 strain: its complete sequence analysis and comparison to the genomes of cardiovirulent coxsackieviruses. Arch Virol 135:115 – 130

Crowell R, Finkelstein SD, Hsu K-HL, Landau BJ, Stahlhandske P, Whittier PS (1988) A murine model for coxsackievirus B3-induced acute myocardial necrosis for study of cellular receptors as determinants of viral tropism. In: Schultheiss HP (ed) New concepts in viral heart disease. Springer, Berlin, Heidelberg, New York, pp 79 – 92

Crowell RL, Landau BJ (1983) Receptors in the initiation of picornavirus infections. In: Fraenkel-Conrat H, Wagner RR (eds) Comprehensive virology, Plenum, New York, 18:1 – 42

Crowell RL, Siak JS (1978) Receptors for the group B coxsackieviruses: characterization and extraction from HeLa cell membranes. In: Pollard M (ed) Perspectives in virology. Raven, New York, pp 39 – 55

Crowell RL, Tomko RP (1994) Receptors for picornaviruses. In: Wimmer E (ed) Cellular receptors for animal viruses. Cold Spring Harbor Laboratory Press, Cold Spring Harbor, New York

Cunningham L, Bowles NE, Lane RJ, Dubowitz V, Archard LC (1990) Persistence of enteroviral RNA in chronic fatigue syndrome is associated with the abnoormal production of equal amounts of positive and negative strands of enteroviral RNA. J Gen Virol 71:1399 – 1402

Cunningham MW, Antone SM, Gulizia JM, McManus B M, Fischetti VA, Gauntt CJ (1992) Cytotoxic and viral neutralizing antibodies crossreact with streptococcal M-protein, enteroviruses, and human cardiac myosin. Proc Natl Acad Sci USA 89:1320 – 1324

D'Alessio DJ (1992) A case control study of group B coxsackievirus immunoglobulin M antibody prevalence and HLA-DR antigens in newly diagnosed cases of insulin dependent diabetes mellitus. Amer J Epid 135:1331 – 1338

Dalldorf G, Sickles GM (1948) An unidentified, filterable agent isolated from the feces of children with paralysis. Science 108:61 – 62

Enders JF, Weller TH, Robbins FC (1949) Cultivation of the Lansing strain of poliomyelitis virus in cultures of various human embryonic tissues. Science 109:85 – 87

Fohlman J, Friman G (1993) Is juvenile diabetes a viral disease? Ann Med 25:569 – 574

Fohlman J, Pauksen K, Morein B, Bjare U, Ilback NG, Friman G (1993) High yield production of an inactivated coxsackie B3 adjuvant vaccine with protective effect against experimental myocarditits. Scand J Inf Dis 88:103 – 108

Freistadt MS, Kaplan G, Racaniello VR (1990) Heterogeneous expression of poliovirus receptor-related proteins in human cells and tissues. Mol Cell Biol 10:5700 – 5706

Frisk G, Diderholm H (1992) Increased frequency of coxsackie B virus IgM in women with spontaneous abortion. J Infect 24:141 – 145

Gauntt CJ (1988) The possible role of viral variants in pathogenesis. In: Bendinelli M, Friedman H (eds) Coxsackieviruses: a general update. Plenum, New York, pp 159 – 179

Gauntt CJ, Arizipe HM, Higdon AL, Wood HJ, Bowers DF, Rozek MM, Crawley R (1995) Molecular mimicry, anti-coxsackievirus B3 neutralizing monoclonal antibodies, and myocarditis. J Immunol 154:2983 – 2995

Gauntt CJ, Gudvangen RJ, Brans VW, Marlin AE (1985) Coxsackievirus group B antibodies in the ventricular fluid of infants with severe anatomic defects in the central nervous system. Pediatrics 76:64 – 68

Gifford R, Dalldorf G (1951) The morbid anatomy of experimental coxsackie virus infection. Am J Pathol 27:1047 – 1064

Gregor GR, Geller SA, Walker GF, Campomanes BA (1975) Coxsackie hepatitis in an adult, with ultrastructural demonstration of the virus. Mount Sinai J Med 42:575 – 580

Haase A, Brahic M, Stowring L, Blum H (1984) Detection of viral nucleic acids by in situ hybridization. In: Maramorosh K, Koprowski H (eds) Methods in Virology. Academic, New York, 7:189 – 226

Heim A, Stille-Seigener M, Kandolf R, Kreuzer H, Figulla HR (1994) Enterovirus-induced myocarditis: hemodynamic deterioration with immunosuppressive therapy and successful application of interferon-alpha. Clin Cardiology 17:563 – 565

Heinz BA, Rueckert RR, Shepard DA, Dutko FJ, McKinlay MA, Fancher M, Rossmann MG, Badger J, Smith TJ (1989) Genetic and molecular analysis of spontaneous mutants of human rhinovirus 14 resistant to an antiviral compound. J Virol 63:2476 – 2485

Holland JJ, McLaren LC, Syverton JT (1959) The mammalian cell-virus relationship. IV: Infection of naturally insusceptible cells with enterovirus ribonucleic acid. J Exp Med 110:65 – 80

Huber SA, Moraska A, Choate M (1992) T cells expressing the gamma delta T-cell receptor potentiate coxsackievirus B3-induced myocarditis. J Virol 66:6541 – 6546

Huber SA, Moraska A, Cunningham M (1994a) Alterations in major histocompatibility complex association of myocarditis induced by coxsackievirus B3 mutants selected with monoclonal antibodies to group A streptococci. Proc Natl Acad Sci USA 91:5543 – 5547

Huber SA, Polgar J, Schultheiss P, Schwimmbeck P (1994b) Augmentation of pathogenesis of coxsackievirus B3 infections in mice by exogenous administration of interleukin-1 and interleukin-2. J Virol 68:195 – 206

Hyypia T, Kallajoki M, Maaronen M, Stanway G, Kandolf R, Auvinen P, Kalimo H (1993) Pathogenetic differences between coxsackie A and B virus infections in newborn mice. Virus Res 27:71 – 78

Iizuka N, Kuge S, Nomoto A. (1987) Complete nucleotide sequence of the genome of coxsackievirus B1. Virology 156:64-73

Iizuka N, Yonekawa H, Nomoto A. (1991) Nucleotide sequences important for translation initiation of enterovirus RNA. J Virol 65:4867-73
International Enterovirus Study Group (1963) Picornavirus group. Virology 19:114 – 116
Jongen PJ, Heessen FW, ter Laak HJ, Galama JM, Gabreels FJ (1994) Coxsackie B1 virus-induced murine myositis: relationship of disease severity to virus dose and antiviral antibody response. Neuromuscular Disorders 4:17 – 23
Kamei S, Hersch SM, Kurata T, Takei Y (1990) Coxsackie B antigen in the central nervous system of a patient with fatal acute encephalitis: immunohistochemical studies of formalin-fixed paraffin-embedded tissue. Acta Neuropath 80:216 – 221
Kandolf R (1988) The impact of recombinant DNA technology on the study of enterovirus heart disease. In: Bendinelli M, Friedman H (eds) Coxsackieviruses: a general update. Plenum, New York, pp 293 – 318
Kandolf R, Hofschneider PH (1985) Molecular cloning of the genome of a cardiotropic coxsackie B3 virus: full-length reverse-transcribed recombinant cDNA generates infectious virus in mammalian cells. Proc Natl Acad Sci USA 82:4818 – 4822
Kandolf R, Klingel K, Zell R, Canu A, Fortmuller U., Hohenadl C, Albrecht M, Reimann BY, Franz WM, Heim A, et al. (1993) Molecular mechanisms in the pathogenesis of enteroviral heart disease: acute and persistent infections. Clin Immunol Immunopathol 68:153 – 158
Kang Y, Chatterjee NK, Nodwell MJ, Yoon JW (1994) Complete nucleotide sequence of a strain of coxsackie B4 virus of human origin that induces diabetes in mice and its comparison with nondiabetogenic coxsackie B4 JBV strain. J Med Virol 44:353 – 361
Kaplan G, Freistadt MS, Racaniello V (1990) Neutralization of poliovirus by cell receptors expressed in insect cells. J Virol 64:4697 – 4702
Keeling PJ, Lukaszyk A, Poloniecki J, Caforio AL, Davies MJ, Booth JC, McKenna WJ (1994) A prospective case-control study of antibodies to coxsackie B virus in idiopathic dilated cardiomyopathy. J Amer Coll Cardiol 23:593 – 598
Klingel K, Hohenadl C, Canu A, Albrecht M, Seemann M, Mall G, Kandolf R (1992) Ongoing enterovirus-induced myocarditis is associated with persistent heart muscle infectionn: quantitative analysis of virus replication, tissue damage, and inflammation. Proc Natl Acad Sci USA 89:314 – 318
Koide H, Kitaura Y, Deguchi H, Ukimura A, Kawamura K, Hirai K (1992) Genomic detection of enteroviruses in the myocardium: studies on animal hearts with coxsackievirus B3 myocarditis and endocardial biopsies from patients with myocarditis and dilated cardiomyopathy. Jap Circul J 56:1081 – 1093
Koike S, Toya C, Kurata T, Abe S, Ise I, Yonekawa H, Nomoto A (1991) Transgenic mice susceptible to poliovirus. Proc Natl Acad Sci 88:951 – 955
Kunin CM, Halmagyi NE (1961) The relative abundance of viral receptors: an explanation of the differential susceptibility of suckling and adult mice to coxsackie B1 infection. J Clin Invest 40:1055 – 1056
Landau BJ, Whittier PS, Finkelstein SD, Alstein B, Grun JB, Schultz M, Crowell RL (1990) Induction of heterotypic virus resistance in adult inbred mice immunized with a variant of coxsackievirus B3. Microb. Pathogen 8:289 – 298
Lane JR, Neumann DA, Lafond-Walker A, Herskowitz A, Rose NR (1992) Interleukin 1 or tumor necrosis factor can promote coxsackie B3-induced myocarditis in resistant B10.A mice. J Exp Med 175:1123 – 1129
Lindberg AM, Crowell RL, Zell R, Kandolf R, Pettersson U. (1992) Mapping of the RD phenotype of the Nancy strain of coxsackievirus B3. Virus Res 24:187-96
Mason JW, O'Connell JB, Herskowitz A, Rose NR, McManus BM, Billingham ME, Moon TE, and the Myocarditis Treatment Investigators (1995) A clinical trial of immunosuppressive therapy for myocarditis. N Eng J Med 333:269 – 275
Melnick JL (1990) Enteroviruses: polioviruses, coxsackieviruses, echoviruses, and the newer enteroviruses. In: Fields BN, Knipe DM, et al. (eds) Virology (2nd edn). Raven, New York, pp 549 – 606
Melnick JL, Dalldorf G, Enders JF, Gelfand HM, Hammon W McD, Huebner RJ, Rosen L, Sabin AB, Syverton JT, Wenner HA (1962) Classification of human enteroviruses. Virology 16:501 – 504
Muckelbauer JK, Kremer M, Minor I, Diana G, Dutko F, Groarke J, Pevear DC, Rossmann MG (1995) The structure of coxsackievirus B3 at 3.5 A resolution. Structure 3:653 – 667
Neu N, Rose NR, Beisel KW, Herskowitz A, Gurri-Glass G, Craig SW (1987) Cardiac myosin induces myocarditis in genetically predisposed mice. J Immunol 139:3630 – 3636

Raab de Verdugo U, Selinka H-C, Huber M, Kramer B, Kellermann J, Hofschneider PH, Kandolf R (1995) Characterization of a 100-kilodalton binding protein for the six serotypes of coxsackie B viruses. J Virol 69:6751 – 6757

Rabausch-Starz I, Schwaiger A, Grunewald K, Muller-Hermelink HK, Neu N (1994) Persistence of virus and viral genome in myocardium after coxsackievirus B3-induced murine myocarditis. Clin Exp Immunol 96:69 – 74

Ramsingh A, Araki H, Bryant S, Hixson A (1992) Identification of candidate sequences that determine virulence in coxsackievirus B4. Virus Res 23:281 – 292

Ramsingh A, Hixson A, Duceman B, Slack J. (1990) Evidence suggesting that virulence maps to the P1 region of the coxsackievirus B4 genome. J Virol 64:3078-81

Ramsingh A, Slack J, Silkworth J, Hixson A (1989) Severity of disease induced by a pancreatropic coxsackie B4 virus correlates with the H-2Kq locus of the major histocompatibility complex. Virus Res 14:347 – 358

Reagan K, Goldberg B, Crowell R (1984) Altered receptor specificity of coxsackievirus B3 after growth in rhabdosarcoma cells. J Virol 49:635 – 640

Ren R, Constantini FC, Gorgacz EJ, Lee JJ, Racaniello VR (1990) Transgenic mice expressing a human poliovirus receptor: a new model for poliomyelitis. Cell 63:353 – 362

Rose NR, Herskowitz A, Neumann DA (1993) Autoimmunity in myocarditis: models and mechanisms. Clin Immunol Immunopathol 68:95 – 99

Rossmann MG, Palmenberg AC (1988) Conservation of the putative receptor attachment site in picornaviruses. Virology 164:373 – 382

Rotbart HA (1992) DNA probes for viral diagnosis. In: Block TM, Jungkind D, Crowell RL, Denison M, Walsh LR (eds) Innovations in antiviral development and detection of virus infections. Plenum, New York, pp 201 – 209

Rueckert RR (1990) Picornaviridae and their replication. In: Fields BN, Knipe DM, et al. (eds) Virology (2nd edn). Raven, New York, pp 507 – 548

Sabin AB (1955) Characteristics and genetic potentialities of experimentally produced and naturally occurring variants of poliomyelitis virus. Ann NY Acad Sci 61:924 – 938

Salk JE (1953) Principles of immunization as applied to poliomyelitis and influenza. Am J Pub Health 43:1384 – 1398

Schultz M, Crowell R (1983) Eclipse of coxsackievirus infectivity: the restrictive event for a non-fusing myogenic cell line. J Gen Virol 64:1725 – 1734

Sexton AM (1967) A chronical of the division of laboratories and research, New York State Department of Health: the first fifty years 1914 – 1964. Stinehour, Lunenburg, VT

Tilles JG (1994) Efficacy of a polyvalent inactivated-virus vaccine in protecting mice from infection with clinical strains of group B coxsackieviruses. Scand J Inf Dis 26:739 – 747

Tilles JG (1995) Pathogenesis of virus-induced diabetes in mice. J Inf Dis 171:1131 – 1138

Tracy S, Chapman NM, Tu, Z (1992) Coxsackievirus B3 from an infectious cDNA copy of the genome is cardiovirulent in mice. Arch Virol 122:399 – 409

Tu Z, Chapman NM, Hufnagel G, Tracy S, Romero JR, Barry WH, Zhao L, Currey K, Shapiro B (1995) The cardiovirulent phenotype of coxsackievirus B3 is determined at a single site in the genomic 5' nontranslated region. J Virol 69:4607 – 4618

Ursing B (1973) Acute pancreatitis in coxsackie B infection. Br Med J 3:524 – 525

Webb SR, Madge GE (1980) The role of host genetics in the pathogenesis of coxsackievirus infection in the pancreas of adult mice. J Inf Dis 141:47 – 54

Wolfgram LJ, Beisel KW, Herskowitz A, Rose NR (1986) Variations in the susceptibility to coxsackievirus B3-induced myocarditis among different strains of mice. J Immunol 136:1846 – 1852

Yoon JW, Austin M, Onodera T, Notkins AL (1979) Virus-induced diabetes mellitus. Isolation of a virus from the pancreas of a child with diabetic ketoacidosis. N Engl J Med 300:1173 – 1179

Zhang G, Wilsden G, Knowles NJ, McCauley JW. (1993) Complete nucleotide sequence of a coxsackie B5 virus and its relationship to swine vesicular disease virus. J Gen Virol 74:845-53

Zibert A, Selinka HC, Elroy-Stein O, Wimmer E (1992) The soluble form of two N-terminal domains of the poliovirus receptor is sufficient for blocking viral infection. Virus Res 25:51 – 61

Coxsackievirus B Epidemiology and Public Health Concerns

M. A. Pallansch

1 Introduction

The coxsackieviruses group B (CVB) are members of the genus *Enterovirus* and share many properties with other members of this group. There are several excellent reviews covering the basic epidemiology and clinical characteristics of enteroviruses and their infections (Modlin 1990; Morens et al. 1991; Melnick 1996), and this material will not be repeated in detail except as it relates to recent studies. Most of the conclusions concerning enteroviruses in general are directly applicable to CVB, including: pronounced seasonality in temperate climates; geographic, age, and sex distribution; high incidence and prevalence; moderately high transmission; and most clinical syndromes.

This chapter will first review the literature published since the previous reviews as it relates to the epidemiologic patterns of CVB infections and disease.

Division of Viral and Rickettsial Diseases, National Center for Infectious Diseases, Centers for Disease Control and Prevention, Atlanta, Georgia 30333, USA

These data comprise individual reports, outbreaks, serological studies, and surveillance data. Second, several diseases will be discussed for which CVB infection is implicated as an etiologic factor, including some "chronic" diseases, but the extent or existence of CVB involvement is not well defined. These areas are currently the focus of most clinical and epidemiologic studies. Third, the design of possible future epidemiologic and laboratory studies of these viruses will be considered, including possible limitations and new techniques.

2 Surveillance

2.1 Viral Isolates

2.1.1 Serotype Variation

Most studies of the occurrence or prevalence of CVB have made use of the reported isolation of virus from clinical specimens, using both active and passive surveillance. The basic sources of surveillance data have been reviewed (MORENS et al. 1991) and are primarily passive. Because of the considerable expense and time involved, surveillance data for enteroviruses is far less common than in the past. Even with these limitations occasional reports do appear (BELL and McCARTNEY 1984; STRIKAS et al. 1986). In addition, occurrences of outbreaks continue to be reported (NAKAYAMA et al. 1989; ALEXANDER et al. 1993; IKEDA et al. 1993; MOREIRA et al. 1995).

In the United States, enterovirus surveillance data are collected and analyzed by the Centers for Disease Control and Prevention. The data have been reported irregularly since the beginning of the program in 1961 (MORENS et al. 1979; MOORE 1982). Like enteroviruses in general, CVB isolations fall into two patterns of incidence, endemic and epidemic. The epidemic pattern, as typified by CVB5, is characterized by distinct outbreaks with few isolates observed in intervening years. During the period 1961–1995 there were four major epidemics of CVB5 in the United States, in 1961, 1967, 1972, and 1983. By contrast, endemic viruses are isolated in nearly every year, with mostly minor yearly differences. This is the pattern seen with CVB2 and CVB4, although larger outbreaks do occur occasionally, as with CVB3 in 1980.

Enterovirus isolates are routinely characterized by serotype using polyclonal animal sera, and by this criteria the six CVB serotypes appear to be quite stable and well defined. However, studies using monoclonal antibodies prepared against CVB3 and CVB4 suggest that there can be considerable antigenic variation within a serotype of CVB (PRABHAKAR et al. 1982; HU et al. 1993). Monoclonal antibodies have been used to study CVB strains, plaque variants of the prototypes, and clinical isolates (PRABHAKAR et al. 1985; WEBB et al. 1986; HU et al. 1993). In all cases, multiple differences between strains were detected at many epitopes, although no correlation with clinical symptoms was made. This antigenic diversity among the multiple viral epitopes has so far yielded very little additional understanding about the biology of CVB infection and pathogenesis.

2.1.2 Genotypic Variation

The genotype of a virus is defined by the primary sequence of its genome and has been studied by several techniques including sequencing, hybridization, and oligonucleotide fingerprinting. As a result of rapid progress in nucleic acid sequencing technology and the considerable power of the epidemiologic inference derived from these type of data, most recent genotypic studies of CVB isolates have used this approach. The use of these type of data has been most comprehensively applied to studies of poliovirus, where the information has proved valuable for supporting the global poliovirus eradication program (KEW et al. 1993, 1995). From these studies it is possible to both determine if an isolated poliovirus is related to vaccine virus and to determine similarities or differences between isolates in an epidemic and from different geographic areas.

Of the CVB, molecular epidemiologic studies using sequencing have been reported for CVB1 and CVB5 (DREBOT et al. 1994; ZOLL et al. 1994; KOPECKA et al. 1995). The study of the CVB5 isolates examined the pattern of genetic changes over three separate outbreaks in the United States. First, as noted above from the surveillance data, CVB5 has an epidemic pattern of occurrence in the United States. The nucleotide sequences from multiple isolates from the epidemics show that each of the epidemics was caused by a single genotype. Second, the genotype of CVB5 observed in the 1967 epidemic shows more similarity to the virus observed in the 1983 epidemic than to viruses isolated during the intervening years. These results suggest discontinuous transmission of the epidemic CVB5 in the United States during the period from 1967 through 1983.

2.2 Serology

Many studies have been performed to examine the level of antibodies to the CVB in defined populations, and the results of these serosurveys have been reviewed (MODLIN 1990; MORENS et al. 1991; MELNICK 1996). Several important conclusions should be reiterated. First, the number of persons with neutralizing antibody to any of the CVBs is high, indicating a high incidence of past infection. A high incidence of recent infection is also suggested by IgM surveys, which typically show at least 4%–6% positivity, and even higher in children. Second, infections with one serotype of CVB can boost the antibody titers to other CVB serotypes as measured by either IgM or neutralization. The pattern of the heterotypic response is not consistent with respect to serotype and varies among persons. The nature of this heterotypic response has been recently explored through the identification of specific epitopes using monoclonal antibodies and peptide antisera (YAGI et al. 1992; HAARMANN et al. 1994; SAMUELSON et al. 1994). Other enteroviruses can also cause an anamnestic response toward CVBs. Third, the prevalence of antibody to each of the serotypes varies by time and location; thus data from different years and locations are not directly comparable. These three points must be considered when designing and interpreting serologic studies of possible associations between enterovirus infection and a disease.

3 Association Between Infection and Clinical Disease

3.1 Review of Previous Findings

Many reports have associated CVB infection with various diseases and syndromes. Most of these associations are based on studies of outbreaks of enteroviral infections. Some are based on one or several clinical case descriptions associated with large numbers of people with similar symptoms who showed evidence of infection with the same serotype of CVB. In outbreaks, it is often possible to determine an etiologic association with confidence. From such studies CVB infection has been ascribed as the cause of aseptic meningitis, encephalitis, pleurodynia, myocarditis, and pericarditis. By far the most common result of CVB infection is asymptomatic infection, an undifferentiated febrile illness, or mild upper respiratory symptoms.

Determining the etiology of single clinical cases is difficult. LERNER and WILSON (1973) proposed guidelines for evaluating individual cases of myocarditis, which are generally applicable to other enteroviral diseases. A high level of association between the virus and the disease may be obtained when the virus is isolated from the tissue affected, such as the cerebrospinal fluid (CSF) during a case of meningitis, or detected directly, such as by polymerase chain reaction (PCR), in situ hybridization, or appropriate immunofluorescence. Isolating virus from throat washings, stool specimens or other secondary site supports only a moderate level of association unless there is also evidence of serological conversion. Detecting antibodies alone indicates only a low level of association. Due to the large number of inapparent infections and frequency of prolonged virus excretion, especially in stool, infection does not prove etiology for the disease. If the isolated virus is known to cause the symptoms and no other agent or cause has been identified, then the isolation provides a plausible explanation for the illness.

3.2 Established Etiologic Associations

3.2.1 Central Nervous System Disease

The most frequent serious manifestation of CVB and enterovirus infection is central nervous system (CNS) disease, most commonly aseptic meningitis. Although this clinical manifestation is not uniquely associated with CVB, outbreaks due to specific CVB continue to be reported (ALEXANDER et al. 1993). The significant contribution of enterovirus and CVB infection to aseptic meningitis in infants less than 2 years of age has been clearly shown in a recent study (BERLIN et al. 1993) which aggressively attempted to establish the etiology for hospitalized children over a 5 year period. There was laboratory evidence of enterovirus infection in 62% of these cases and nearly half of laboratory diagnosed cases had evidence of CVB infection. More significant cases of CNS disease have also been associated with CVB infection. These include reports of encephalitis and paralysis (KAMEI et al. 1990; YUI and GLEDHILL 1991). The associations are not always confirmed by viral isolation from CSF, but rely on serologic evidence, detection or isolation of CVB from other specimens. Improvements in technology, as described in Sect. 4.1, offers the ability to detect virus in the CNS when it cannot be isolated and thus extends diagnostic capabilities.

3.2.2 Carditis

Another recognized clinical syndrome associated with CVB infection is acute myocarditis and pericarditis. While there continue to be numerous clinical reports of myocarditis and pericarditis associated with CVB infection and specific studies addressing the involvement of CVB in myocarditis (TRACY et al. 1990), the fraction of cases of acute myocarditis that may be due to CVB infection is currently unresolved. These and other studies described elsewhere in this volume suggest, but do not clearly prove, that CVB infection may be associated with a large fraction of cases of acute myocarditis. Another obstacle for fully characterizing the public health significance of viral myocarditis is that the total disease burden from myocarditis of any etiology is poorly described on a population basis.

3.2.3 Neonatal Infections

Coxsackieviruses group B are associated with severe illness in neonates, including sepsis and generalized disseminated infection. One of the most systematic studies of newborn infection was reported by KAPLAN et al. (1983), covering records over a 10 year period in Nassau County, New York. Considering only cases in which infection was documented by CVB isolation, there were 77 cases and six fatalities in over 153 000 live births. This rate of infection resulting in hospitalization, one for every 2000 births during the first 3 months of life, is a conservative estimate because of the recognized insensitivity of viral isolation in detecting viral infection.

It is possible that the neonate can be infected in utero, from its mother in the peripartum period, and through postpartum exposure, including nosocomial infection. The consequences of neonatal infection can be quite severe, including fatal multi-organ disease. Several recent studies have documented the varied outcomes of meningitis, myocarditis, and fatal systemic infection in the CVB-infected neonate (DRUYTS-VOETS et al. 1993; HADDAD et al. 1993; ISACSOHN et al. 1994). One report has described the isolation of CVB1 from amniotic fluid obtained through transabdominal amniocentesis, demonstrating intrauterine infection at 34 weeks gestation (STRONG and YOUNG 1995). In another study, CVB5 was isolated from the cervixes of four women with febrile illness during the third trimester of pregnancy (REYES et al. 1986), supporting one of the possible mechanisms by which neonatal infections may occur.

3.3 Suspected Coxsackievirus B Disease

In addition to recognized associations with acute disease, several studies have implicated CVB infection with several chronic diseases, such as diabetes mellitus, chronic heart disease, and other neurologic or muscle disease. While the association of CVB infection with these individual diseases are certainly not proven, in some cases considerable evidence is accumulating which gives at least a strong implication of some involvement.

3.3.1 Diabetes Mellitus

The most significant potential association of CVB infection with chronic disease is with juvenile onset insulin-dependent diabetes mellitus (IDDM). This topic, including most of the published studies, has been reviewed previously (BARRETT-CONNER 1985; FOHLMAN and FRIMAN 1993). To date, the epidemiologic studies have failed to prove or disprove the suggested disease association and often provide conflicting conclusions from very similar study designs. Despite these difficulties, several lines of evidence have maintained interest in this possible association. Continued studies in murine models, as described elsewhere in this volume, in which CVB infection of certain inbred strains of mice produce diabetes, contribute to the continued plausibility of CVB virus-induced IDDM. A circumstantial association between an outbreak of CVB5 and a subsequent increase in the incidence of IDDM (Wagenknecht et al. 1991) provides the latest example of a temporal association between a CVB outbreak and increased disease incidence in the general population. Finally, serologic case/control studies and several individual case reports continue to support an association between CVB infection and IDDM.

Although some studies have failed to observe an association between IgM positivity to CVB and IDDM (TUVEMO et al. 1989) and others have been inconclusive (SCHERBAUM et al. 1991), there are many studies which continue to show a higher rate of IgM positivity among recent onset cases of IDDM when compared to a healthy control group (FRISK et al. 1992; HYÖTY et al. 1995). Another study observed a difference between antibody levels in mothers of children who developed IDDM compared to mothers whose children were born at the same time and the same hospital who did not develop disease (DAHLQUIST et al. 1995). Despite these apparently contradictory results, two studies have specifically identified factors which may be important for the interpretation of all serologic studies (D'ALESSIO 1992; HELFAND et al. 1995). In both studies, the difference between IDDM cases and controls was focused in the adolescent age group and not in younger children. In addition, in one of these studies (D'ALESSIO 1992), in which genetic risk markers were determined, almost all of the differences between the study groups were found only in the genetic risk group that was HLA-DR3-positive.

In summary, these recent studies provide further support for the association between CVB infection and IDDM. The complexity of current concepts about the pathogenesis of this disease can still include a potential role for CVB infection in the development or onset of IDDM. Several points, however, should be stressed. Because it is clear that the risk for disease includes both a genetic risk and some environmental exposure, possibly viral, and that the development of clinical symptoms can take many years, serologic studies must be very carefully designed and interpreted. Studies must take into account the choice of controls, the age and genetics of both cases and controls, timing of serum collection, sufficient study duration to include one or more virus seasons, and the sensitivity and specificity of the antibody detection system. Despite these limitations, the evidence clearly warrants further investigation into this potential etiologic association with an important disease.

3.3.2 Inflammatory Heart Disease

Although there is widespread acceptance of the etiologic association between CVB infection and myocarditis, only recently has CVB involvement in idiopathic dilated cardiomyopathy (DCM) in general been systematically examined. This is partly a result of the realization that at least some cases of DCM may result from continued pathology initiated by clinical or subclinical myocarditis. It is possible that the spectrum of dilated heart disease may include both acute myocarditis and DCM and that both may have a least some fraction of these cases attributable to CVB infection. Several aspects of this potential association have been reviewed (TRACY et al. 1992, 1996).

There are several studies which now have detected evidence of CVB infection associated with DCM. This association has been observed both with serologic studies (MUIR et al. 1989; WESSLÉN et al. 1993; KEELING et al. 1994) and by detection of enterovirus RNA in heart tissue (BOWLES et al. 1989; COCHRANE et al. 1991; SCHWAIGER et al. 1993; SATOH et al. 1994). The remarkable consistency of the results from these studies have provided preliminary evidence which, despite some study design and methodology issues, warrant further investigation of this potential association.

3.3.3 Other Chronic Diseases

Several case reports and studies have suggested links between CVB infection and other clinical conditions, including amyotrophic lateral sclerosis (ALS), chronic fatigue syndrome (CFS), chronic myopathies, and fetal demise. Two studies have detected evidence of enterovirus infection among patients with ALS. The first study, using PCR, detected enterovirus nucleic acid in spinal cord specimens (WOODALL et al. 1994). The nucleic acid sequences of these PCR products were also consistent with enterovirus and CVB genome data available. Lending further credibility to the observation, the sequences of the PCR products were similar in different sections of spinal cord from the same patient, but were clearly different when compared between PCR products from different patients. In a second study (BARTFELD et al. 1989), circulating immune complexes from ALS patients contained antigen that appeared to be enterovirus related.

By contrast, the association of CVB infection with CFS continues to generate conflicting results in various studies. It still seems that for every study that shows a difference in CVB antibody among CFS cases and controls, there are others which fail to find any difference (MILLER et al. 1991). There are some studies which do detect enterovirus nucleic acid in specimens from CFS patients. In one study, muscle biopsy specimens were examined using strand-specific hybridization probes (CUNNINGHAM et al. 1990). These results suggested that enteroviruses had persistently infected muscle tissue in CFS patients. In a separate study (NAIRN et al. 1995), using PCR and serum specimens, enterovirus nucleic acid was detected in 42% of CFS patients and only 9% of the comparison group. These results did not correlate with the presence of CVB neutralizing antibodies in the same specimens. Because of numerous differences among studies, including case definition, choice of controls, and viral detection methodologies, as well as inconsistent results from seemingly similar studies, the association of CVB infection with CFS remains unestablished.

A similar series of studies have attempted to examine the association of enterovirus and CVB infection with various inflammatory muscle disease. Despite several early reports of antibody evidence of CVB infection among cases of juvenile dermatomyositis, subsequent studies have not been able to confirm this association. Attempts to detect virus directly in muscle tissue have not been consistently able to show evidence of virus at higher rates than in specimens from controls (HILTON et al. 1994). By contrast, a serologic study looking at neutralizing antibodies in patients with adult polymyositis and adult dermatomyositis shows some evidence of enterovirus infection (NISHIKAI 1994).

Two recent studies contribute additional evidence that enterovirus infection may be responsible for cases of fetal demise. In the first study (FRISK and DIDERHOLM 1992), IgM antibody to CVB were measured from women who had experienced either spontaneous abortion or stillbirths. Among women with spontaneous abortions, 33% were CVB IgM-positive compared to 8% for healthy pregnant women. In a separate study of miscarriage before the 13th week of gestation (AXELSSON et al. 1993), 42% of women with miscarriage were CVB IgM-positive compared to 18% positive for control women. These studies, as well as several previous reports, suggest that the role of CVB and enterovirus infection in fetal demise should continue to be examined.

4 Methodologies and Studies

4.1 Methods for Detection of Coxsackievirus B Infection

The techniques available for laboratory support of epidemiologic investigations of CVB have changed markedly in the last 10 years. The traditional techniques for detecting and characterizing enteroviruses rely on the time-consuming and expensive procedures of viral isolation in tissue culture and neutralization by reference antisera (GRANDIEN et al. 1989). Isolating CVB from specimens using appropriate tissue culture cell lines is possible within 2 or 3 days in the majority of cases and remains a very sensitive method for detecting these viruses. Several technical advances, however, have allowed the development of alternative procedures for detecting and characterizing CVB and other enteroviruses, including nonviable virus.

Nucleic acid probes were the first application of molecular techniques developed for detecting enterovirus genomes. These early probes were made by synthesizing complimentary DNA (cDNA) from CVB viral RNA. The cDNA can be cloned into bacteria, amplified, and radiochemically labeled to provide a reagent for detecting enterovirus genomes by nucleic acid hybridization. Further refinement of this technique included the expression of RNA probes to improve sensitivity and the application of this to in situ hybridization in myocardial tissue (HOHENADL et al. 1991). This technique has been particular useful for understanding the cytopathology of enterovirus infection, particularly in studies of myocarditis. The technique, however, cannot be used to identify the serotype of the enterovirus in a specimen.

By far the most important technical development in CVB studies and most of molecular biology, in general, has been the PCR. Although this approach has been extensively described as applied to enterovirus detection, numerous modifications of this technique have been reported to improve detection from clinical specimens (WEISS et al. 1991; SHIMIZU et al. 1994; SEVERINI et al. 1993; GOW et al. 1995). A few important points about enterovirus PCR need to be noted. First, no existing PCR approach is capable of uniquely differentiating all enterovirus serotypes while retaining detection of all isolates. PCR can be used to confirm the presence of an enterovirus, but not which serotype of an enterovirus. Second, detection of enteroviruses from clinical specimens is not reliably as sensitive as from purified or tissue culture grown virus. Although significant improvements in this area have been achieved, there is an urgent need for a simple, effective, reproducible and highly sensitive methods for processing clinical specimens. Third, many reported studies fail to use or document procedures to avoid artifactual PCR results. This is important not only for the assay methodology itself, but also at the level of blinded and blank specimens. Inclusion of this documentation is critical to the acceptance of results from major studies of CVB infection and disease using this technique.

Many serologic studies rely upon the detection of IgM antibody as evidence of recent CVB infection and this is now widely used as an alternative to the neutralization and complement fixation test. In the last several years, several groups have developed an enzyme-linked immunosorbent assay (ELISA) for CVB-specific IgM. These tests were positive for nearly 90% of culture-confirmed CVB infections and are rapid to perform. The ELISAs have been successfully applied for epidemiologic investigations of outbreaks (GOLDWATER 1995) and in specific diagnostic use (DAY et al. 1989).

In most cases, however, the test is not completely serotype-specific. Depending on the configuration and sensitivity of the test, from 10% to nearly 70% of serum samples show heterotypic responses due to other enterovirus infections. This heterotypic response has been exploited to measure broadly reactive antibody and the assay used to detect enterovirus infection generically (BOMAN et al. 1992; SWANINK et al. 1993). In attempting to characterize the exact nature of the response using different antigens, it is clear that the human immune response to enterovirus infection includes antibodies that react with both serotype-specific epitopes and shared epitopes (FRISK et al. 1989). Despite this problem, which is inherently biological, there remains a fairly high concordance of results between assays of different configurations (HODGSON et al. 1995). To summarize, the IgM assays that are generally used in epidemiologic studies have very good sensitivity. These assays detect heterotypic antibodies resulting from other enterovirus infection and therefore cannot be considered serotype-specific. The assays, however, appear to be very specific for enterovirus infection. Attempts to improve the serotype specificity of these assays have, as expected, either reduced the sensitivity of the assay or decreased the convenience.

4.2 Limitations and Study Design Considerations

4.2.1 Considerations Due to the Viruses

In epidemiologic studies of CVB viruses, several important properties of the viruses are critical to study design and data interpretation. First, the grouping of these six enterovirus serotypes is based principally upon growth characteristics in newborn mice; they produce a focal myositis and a spastic paralysis (MORENS et al. 1991; MELNICK 1996). This growth property in mice has never been uniquely correlated with any human disease. While CVB viruses are more often associated with myocarditis and pleurodynia, other enteroviruses are frequently associated with these illnesses as well.

Second, the sequences responsible for the pathogenic potential of a virus are not necessarily the same as the genes that code for serotype (TU et al. 1995). Since recombination occurs often with poliovirus, it almost certainly occurs with the other enteroviruses. Therefore, genes for pathogenic potential found in one serotype may recombine with virus of a different serotype and spread this gene among several serotypes. The number of serotypes is limited only by the range of potential recombination partners.

Another factor which has the potential to significantly affect the interpretation of epidemiologic studies is the recognition that even single virus isolates are actually a collection of genetic variants. This is particularly important since several studies indicate that these variants may have different pathogenic potential in animal models. While these variants have been observed for both murine diabetogenic (SZOPA et al. 1990) and cardiotropic (GOMEZ et al. 1993) potential, there is no reason to suspect that this is not a generic property for any viral genetic phenotype. Since there is no way to know if these variants have some selective advantage in the human population, the prevalence of these pathogenic variants may vary widely over time and geography. This could lead to variable results in epidemiologic studies that are highly focused in temporal or geographic scope, since at any point in time the circulating viruses may have a higher or lower frequency of these variants.

Fourth, enterovirus infections are very common. It has been conservatively estimated that there are more than 20 million enterovirus infections yearly in the United States, and of these between one quarter and one half may be due to CVB. Serosurveys of IgM antibodies in most study control groups typically show at least 5% positive for CVB IgM. These numbers support a high incidence of enterovirus infection and are consistent with the high prevalence of neutralizing antibody to CVB.

Finally, the prevalence of any given serotype in any given year can vary dramatically. There can also be considerable variation during the same year between different locations within the United States (STRIKAS et al. 1986). Studies attempting to correlate virus infection with a given disease must consider what serotypes are prevalent in a particular location at a particular time in designing the study. Alternatively, study design should not be premised on the presence of particular serotypes.

4.2.2 Considerations Due to the Nature of Enterovirus Infections

In addition to considerations regarding the viruses, several characteristics of enterovirus infections should be reiterated. First, most infections are either subclinical or associated with minor illness. The ratio of infected individuals to reported clinical cases for any serious enterovirus disease is typically quite high (MODLIN 1990; MORENS et al. 1991), for example, around 1000:1 or greater for enteroviral aseptic meningitis. This ratio can be significantly lower in certain outbreaks. These large ratios make prospective studies difficult because of the large at-risk population. Since many of the factors responsible for the variation in clinical manifestation are not known, it is not possible to refine the definition of the at-risk population and increase the chance of finding an association. Certain populations, such as neonates, seem to develop serious disease at a higher rate, and other factors such as physical stress have been suggested to be associated with more severe disease

(Modlin 1990; Morens et al. 1991). For chronic disease in animal models, the genetic makeup of the host is clearly important, as described elsewhere in this volume. The association of certain genetic markers with risk for IDDM may be parallel to this genetically defined susceptibility in humans. When host factors are identified which increase risk for a particular disease, a prospective study could be designed to achieve lower numbers in the at-risk population per case.

A second host consideration, which is important for serologic studies, is the variability in the immune response. Heterotypic immune response, as measured by neutralization or IgM ELISA, is widely recognized (Muir and Banatvala 1990) and seems to be age-dependent. It is not clear what factors other than age lead to variation in this parameter. If, for example, prior number of enterovirus infections is correlated with heterotypic response, then control groups in serologic studies should be matched by family size, socioeconomic status, and other factors correlated with increased rates of enterovirus infection.

Recent reports have highlighted the possibility that the pathogenic potential of a virus may be modified by additional co-factors which may act on the host. In particular, selenium deficiency resulted in increased virulence of a myocarditic virus strain of CVB3 (Beck et al. 1994b). Vitamin E deficiency also increased the pathology observed with the myocarditic strain, but also resulted in significant pathology from infection with an amyocarditic strain of CVB3 (Beck et al. 1994a). These observations stress the importance of considering additional factors affecting pathogenesis beyond simple genetic risk factors of the host and are described more completely elsewhere in this volume.

Included in some concepts of pathogenic mechanisms for several diseases that have observed and proposed associations with CVB infection are elements of viral persistence. To date, several examples have been reported of equilibrium between a low rate of viral replication matched to dividing cells in culture, resulting in persistently infected cell cultures (Argo et al. 1992; Tonew et al. 1995). Although it is very clear that virus genome can be detected in myocytes and other nondividing cell types in animals, weeks and maybe longer after infectious virus is no longer detectable (Tam et al. 1991; Rabausch-Starz et al. 1994), the nature and duration of this persistence is not clear. Another study which followed animals with chronic myositis further failed to detect viral genome after more than 4 weeks, despite continued pathology for at least 16 weeks (Zoll et al. 1993). Although there is still much to learn about viral replication, and even more to learn about the survival of viral RNA in double-stranded forms in nonreplicating cells in vivo, it seems unlikely that nonreplicating RNA is stable indefinitely in cells. Slow replication, however, is a distinct possibility and may be functionally equivalent in terms of pathogenic mechanisms.

A key area for epidemiologic study of CVB infections is their role in diseases such as diabetes mellitus. Studies of this topic, if they are to be successful, must take into account the characteristics of the viruses and their infections as noted above. For example, if an IgM serologic study had been conducted in Bulgaria in 1975 to determine if poliovirus was associated with paralysis in a group of nearly

140 patients with recent onset of flaccid paralysis, the results would have shown no significant differences between patients and controls for all three serotypes of poliovirus. However, Bulgaria in that year had an outbreak of enterovirus 71 that proved to be the etiologic agent of acute flaccid paralysis for that outbreak. The erroneous implied conclusion of the study would have been that poliovirus is not a major cause of acute flaccid paralysis. Various studies may demonstrate an association between virus infection and disease, but there can be numerous reasons why a study can fail to show an association, even when one exists.

Therefore, in serologic studies controls must be matched for factors related to prevalence of enterovirus antibodies such as age, sex, serum collection timing, and socioeconomic status. These studies probably should consider any enterovirus as a possible etiologic agent and should include the circulating enteroviruses in the subset to be examined.

By careful consideration of the unique features of enteroviruses and their infections and how they affect study design and data interpretation, as well as use of improved laboratory techniques, it should be possible in the near future to better understand correlations between CVB or enterovirus infection and several chronic diseases.

4.3 Prospects for Intervention

Even with considerable progress in understanding human disease resulting from CVB infection, it is necessary that the ultimate goal of such studies is to achieve control and/or prevention of disease. If some fraction of the diseases mentioned above are attributable to infection with CVB or other enteroviruses, then control may be achieved by preventing these infections or by successful treatment. The prospects for progress in this area are improving rapidly through several approaches.

4.3.1 Treatment

Although no antiviral therapy is presently available for enterovirus infections and current treatment is directed toward alleviating symptoms, candidate drugs have been identified that exhibit antiviral activity against several enteroviruses in tissue culture and experimental animals (Rozhon et al. 1993). As evaluation of some of these drugs progress, they have been tested in the treatment of disease in animal model systems of both myocarditis and diabetes (Ilbäck et al. 1993; Pauksen et al. 1993; See and Tilles 1993). Although efficacy was demonstrable in these studies, some parameters affecting efficacy, including post-infection timing, have not been systematically determined.

In patients with agammaglobulinemia, chronic enterovirus infections have been treated with γ-globulin, and in some cases this has controlled the infection. Use of γ-globulin in other clinical illness has not been systematically evaluated. In murine animal models, however, the use of immunoglobulin has been successfully used to suppress myocarditis and protect against cardiac damage (WELLER et al. 1992; TAKADA et al. 1995). The efficacy of this treatment needs to be evaluated in human clinical trials.

4.3.2 Prevention

There are no vaccines available for nonpolio enteroviruses. Many investigators, however, have demonstrated the potential usefulness of vaccination in the prevention of acute infection or myocarditis in animal models (BECK et al. 1990; FOHLMAN et al. 1990, 1993; SEE and TILLES 1994). Caution may be necessary, because it has also been shown that some monoclonal antibodies to CVB potentiate pathogenesis and cardiac damage in similar animal models (GAUNTT et al. 1995).

References

Alexander JP Jr, Chapman LE, Pallansch MA, Stephenson WT, Török TJ, Anderson LJ (1993) Coxsackievirus B2 infection and aseptic meningitis: a focal outbreak among members of a high school football team. J Infect Dis 167:1201–1205

Argo E, Gimenez B, Cash P (1992) Non-cytopathic infection of rhabdomyosarcoma cells by coxsackie B5 virus. Arch Virol 126:215–229

Axelsson C, Bondestam K, Frisk G, Bergström S, Diderholm H (1993) Coxsackie B virus infections in women with miscarriage. J Med Virol 39:282–285

Barrett-Conner E (1985) Is insulin-dependent diabetes mellitus caused by coxsackievirus B infection? A review of the epidemiologic evidence. Rev Infect Dis 7:207–215

Bartfeld H, Dham C, Donnenfeld H, Ollar RA, deMasi MT, Kascsak R (1989) Enteroviral-related antigen in circulating immune complexes of amyotrophic lateral sclerosis patients. Intervirology 30:202–212

Beck MA, Chapman NM, McManus BM, Mullican JC, Tracy S (1990) Secondary enterovirus infection in the murine model of myocarditis. Pathologic and immunologic aspects. Am J Pathol 136:669–681

Beck MA, Kolbeck PC, Rohr LH, Shi Q, Morris VC, Levander OA (1994a) Vitamin E deficiency intensifies the myocardial injury of coxsackievirus B3 infection of mice. J Nutr 124:345–358

Beck MA, Kolbeck PC, Shi Q, Rohr LH, Morris VC, Levander OA (1994b) Increased virulence of a human enterovirus (coxsackievirus B3) in selenium-deficient mice. J Infect Dis 170:351–357

Bell EJ, McCartney RA (1984) A study of coxsackie B virus infections, 1972–1983. J Hyg (Lond) 93:197–204

Berlin LE, Rorabaugh ML, Heldrich F, Roberts K, Doran T, Modlin JF (1993) Aseptic meningitis in infants < 2 years of age: diagnosis and etiology. J Infect Dis 168:888–892

Boman J, Nilsson B, Juto P (1992) Serum IgA, IgG, and IgM responses to different enteroviruses as measured by a coxsackie B5-based indirect ELISA. J Med Virol 38:32–35

Bowles NE, Rose ML, Taylor P, Banner NR, Morgan-Capner P, Cunningham L, Archard LC, Yacoub MH (1989) End-stage dilated cardiomyopathy. Persistence of enterovirus RNA in myocardium at cardiac transplantation and lack of immune response. Circulation 80:1128–1136

Cochrane HR, May FE, Ashcroft T, Dark JH (1991) Enteroviruses and idiopathic dilated cardiomyopathy. J Pathol 163:129–131

Cunningham L, Bowles NE, Lane RJ, Dubowitz V, Archard LC (1990) Persistence of enteroviral RNA in chronic fatigue syndrome is associated with the abnormal production of equal amounts of positive and negative strands of enteroviral RNA. J Gen Virol 71:1399–1402

D'Alessio DJ (1992) A case-control study of group B Coxsackievirus immunoglobulin M antibody prevalence and HLA-DR antigens in newly diagnosed cases of insulin-dependent diabetes mellitus. Am J Epidemiol 135:1331–1338

Dahlquist G, Frisk G, Ivarsson SA, Svanberg L, Forsgren M, Diderholm H (1995) Indications that maternal coxsackie B virus infection during pregnancy is a risk factor for childhood-onset IDDM. Diabetologia 38:1371–1373

Day C, Cumming H, Walker J (1989) Enterovirus-specific IgM in the diagnosis of meningitis. J Infect 19:219–228

Drebot MA, Nguan CY, Campbell JJ, Lee SH, Forward KR (1994) Molecular epidemiology of enterovirus outbreaks in Canada during 1991-1992: identification of echovirus 30 and coxsackievirus B1 strains by amplicon sequencing. J Med Virol 44:340–347

Druyts-Voets E, Van Renterghem L, Gerniers S (1993) Coxsackie B virus epidemiology and neonatal infection in Belgium. J Infect 27:311–316

Fohlman J, Friman G (1993) Is juvenile diabetes a viral disease? Ann Med 25:569–574

Fohlman J, Ilbäck NG, Friman G, Morein B (1990) Vaccination of Balb/c mice against enteroviral mediated myocarditis. Vaccine 8:381–384

Fohlman J, Pauksen K, Morein B, Bjare U, Ilbäck NG, Friman G (1993) High yield production of an inactivated coxsackie B3 adjuvant vaccine with protective effect against experimental myocarditis. Scand J Infect Dis Suppl 88:103–108

Frisk G, Diderholm H (1992) Increased frequency of coxsackie B virus IgM in women with spontaneous abortion. J Infect 24:141–145

Frisk G, Friman G, Tuvemo T, Fohlman J, Diderholm H (1992) Coxsackie B virus IgM in children at onset of type 1 (insulin-dependent) diabetes mellitus: evidence for IgM induction by a recent or current infection. Diabetologia 35:249–253

Frisk G, Nilsson E, Ehrnst A, Diderholm H (1989) Enterovirus IgM detection: specificity of mu-antibody-capture radioimmunoassays using virions and procapsids of Coxsackie B virus. J Virol Methods 24:191–202

Gauntt CJ, Arizpe HM, Higdon AL, Wood HJ, Bowers DF, Rozek MM, Crawley R (1995) Molecular mimicry, anti-coxsackievirus B3 neutralizing monoclonal antibodies, and myocarditis. J Immunol 154:2983–2995

Goldwater PN (1995) Immunoglobulin M capture immunoassay in investigation of coxsackievirus B5 and B6 outbreaks in South Australia. J Clin Microbiol 33:1628–1631

Gómez RM, Cui X, Castagnino CG, Berría MI (1993) Differential behaviour in pancreas and heart of two coxsackievirus B3 variants. Intervirology 36:153–160

Gow J, Cash P, Behan W, McGarry F, Simpson K, Behan P (1995) Detection of picornavirus genomic and template RNA strands by a novel semi-nested polymerase chain reaction-technique and agarose-gel electrophoresis. Electrophoresis 16:338–340

Grandien M, Forsgren M, Ehrnst A (1989) Enteroviruses and reoviruses. In: Schmidt NJ, Emmons RW (eds), Diagnostic procedures for viral, rickettsial and chlamydial infections. American Public Health Association, Washington, DC, pp 513–569

Haarmann CM, Schwimmbeck PL, Mertens T, Schultheiss HP, Strauer BE (1994) Identification of serotype-specific and nonserotype-specific B-cell epitopes of coxsackie B virus using synthetic peptides. Virology 200:381–389

Haddad J, Gut JP, Wendling MJ, Astruc D, Jernite M, Obert G, Messer J (1993) Enterovirus infections in neonates. A retrospective study of 21 cases. Eur J Med 2:209–214

Helfand RF, Gary HE Jr, Freeman CY, Anderson LJ, Pallansch MA (1995) Serologic evidence of an association between enteroviruses and the onset of type 1 diabetes mellitus. Pittsburgh Diabetes Research Group. J Infect Dis 172:1206–1211

Hilton DA, Fletcher A, Pringle JH (1994) Absence of coxsackie viruses in idiopathic inflammatory muscle disease by in situ hybridization. Neuropathol Appl Neurobiol 20:238–242

Hodgson J, Bendig J, Keeling P, Booth JC (1995) Comparison of two immunoassay procedures for detecting enterovirus IgM. J Med Virol 47:29–34

Hohenadl C, Klingel K, Mertsching J, Hofschneider PH, Kandolf R (1991) Strand-specific detection of enteroviral RNA in myocardial tissue by in situ hybridization. Mol Cell Probes 5:11–20

Hu H, Dong D, Tang E, Zhang M, Cao Y (1993) Antigenic analysis of coxsackievirus B3 with monoclonal antibodies. Jpn J Med Sci Biol 46:87–93

Hyöty H, Hiltunen M, Knip M, Laakkonen M, Vähäsalo P, Karjalainen J, Koskela P, Roivainen M, Leinikki P, Hovi T, Åkerblom HK, Childhood Diabetes in Finland (DiMe) Study Group (1995) A prospective study of the role of coxsackie B and other enterovirus infections in the pathogenesis of IDDM. Childhood Diabetes in Finland (DiMe) Study Group. Diabetes 44:652–657
Ikeda RM, Kondracki SF, Drabkin PD, Birkhead GS, Morse DL (1993) Pleurodynia among football players at a high school. An outbreak associated with coxsackievirus B1. JAMA 270:2205–2206
Ilbäck NG, Wesslén L, Pauksen K, Stålhandske T, Friman G, Fohlman J (1993) Effects of the antiviral WIN 54954 and the immune modulator LS 2616 on cachectin/TNF and gamma-interferon responses during viral heart disease. Scand J Infect Dis Suppl 88:117–123
Isacsohn M, Eidelman AI, Kaplan M, Goren A, Rudensky B, Handsher R, Barak Y (1994) Neonatal coxsackievirus group B infections: experience of a single department of neonatology. Isr J Med Sci 30:371–374
Kamei S, Hersch SM, Kurata T, Takei Y (1990) Coxsackie B antigen in the central nervous system of a patient with fatal acute encephalitis: immunohistochemical studies of formalin-fixed paraffin-embedded tissue. Acta Neuropathol (Berl) 80:216–221
Kaplan MH, Klein SW, McPhee J, Harper RG (1983) Group B coxsackievirus infections in infants younger than three months of age: a serious childhood illness. Rev Infect Dis 5:1019–1032
Keeling PJ, Lukaszyk A, Poloniecki J, Caforio AL, Davies MJ, Booth JC, McKenna WJ (1994) A prospective case-control study of antibodies to coxsackie B virus in idiopathic dilated cardiomyopathy. J Am Coll Cardiol 23:593–598
Kew O, De L, Yang CF, Nottay B, da Silva E, Pallansch M (1993) The role of virological surveillance in the global initiative to eradicate poliomyelitis. In: Kurstak E (ed), Control of virus diseases. Marcel Dekker, New York, pp 215–246
Kew OM, Mulders MN, Lipskaya GY, da Silva EE, Pallansch MA (1995) Molecular epidemiology of polioviruses. In: Kew O, Nathanson N (eds), Molecular epidemiology. Seminars in Virology, vol 6. Academic, New York, pp 401–414
Kopecka H, Brown B, Pallansch M (1995) Genotypic variation in coxsackievirus B5 isolates from three different outbreaks in the United States. Virus Res 38:125–136
Landau BJ, Whittier PS, Finkelstein SD, Alstein B, Grun JB, Schultz M, Crowell RL (1990) Induction of heterotypic virus resistance in adult inbred mice immunized with a variant of Coxsackievirus B3. Microb Pathog 8:289–298
Lerner AM, Wilson FM (1973) Virus myocardiopathy. Prog Med Virol 15:63–91
Melnick JL (1996) Enteroviruses: polioviruses, coxsackieviruses, echoviruses, and newer enteroviruses. In: Fields BN, Knipe DM, Howley PM, Chanock RM, Melnick JL, Monath TP, Roizman B, Straus SE (eds), Virology. Lippincott-Raven, Philadelphia, pp 655–712
Miller NA, Carmichael HA, Calder BD, Behan PO, Bell EJ, McCartney RA, Hall FC (1991) Antibody to Coxsackie B virus in diagnosing postviral fatigue syndrome. BMJ 302(6769):140–143
Modlin JF (1990) Coxsackieviruses, echoviruses, and newer enteroviruses. In: Mandell GL, Douglas RG Jr, Bennett JE (eds), Principles and practices of infectious diseases. Churchill Livingstone, New York, pp 1367–1383
Moreira RC, Castrignano SB, Carmona R de C. Gomes FM, Saes SG, Oliveira RS, Souza DF, Takimoto S, Costa MC, Waldman EA (1995) An exanthematic disease epidemic associated with coxsackievirus B3 infection in a day care center. Rev Inst Med Trop São Paulo 37:235–238
Moore M (1982) Enteroviral disease in the United States, 1970–1979. J Infect Dis 146:103–108
Morens DM, Zweighaft RM, Bryan JM (1979) Non-polio enterovirus disease in the United States, 1971–1975. Int J Epidemiol 8:49–54
Morens DM, Pallansch MA, Moore M (1991) Polioviruses and other enteroviruses. In: Belshe RB (ed), Textbook of Human Virology. Mosby Year Book, St Louis, Missouri, pp 427–497
Muir P, Banatvala JE (1990) Reactivity of enterovirus-specific IgM with infective and defective coxsackie B virions in patients with monotypic and multitypic IgM responses. J Virol Methods 29:209–224
Muir P, Nicholson F, Tilzey AJ, Signy M, English TA, Banatvala JE (1989) Chronic relapsing pericarditis and dilated cardiomyopathy: serological evidence of persistent enterovirus infection. Lancet 1(8642):804–807
Nairn C, Galbraith DN, Clements GB (1995) Comparison of coxsackie B neutralisation and enteroviral PCR in chronic fatigue patients. J Med Virol 46:310–313

Nakayama T, Urano T, Osano M, Hayashi Y, Sekine S, Ando T, Makinom S (1989) Outbreak of herpangina associated with coxsackievirus B3 infection. Pediatr Infect Dis J 8:495–498

Nishikai M (1994) Coxsackievirus infection and the development of polymyositis/dermatomyositis. Rheumatol Int 14:43–46

Pauksen K, Ilbäck NG, Friman G, Fohlman J (1993) Therapy of coxsackie virus B3-induced myocarditis with WIN 54954 in different formulations. Scand J Infect Dis Suppl 88:125–130

Prabhakar BS, Haspel MV, McClintock PR, Notkins AL (1982) High frequency of antigenic variants among naturally occurring human coxsackie B4 virus isolates identified by monoclonal antibodies. Nature 300:374–376

Prabhakar BS, Menegus MA, Notkins AL (1985) Detection of conserved and nonconserved epitopes on Coxsackievirus B4: frequency of antigenic change. Virology 146:302–306

Rabausch-Starz I, Schwaiger A, Grünewald K, Müller-Hermelink HK, Neu N (1994) Persistence of virus and viral genome in myocardium after coxsackievirus B3-induced murine myocarditis. Clin Exp Immunol 96:69–74

Reyes MP, Zalenski D, Smith F, Wilson FM, Lerner AM (1986) Coxsackievirus-positive cervices in women with febrile illnesses during the third trimester in pregnancy. Am J Obstet Gynecol 155:159–161

Rozhon E, Cox S, Buontempo P, O'Connell J, Slater W, De Martino J, Schwartz J, Miller G, Arnold E, Zhang A., Morrow C, Jablonski S, Pinto P, Versace R, Duelfer T, Girijavallabhan V (1993) SCH 38057: a picornavirus capsid-binding molecule with antiviral activity after the initial stage of viral uncoating. Antiviral Res 21:15–35

Samuelson A, Forsgren M, Johansson B, Wahren B, Sällberg M (1994) Molecular basis for serological cross-reactivity between enteroviruses. Clin Diagn Lab Immunol 1:336–341

Satoh M, Tamura G, Segawa I, Hiramori K, Satodate R (1994) Enteroviral RNA in dilated cardiomyopathy. Eur Heart J 15:934–939

Scherbaum WA, Hampl W, Muir P, Glück M, Seißler J, Egle H, Hauner H, Boehm BO, Heinze E, Banatvala JE, Pfeiffer EF (1991) No association between islet cell antibodies and Coxsackie B, mumps, rubella and cytomegalovirus antibodies in non-diabetic individuals aged 7-19 years. Diabetologia 34:835–838

Schwaiger A, Umlauft F, Weyrer K, Larcher C, Lyons J, Muhlberger V, Dietze O, Grünewald K (1993) Detection of enteroviral ribonucleic acid in myocardial biopsies from patients with idiopathic dilated cardiomyopathy by polymerase chain reaction. Am Heart J 126:406–410

See DM, Tilles JG (1993) WIN 54954 treatment of mice infected with a diabetogenic strain of group B coxsackievirus. Antimicrob Agents Chemother 37:1593–1598

See DM, Tilles JG (1994) Efficacy of a polyvalent inactivated-virus vaccine in protecting mice from infection with clinical strains of group B coxsackieviruses. Scand J Infect Dis 26:739–747

Severini GM, Mestroni L, Falaschi A, Camerini F, Giacca M (1993) Nested polymerase chain reaction for high-sensitivity detection of enteroviral RNA in biological samples. J Clin Microbiol 31:1345–1349

Shimizu H, Schnurr DP, Burns JC (1994) Comparison of methods to detect enteroviral genome in frozen and fixed myocardium by polymerase chain reaction. Lab Invest 71:612–616

Strikas RA, Anderson LJ, Parker RA (1986) Temporal and geographic patterns of isolates of nonpolio enterovirus in the United States, 1970 –1983. J Infect Dis 153:346–351

Strong BS, Young SA (1995) Intrauterine coxsackie virus, group B type 1, infection: viral cultivation from amniotic fluid in the third trimester. Am J Perinatol 12:78–79

Swanink CM, Veenstra L, Poort YA, Kaan JA, Galama JM (1993) Coxsackievirus B1-based antibody-capture enzyme-linked immunosorbent assay for detection of immunoglobulin G (IgG), IgM, and IgA with broad specificity for enteroviruses. J Clin Microbiol 31:3240–3246

Szopa TM, Ward T, Dronfield DM, Portwood ND, Taylor KW (1990) Coxsackie B4 viruses with the potential to damage beta cells of the islets are present in clinical isolates. Diabetologia 33:325–328

Takada H, Kishimoto C, Hiraoka Y (1995) Therapy with immunoglobulin suppresses myocarditis in a murine coxsackievirus B3 model. Antiviral and anti-inflammatory effects. Circulation 92:1604–1611

Tam PE, Schmidt AM, Ytterberg SR, Messner RP (1991) Viral persistence during the developmental phase of Coxsackievirus B1-induced murine polymyositis. J Virol 65:6654–6660

Tonew M, Hartmann M, Schmidtke M, Stelzner A (1995) Replication and persistence of coxsackieviruses B3 in human fibroblasts. Int J Med Microbiol Virol Parasitol Infect Dis 282:92–101

Tracy S, Chapman N, Romero J, Ramsingh A (1996) Coxsackievirus B3 cardiovirulence determinants and their role in inflammatory heart muscle disease. Trends in Microbiology 4:175–178

Tracy S, Hufnagel G, Chapman N (1992) Interesting problems in enteroviral inflammatory heart disease. Herz 17:79–84

Tracy S, Chapman NM, McManus BM, Pallansch MA, Beck MA, Carstens J (1990) A molecular and serologic evaluation of enteroviral involvement in human myocarditis. J Mol Cell Cardiol 22:403–414

Tu Z, Chapman NM, Hufnagel G, Tracy S, Romero JR, Barry WH, Zhao L, Currey K, Shapiro B (1995) The cardiovirulent phenotype of coxsackievirus B3 is determined at a single site in the genomic 5' nontranslated region. J Virol 69:4607–4618

Tuvemo T, Dahlquist G, Frisk G, Blom L, Friman G, Landin-Olsson M, Diderholm H (1989) The Swedish childhood diabetes study III: IgM against coxsackie B viruses in newly diagnosed type 1 (insulin-dependent) diabetic children – no evidence of increased antibody frequency. Diabetologia 32:745–747

Wagenknecht LE, Roseman JM, Herman WH (1991) Increased incidence of insulin-dependent diabetes mellitus following an epidemic of Coxsackievirus B5. Am J Epidemiol 133:1024–1031

Webb SR, Kearse KP, Foulke CL, Hartig PC, Prabhakar BS (1986) Neutralization epitope diversity of Coxsackievirus B4 isolates detected by monoclonal antibodies. J Med Virol 20:9–15

Weiss LM, Movahed LA, Billingham ME, Cleary ML (1991) Detection of Coxsackievirus B3 RNA in myocardial tissues by the polymerase chain reaction. Am J Pathol 138:497–503

Weller AH, Hall M, Huber SA (1992) Polyclonal immunoglobulin therapy protects against cardiac damage in experimental coxsackievirus-induced myocarditis. Eur Heart J 13:115–119

Wesslén L, Waldenström A, Lindblom B, Høyer S, Friman G, Fohlman J (1993) Genotypic and serotypic profile in dilated cardiomyopathy. Scand J Infect Dis Suppl 88:87–91

Woodall CJ, Riding MH, Graham DI, Clements GB (1994) Sequences specific for enterovirus detected in spinal cord from patients with motor neurone disease. BMJ 308:1541–1543

Yagi S, Schnurr D, Lin J (1992) Spectrum of monoclonal antibodies to coxsackievirus B-3 includes type- and group-specific antibodies. J Clin Microbiol 30:2498–2501

Yui LA, Gledhill RF (1991) Limb paralysis as a manifestation of Coxsackie B virus infection. Dev Med Child Neurol 33:427–438

Zoll J, Galama J, Melchers W (1994) Intratypic genome variability of the coxsackievirus B1 2A protease region. J Gen Virol 75:687–692

Zoll J, Jongen P, Galama J, van Kuppeveld F, Melchers W (1993) Coxsackievirus B1-induced murine myositis: no evidence for viral persistence. J Gen Virol 74:2071–2076

Coxsackie B Viruses and Human Heart Disease

C. BABOONIAN[1], M. J. DAVIES[1], J. C. BOOTH[2], AND W. J. MCKENNA[1]

1 Coxsackie B Viruses and Human Heart Disease

A large number of viruses can cause infections of the heart. Among these are the enteroviruses, adenoviruses, retroviruses and even orthomyxoviruses (CLEMENTS 1993). Human viruses causing systemic infections have a viremic stage during which the heart may be exposed to infectious agents. Thus the presence of virus in the myocytes or the infiltrating immune cells is likely to be a more common event than suspected. However, clinically overt cases of heart disease following common viral infections are rare and may represent a small percentage of all viral heart

[1]Department of Cardiological Sciences, St George's Hospital Medical School, Cranmer Terrace, Tooting, London SW17 0RE, UK

[2]Department of Medical Microbiology, St George's Hospital Medical School, Cranmer Terrace, Tooting, London SW17 0RE, UK

infections. Of the numerous agents commonly infecting humans, enteroviruses and in particular coxsackie B viruses (CVB) have been the most closely scrutinised agents implicated in human heart disease. The aim of this review is to discuss the evidence for an etiologic link between enteroviruses and in particular CVB and human heart disease.

2 General Characteristics of Coxsackie B Viruses

Coxsackie B viruses belong to the enterovirus group and are members of the Picornaviridae. There are six serotypes of these small nonenveloped, single-stranded RNA viruses (CBV1-6). The virus particle has an icosahedral symmetry and the nucleocapsid consists of four capsid proteins (VP1-4). The genome is positive-stranded RNA of 7000–7500 bases which is prone to mutations resulting in antigenic variations in the virus proteins (Rueckert 1996). Replication of the virus takes place in the cytoplasm of infected cells generating a negative sense RNA which then acts as template for synthesis of viral genomic RNA. The viral genome is polyadenylated at the 3' end and acts as messenger RNA for synthesis of a large polyprotein that is autocatalysed to generate the structural and nonstructural polypeptides of the virion. At either end of the open reading frame there are nontranslated regions of approximately 600 and 150 bases (5' and 3' ends respectively) which are highly conserved among different enteroviruses and are often used for probing in situ hybridization assays for detection of all enteroviruses. Regions of the viral RNA coding for structural polypeptides such as VP1 show less homology between serotypes.

Coxsackie viruses are ubiquitous agents and can spread rapidly within the community causing small epidemics. Although a high proportion of infections with these viruses are subclinical, mild presentation may range from mild, undifferentiated febrile illness or upper respiratory tract infection to a severe, systematic and sometimes fatal disease of neonates. The site of entry of enteroviruses is believed to be the alimentary tract, but in communities where higher standards of hygiene and sanitation result in a reduction in faecal-oral spread of these agents, aerosol transmission acquires significance. The incubation period is approximately one to 2 weeks but longer periods of incubation of 35 days have been noted. Virus multiplication at the portal of entry may be followed by viremia and further proliferation of the virus in the reticuloendothelial system resulting in the dissemination of virus and involvement of target organs (Melnick 1996).

3 Laboratory Diagnosis

3.1 Virus Culture and Antigen Detection

Coxsackie B viruses can grow in primary and passaged cultures of monkey kidney cells and human cell cultures such as WI 38, human embryo lung fibroblasts and human foetal diploid kidney cells. Infected cells round up and become refractile and show a characteristic cytopathic effect. Identification of the viral serotype can be carried out by neutralisation assays using a panel of antisera (MELNICK and WIMBERLY 1985).

Coxsackie B viruses are present in the pharyngeal washings of infected individuals for 1–2 weeks after infection and for a prolonged period of several weeks in the faeces. As these viruses are common infectious agents, even during nonepidemic periods, a relatively large number of healthy individuals particularly children may be excreting CVBs (PARKS et al. 1967; JENISTA et al. 1984). The presence of virus in the pharynx and alimentary tract is therefore not of diagnostic significance in relation to heart disease, and for an evaluation of the role of these viruses in myocardial disease, detection of virus in the affected target organ is important. Although historically isolation of CVBs from heart tissue was the first indication that this group of pathogens may be involved in myocarditis (JAVETT et al. 1956), in practice isolation of infectious virus from the myocardium is rarely successful. One of the reasons is that symptoms of cardiac disease may not appear immediately following infection even if disease is a direct consequence of the cytolytic action of the virus. The period of time that CVBs are resident in the heart is not known and virus may be eliminated from the tissues by the time a biopsy sample is taken for examination.

Detection of viral antigens using immunofluorescence staining of histological samples has proved valuable for quick diagnosis of many infectious diseases. However this method has not been frequently used either for diagnostic or epidemiological studies of CVB infections of the myocardium. The sensitivity of antigen detection techniques does not compare well with methods such as polymerase chain reaction (PCR) for nucleic acid amplification. The polyclonal sera available for serotyping of CVBs in cell culture may not be suitable for use on heart tissue as highly specific reagents with no binding to tissue sections are needed for such assays. Such reagents are becoming available, but it is important to remember that antigen detection strategies will also encounter problems similar to virus isolation in that a negative finding may not disprove CVBs as etiologic agents of heart disease. Perhaps one major advantage of detecting viral proteins as opposed to live virus is that defective particles, which would not replicate in cell culture and which express only some CVB antigens, may be seen by immunofluorescence.

3.2 Serological Diagnosis

The humoral response to CVBs is an important defence mechanism against disseminated infection. As CVBs circulate in the community, by adulthood a majority of people would have experienced infection with some of these viruses and consequently have developed neutralising antibodies. Traditionally serodiagnosis of a recent CVB infection was carried out by testing acute and convalescent sera in a virus neutralisation assay. A fourfold rise in antibody titre was taken to be indicative of a recent infection. The interpretation of data obtained from these assays is particularly difficult when patients suffering from myocardial disease are being investigated. By the time symptoms of cardiac malfunction appear, antibody levels to CVBs may have already reached their highest levels. Persistence of high levels of antibody from previous infections and occurrence of a heterotypic response following infection with other enteroviruses can also confuse diagnosis of CVB related heart disease.

Recent infection can be more easily identified using a virus specific IgM assay. The presence of IgM class antibodies generally indicates infection with CVBs in the past 1–3 months. However some of the difficulties associated with the neutralisation assay also apply to the detection of IgM. Presence of IgM class antibodies does not prove an etiologic relationship, as antibodies may persist for long periods in some individuals and heterotypic response particularly in older individuals can be problematic (Minor et al. 1990). By contrast, the absence of IgM antibodies may mean that the patient presented to the clinic months or even years after infection and after the decline of virus specific IgM antibodies to undetectable levels. There are other technical difficulties. To detect CVB specific antibodies, gradient purified virus or synthetic peptides must be used. Cell lysates used as antigen will detect not only CVB specific antibodies but also enterovirus group specific immunoglobulins. Even with pure virus used in antibody capture assays, an anamnestic response can interfere with evaluation of the data. Apart from technical problems in interpreting results, population studies in which patient and control groups are compared also need to take into account the transmission of these viruses in families and communities in close contact and therefore occurrence of localized small epidemics. Control groups not matched to the patient population can falsely implicate CVBs in cardiac disease.

3.3 Detection of Viral RNA

There are several molecular methods currently available for detection of coxsackie viral RNA in myocardial tissue. These include dot or slot blot and northern blot hybridization, in situ hybridization, PCR and, more recently, in situ PCR. Hybridization methods rely on the fact that viral RNA can form double stranded molecules with complementary probes which were initially developed following cloning of the RNA genome of CVB3 (Tracy 1984, 1985; Kandolf and Hofschneider

1985; TRACY et al. 1985a,b; ROBART et al. 1985). The cDNA copy of the viral genome with sequence homology to a broad spectrum of enteroviruses has been used as a probe for detecting virus specific sequences in myocardial tissue. A large number of early studies suggesting an association between enteroviruses and heart disease have used the dot or slot blot hybridization techniques (BOWLES et al. 1986). RNA is usually extracted from biopsy samples and applied to the hybridization membrane followed by the probing of samples using radiolabelled probes. One of the problems associated with this technique is that binding to uninfected cells is sometimes observed making the interpretation of the results difficult. A more time consuming option is using northern blotting followed by hybridization (OKADA et al. 1990). This method will by nature produce less nonspecific binding as the samples have to be run on a gel prior to blotting and hybridization. This makes it unlikely that a host nucleic acid corresponding to the precise molecular weight of the viral RNA will bind to the probe.

In situ hybridization is another system frequently used. It is a technically demanding method which relies on an adequate number of RNA molecules being present inside each cell, to be detected by a labelled probe. Although if appropriate hybridization conditions are used as few as 20 genome copies of CVB3 RNA can be detected in laboratory infected cultured cells, a general comparison of data published using slot or dot blot and in situ hybridization techniques suggests that the number of positive samples obtained in the former group is greater (BOWLES et al. 1986) than those seen in the latter (FIGULLA et al. 1995). It is possible that either the higher specificity of the in situ hybridization is reducing the number of false positive results or by gaining the advantage of observing cellular localization of viral RNA, some sensitivity is lost. It is worth noting that most studies carried out to date have used probes covering the whole viral genome (FIGULLA et al. 1995) or probes to conserved regions of enteroviral RNA (BOWLES et al. 1989; TRACY 1990). Such probes are not serotype specific and would hybridize to most enteroviruses, making it difficult to analyze the specific role of CVBs in cardiac disease.

A simple yet specific and highly sensitive detection system that is now one of the commonest means of detection of viral genomic material is PCR. Enteroviral RNA isolated from tissue sections can be reverse transcribed and amplified using this technique. Detection is carried out by gel electrophoresis and specificity is confirmed by either hybridization using a probe which anneals to the amplified material or a nested PCR with primers that anneal to the first PCR product. In most studies the primers have been selected to anneal to the generally well-conserved 5' nontranslated portion of the enteroviral genome. Several groups have now investigated the presence of CVB sequences in heart tissue (Tables 2, 4). Gene amplification by PCR can be used directly on permeabilized tissue sections to demonstrate the presence of viral sequences in cells. The in situ PCR technique combines the advantages of the in situ hybridization method with the sensitivity of PCR. As the technique is relatively new no data is available yet on the examination of heart tissue using this method.

4 The Association of Coxsackie B Viruses with Myocardial Disease

4.1 Clinico-Pathological Correlations in Myocarditis

Acute nonspecific myocarditis is characterised by myocyte death associated with a heavy chronic inflammatory cell infiltrate predominantly T lymphocytes. The clinical picture includes a febrile illness associated with the sudden onset of arrhythmias and/or heart failure. Severe cases may develop A-V block and a small proportion of cases have concomitant acute pericarditis. A clinical diagnosis of acute myocarditis is made far more frequently than positive cardiac biopsy confirmation achieves. The reported rates of detection of positive biopsies vary widely but are seldom more than 10%. In order to standardise histological reporting a group of pathologists met in Dallas and laid down the firm morphological criteria, for use on conventionally stained histological material, that both myocyte necrosis and an interstitial T cell infiltrate had to be present. Myocardial fibrosis must be absent because it implies a long standing disease process (ARETZ et al. 1987). Many explanations exist for why the positive rates are so low. A significant proportion of patients who present clinically with apparent sudden onset of heart failure are shown on biopsy to have fibrosis and are therefore examples of a dilated cardiomyopathy (see below). There is also an argument that the Dallas histological criteria are too stringent. One group advocates broadening the morphological criteria for acute myocarditis based on immunohistochemistry. If more than two T lymphocytes per high power field (x40) are present in association with intercellular adhesion molecules and MHC class II expression on endothelial and interstitial cells this is called acute myocarditis (KUHL et al. 1996). This is a very different definition from that used by the Dallas group. This review is concerned with acute myocarditis as defined by the Dallas criteria.

A number of infectious agents can cause an acute myocarditis picture including viruses and protozoa such as *Toxoplasma gondii* but the great majority of human cases do not have an easily identifiable cause. The similarity of the human disease to known viral infections such as coxsackie infection in mice or parvovirus infection in dogs has been a major reason for believing viruses are responsible for the majority of human acute nonspecific myocarditis.

4.2 Coxsackie B Virus Heart Disease in Children

Viral myocarditis in children is a significant cause of morbidity and mortality. As young children are immunologically naive and exposed to a wide range of infectious agents, they are prime targets for enteroviral infections. Clinical symptoms associated with CVB in children may be localized to the heart or more commonly involve multiple organs. Initial report of such cases was by JAVETT et al. (1956)

describing an outbreak of generalised disease of the newborn in a Johannesburg maternity home in which ten newborn babies became ill. The babies were all febrile and developed signs of acute myocarditis accompanied, in some cases, with meningoencephalitis. Post-mortem examination carried out on three of the six infants that died revealed the cause of death to be acute myocarditis. Although virus isolation from the myocardium was not attempted, CVB was demonstrated in peripheral sites in four of the ten cases. One of the four isolates was serotyped and found to be CVB4. MONTGOMERY et al. (1955) also reported an outbreak of neonatal illness in a maternity unit in Rhodesia where one out of three babies infected died. CVB4 was found in the faeces of the dead baby and one of the live infants. In a more recent study HADDAD et al. (1993) retrospectively reviewed 21 cases of multiorgan disease in newborn babies occurring between 1987 and 1991 in France and diagnosed enteroviral infection by virus isolation from peripheral sites and detection of IgM class antibodies. Although the above studies were highly suggestive of the role CVBs played in myocardial disease direct evidence of an etiologic relationship was first provided by VAN CREVELD and DE JAGER (1956), who isolated CVB4 from the heart muscle of a newborn baby with fatal myocarditis. In a larger study in Sarajevo (ZVIZDIC et al. 1990), virus was isolated from the myocardial tissue of six out of seven autopsied cases of newborn babies that died in an outbreak of CVB infection.

Evidence that CVB infection may occur in utero and cause damage to the foetus was provided by several studies (KIBRICK and BENIRSCHKE 1958; BURCH et al. 1968; BASSO et al. 1990; DOMMERGUES et al. 1994) that either isolated virus from multiple foetal organs or demonstrated viral antigens in autopsy samples taken from the myocardium. A prospective study of nearly 23 000 women (BROWN and KARUNAS 1972) using paired sera also found that there were significantly more infections with CVB3 and B4 in mothers of infants with cardiovascular anomalies than in normal controls. Of 139 such mothers 10% had evidence of CVB3 infection compared to 2.7% of 262 controls, and 19.4% had evidence of CVB4 infection compared to 11.1% of controls. KORO'LKOVA et al. (1989) confirmed this work and using immunofluorescence found CVB in the urinary sediments of 84.4% of 62 children with congenital heart disease and only 7.7% of controls. The predominant serotype in this study was CVB3. The studies reviewed here provide substantial evidence that CVBs are involved in childhood myocarditis. Technological problems of the traditional virological methods however, limit systematic analysis of the data. Virus isolation from the heart tissue of fatal cases of myocarditis does not provide an indication of the prevalence of these agents in children and the rate at which they cause myocardial disease. Furthermore, large scale serological analysis of children or detection of virus from peripheral sites does not prove an aetiological involvement in myocarditis. These limitations of virus culture and serology have been partially overcome by molecular biology techniques established in recent years. HILTON et al. (1993) used in situ hybridization and PCR to investigate the presence of coxsackie virus RNA in childhood myocarditis using formalin fixed tissue. Two out of ten cases studied were positive. Viral RNA was predominantly associated with areas showing inflammatory cell infiltrates and myofibre necrosis suggesting a lytic action of the virus. MARTIN et al. (1994) used PCR in the rapid diagnosis of viral infection using 38 myocardial tissues from 34 patients with clinical evidence of myocarditis. Eight of the samples tested were positive for enteroviral sequences. None of the 17 control samples studied had viral sequences detected by hybridization following PCR. Recent work by the same group (TOWBIN 1996) has strengthened the belief that enteroviruses are one of the agents involved in myocardial disease. In a detailed study of 135 paediatric patients with myocarditis Towbin and coworkers workers found 23 to be PCR positive for enteroviruses from the myocardium. In this study a diverse group of control patients, who yielded 2% positivity rates, were included. The studies cited here and a large number of other case reports (TANIMURA et al. 1980; IWASAKI et al. 1985; GAREL et al. 1986; WELLS et al. 1986; LAU 1994) have used serology, virus isolation and molecular techniques in conjunction with clinical data, suggestive of an infectious agent, as criteria for a causal relationship. Although in some studies there is a paucity of data on control subjects, overall there is enough evidence to link enteroviruses to multiorgan disease and cardiac manifestations in children. It still remains to be established however, if some enteroviruses, in particular some strains of CVB, are more cardiotropic in humans than others. The relative predominance of different CVB serotypes in acute myocarditis in children and their possible significance in the clinical course of the disease has only been addressed in part. The proportion of children excreting a particular serotype of CVB seems very much dependent on the population studied (HERDY et al. 1991; GOWRISHANKAR and RAJAJEE 1994). The use of primers and probes specific for CVBs will now allow more sensitive and detailed study of the involvement of individual serotypes of CVB in disease. Such studies have been attempted using virus neutralisation assays, but molecular techniques can provide a more sensitive and accurate diagnosis.

4.3 Coxsackie B viral Myocarditis in Adults

The association between CVB infection and heart disease in adults was not suspected, despite an early observation that pericarditis was seen in some cases of Bornholm's disease (BING 1933), until FLETCHER and BRENAN (1957, 1958) described pericarditis in two patients with raised levels of neutralising antibody to CVB4 and CVB3. These reports were followed by more studies in which pericarditis and myocarditis in adults were associated with a rise in CVB antibody levels (JOHNSON et al. 1961; SWAN 1961; POLLEN 1963) or isolation of virus from the faeces combined with serological evidence of infection (BELL and MEIS 1959; GILLET 1959; GORDON et al. 1959; LEWES and LANE 1961). While evidence for an etiologic link between CVBs and neonatal heart disease was accumulating, such data for adult cardiac involvement were less forthcoming. This was probably due to the fact that neonatal disease is often more severe and has a higher mortality rate than that observed in adults. Moreover, while virus can be isolated from the heart, in severe cases of childhood CVB infection, reports of virus isolation from the myocardium (LONGSON et al. 1969) and pericardial fluid in adults (MONALDI et al. 1963; SOUTAR 1971) are scarce. Also the presence of high levels of specific antibodies to CVB in neonatal serum is a more convincing indicator of recent infection than in adults, in whom an anamnestic response may or may not arise as a consequence of infection with a strain of virus similar to one experienced previously. Despite such problems SAINANI et al. (1968) described 22 cases of heart disease, ten with pericarditis, ten with myocarditis and two with both these conditions who showed either serological evidence of CVB infection or were excreting virus. The serotypes involved were B4 (nine cases), B2 (six cases), B3 and B5 (three cases each) and B1 (one case). HELIN et al. (1968) and SAINANI et al. (1975) also reported 16 and 19 patients with acute myopericarditis who had respectively, either serological or virological evidence of CVB infection. The serotypes involved in the second study were B4 (eight cases), B5 (five cases), B3 (four cases), and B2 (two cases).

Table 1. Serological evidence for association of enteroviruses and acute idiopathic myocarditis

Reference	Method used	Percentage positive	
		Patient	Control
EL-HAGRASSY et al. 1980	Enterovirus-specific IgM	36% (14/38)	10.3% (6/58)
BELL and MCCARTNEY 1984	Rise in CVB neutralising antibody	33% (46/139)	4% (34/950)
FRISK et al. 1984	CVB-specific IgM (RIA)	49% (18/37)	0%[a]
MCCARTNEY et al. 1986	Enterovirus-specific IgM (EIA)	33% (116/352)	5% (4/87)
JOY et al. 1990	Rise in CVB neutralising antibody	34% (11/32)	ND
RAMAMURTHY et al. 1993	Rise in CVB3 neutralising antibody	5% (1/20)	ND
KANDA et al. 1995	Rise in CVB neutralising antibody	44% (8/18)	ND
Summary		34% (214/636)	4% (44/1139)

ND, not done
[a] Small numbers of positives were observed, but numbers were not specified

Serological studies on patients with acute myocardial or pericardial disease have continued through the last four decades (Table 1). BELL and MCCARTNEY (1984) reported high or rising levels of neutralising antibodies to CVB in 33% of patients with acute myopericarditis and 4% of normal volunteers. This was in close agreement with the 34% reported by JOY et al. (1990) who also studied serotype specificity of the patients' antibody response and found that most were infected with either CVB2 or CVB4. Although much higher (KANDA et al. 1995) and lower (RAMAMURTHY et al. 1993) rates of infection have been presented, the predominant number of reports reviewed here have shown increased or rising titres of antibody in acute myocarditis. The neutralisation assays used in the older studies offer the advantage of distinguishing between CVB strains, but the test does not differentiate recent infection and long standing immunity unless serial samples are analyzed. The detection of IgM class antibodies as a marker of recent infection bypasses this problem. Several studies have used this method advantageously (Table 1), but at present group reactive antibodies are all that can be measured with relative ease. The difficulty with IgM measurements is that antigenic epitopes of CVBs that can be used for specific diagnosis of infection have not been characterised adequately. HAARMAN et al. (1994) studied synthetic overlapping peptides and found one antigenic determinant representing residues 1–15 of VP1 to be specific to CVB3. More data are needed on the type specific and group specific epitopes of CVB before reliable serological diagnosis can be made. Not only is the specificity of the IgM tests in question but HODGSON et al. (1995) found qualitative differences between two commonly performed methods of detection of enteroviral IgM and cautioned against sole reliance on serology for diagnosis of enteroviral infections. It is with such problems in mind that the studies using IgM detection systems have to be considered. The reported prevalence of enterovirus specific IgM antibodies in patients with acute myocarditis varies between 33% and 49% (Table 1). These studies were carried out mainly on patients without biopsy proven myocarditis and may therefore include patients with long standing myocardial disease presenting as acute illness as a consequence of other nonvirological factors affecting heart function. Another variable that may have influenced the rate of IgM seropositivity is the occurrence of local epidemics in the communities. When seropositivity rates in control groups are compared there is no suggestion that the studies reporting a higher prevalence of infection in the test population were experiencing epidemics. The control groups studied in these reports however, were often age matched healthy individuals or patients with other nonviral cardiac diseases. A more appropriate control population may have been household or workplace contacts of patients with acute myocarditis.

Table 2. Evidence for association of enteroviruses with acute idiopathic myocarditis

Reference	Method used	Percentage positive Patient	Control
Bowles et al. 1986	Slot blot (enterovirus specific)	50% (4/8)	0% (0/4)
Jin et al. 1990	RT PCR (enterovirus specific)	7% (2/28)	0% (0/9)
Tracy et al. 1990	In situ hybridization (enterovirus specific)	18% (3/17)	ND
Weiss et al. 1991	RT PCR (CVB3-specific)	20% (1/5)	0% (0/21)
Petitjean et al. 1992	RT PCR/hybridization (enterovirus specific)	10% (1/10) positive	34.8% (8/23)
		20% (2/10) equivocal	4.3% (1/23)
Koide et al. 1992	RT PCR (enterovirus specific)	33% (3/9)	0% (0/9)
Giacca et al. 1994	RT nested PCR (enterovirus specific)	33% (1/3)	3.6% (1/28)
Satoh et al. 1994	RT nested PCR (enterovirus specific)	33% (12/36)	0% (0/10)
Seko et al. 1995	RT PCR/hybridization (enterovirus specific)	0% (0/2)	ND
Ueno et al. 1995	RT PCR (enterovirus specific)	80% (4/5)	11% (3/27)
Summary		25% (33/133)	10% (13/131)

ND, not done

The ambiguity of results produced from virus culture and serological assays has forced a change in the detection methods used. Molecular biological techniques have now gained prominence as tools for diagnosis of enteroviral infection of the myocardium, but the hope that these sensitive detection systems would provide conclusive evidence of viral aetiology has not been entirely fulfilled. BOWLES et al. (1986) were one of the first groups to report the detection of CVB RNA in myocardial tissue of 53% of patients with myocarditis. Although the detection system used did not discriminate between enteroviruses as a group from the CVB, nucleotide sequencing of PCR products later confirmed the high rate of positivity and established that the detected sequences were similar to CVB3 (KHAN et al. 1994). WEISS et al. (1991) confirmed the presence of viral RNA in the myocardium using reverse transcriptase (RT) PCR, sensitive enough to detect 1–100 plaque forming units of virus per gram of tissue and detected CVB3 in one out of five patients with myocarditis. No positive signals were obtained in 21 specimens from a wide variety of other cardiac disorders. This 20% positivity rate compared to the 53% reported by BOWLES et al. (1986) may have been due to the specificity of the assay to CVB3. The authors also argued that the potential for nonspecific hybridization was higher for the slot blot technique than for PCR. Later, PETITJEAN et al. (1992) studied ten patients with myocarditis and obtained a clear positive signal in one sample and borderline findings in two patients. Nine of 23 patients with other cardiac abnormalities were also positive by PCR. Although a link between viral infection and cardiomyopathy was suggested by Petitjean et al. (1992) the results were not supportive of such an association. The differences in reported positivity rates that troubled serological assays resurfaced in studies using molecular techniques. Table 2 shows a summary of recent reports. UENO et al. (1995), studied five patients with active myocarditis and demonstrated enteroviral RNA by RT PCR in four of these, a detection rate of 80%, three of 27 noncardiac autopsy cases carried viral RNA in the myocardium. KOIDE et al. (1992) studied nine patients with clinical myocarditis and compared these with patients with hypertrophic cardiomyopathy and other nonvirological heart diseases and recorded 33% positivity by enterovirus PCR in the myocarditis population and none in the control group. Clinical features of patients who were PCR positive were compared with those who were negative but no convincing differences could be seen. GIACCA et al. (1994) and SATOH et al. (1994) reported similar findings. The first group carried out enterovirus specific, RT nested PCR and found one positive sample in three cases of idiopathic myocarditis compared to one in 28 controls. The second group examined 44 tissues obtained from 36 patients with myocarditis as well as ten patients with nonviral cardiac diseases and detected enteroviral RNA in 12 of the myocarditis patients using a nested RT PCR. None of the ten control samples had viral RNA. An overall analysis of recent reports (Table 2) shows an average of 25% enteroviral RNA detection rate in idiopathic myocarditis patients compared with 10% in the control group. Not enough work has been done on CVBs to enable conclusions to be drawn as to the role of these agents in myocarditis but twice as many patients carrying enteroviral sequences in the target organ points to the significance of these viruses in myocardial disease. When the heart tissue is being studied, it is very difficult if not impossible to choose appropriate control groups for comparative purposes. Small epidemics of CVB may have occurred in communities and artificially boosted the numbers of biopsy positive patients. Due to ethical restrictions the control groups cannot be family members or work associates. Matching patients with controls living in the same area around the time disease symptoms were observed in the patients is difficult. At best the controls can be age and sex matched individuals attending the hospital for noninfectious cardiac disorders. For a group of viruses with the transmission characteristics of enteroviruses this strategy may result in misleading information.

4.4 Dilated Cardiomyopathy

Patients with congestive cardiac failure and a dilated left ventricle with impaired systolic function in the absence of an ischaemic, valvular or hypertensive cause are usually classified as dilated cardiomyopathy (DCM). The entity is based on a common pathophysiology rather than a common pathogenesis and the condition is heterogeneous in origin. When cases due to excess use of alcohol or due to genetic defects including mitochondrial abnormalities are excluded the largest group among patients with DCM is idiopathic in origin. The morphological features include interstitial fibrosis and an increased number of interstitial cells including macrophages and lymphocytes. Individual myocyte death occurs at varying rates as evidenced by the presence of single necrotic cells. These histological features bear some similarity to acute viral myocarditis and the view has developed that DCM is a low grade chronic myocarditis due to persistent viral infection. An analogy can be drawn with persistent piecemeal necrosis of liver cells in chronic hepatitis. The role of virus in DCM may either be that infectious virus replication continues at low levels or that persistence of viral RNA in the absence of productive infection occurs. In such a state of quiescence some viral antigens may still be presented on the cell surface directing the immune response to these cells. There is also the possibility that viral infection may trigger autoimmune mediated myocardial damage.

4.5 Association of Enteroviruses with Dilated Cardiomyopathy

The serological and molecular techniques applied to the study of viral aetiology in acute myocarditis have also been used in the study of DCM. One of the earliest reports suggesting a link between enteroviruses and DCM was carried out in 1979 by CAMBRIDGE et al. who found high levels of CVB specific IgG antibodies in 30% of patients with DCM and only 2% of the controls. High titre antibodies were also found by JIANG (1992) in 65.7% of patients and only 25.7% of controls. Measurement of IgM class antibodies to enteroviruses confirmed these findings. MUIR et al. (1989) and BOWLES et al. (1989) reported a higher percentage of IgM positive sera in the DCM group than in controls. The percentages of IgM positivity reported by

Table 3. Serological evidence for association of enteroviruses and idiopathic dilated cardiomyopathy

Reference	Method used	Percentage positive Patient	Control
CAMBRIDGE et al. 1979	High levels of IgG CVB-specific	30% (15/50)	2% (1/50)
MUIR et al. 1989	Enterovirus-specific IgM	33% (28/86)	12% (10/84)
BOWLES et al. 1989	Enterovirus-specific IgM	4.8% (1/21)	0% (0/19)
LILJEQVIST et al. 1993	Enterovirus-specific IgM	1.5% (1/63)	10% (6/63)
WESSELEN et al. 1993	IgM CVB-specific	75% (6/8)	0% (0/7)
KEELING et al. 1994	CVB-specific IgM	37.5% (12/32)	household controls 28% (9/32)
Summary		24% (63/260)	10% (26/255)

the two groups however, were very different. While MUIR et al. (1989) found 33% of DCM patients had enterovirus specific IgM, BOWLES et al. only found a 4.8% IgM positivity. This may be a consequence of a higher incidence of infection in the population studied by MUIR et al. (1989), as 12% of the control group also had recent enteroviral infection. One of the highest recorded frequencies of IgM detection was found in a Swedish population. WESSELEN et al. (1993) studied a small group of eight patients and found six to have CVB IgM. None of the control group were IgM positive. Although the numbers were small it was argued that CVB infection in the normal healthy population in Sweden is uncommon and an infection rate of 75% in this group is highly significant. The choice of the control population in serological studies was addressed by KEELING et al. (1994) who conducted CVB specific IgM measurements in 114 patients with DCM and compared them with 94 unmatched controls, 41 control subjects that attended the same general practice as the patients and 32 members of the same household. They showed a high frequency of 33% positivity for CVB IgM in the total patient population and only 5% in the unmatched controls. The matched controls had a 27% IgM seropositivity and the household contacts of the patients were 28% seropositive. Moreover, they showed that there was a trend in the patients and their contacts to have IgM antibodies to the same serotype of CVB. In the light of this report comparison of DCM cases with controls (Table 3) is difficult. Taking the epidemiological data at face value there is an excess of IgM antibody detection rate in the DCM group. It can be argued however, that the controls in a large number of these studies are not appropriately chosen to take into account the transmission mode of enteroviruses.

Attempts to isolate virus from the heart or detect viral proteins in the myocardium have so far been unsuccessful and direct evidence implicating enteroviruses in DCM is limited to the detection of viral nucleic acids in the heart tissue. BOWLES et al. (1986) were the first group to use the slot blot hybridization technique and to show enteroviral RNA in the myocardium of five of nine patients with DCM. In their second study they reported 28% of the DCM samples to be positive (BOWLES et al. 1989). WERNER et al. (1993) and FIGULLA et al. (1995) confirmed these findings and using in situ hybridization found a large number of patients with DCM to be carriers of enteroviral RNA. Most other studies reported to date have used an RT PCR either alone or combined with blotting and hybridization as the detection strategy. Table 4 shows a summary of recent reports and the outcome of these

Table 4. Evidence for association of enteroviruses with dilated cardiomyopathy

Reference	Method used	Percentage positive Patient	Control
BOWLES et al. 1986	Slot-blot (enterovirus specific)	55% (5/9)	0% (0/4)
BOWLES et al. 1989	Slot-blot (enterovirus specific)	28% (6/21)	5% (1/19)
WIEGAND et al. 1990	In situ hybridization (enterovirus specific)	17% (1/6)	0% (0/8)
JIN et al. 1990	RT PCR (enterovirus specific)	15% (3/20)	0% (0/9)
WEISS et al. 1991	RT PCR CVB3	30% (0/11)	0% (0/21)
PETITJEAN et al. 1992	RT PCR/hybridization (enterovirus specific)	49% (22/45) positive	34.8% (8/23)
		18% (8/45) equivocal	4.3% (1/23)
GRASSO et al. 1992	RT PCR (enterovirus specific) and CVB3	10% (0/21)	0% (0/60)
CHIANG et al. 1992	RT PCR/hybridization (enterovirus specific)	75% (3/4)	0% (0/6)
KEELING et al. 1992	RT PCR/hybridization (enterovirus specific)	12% (6/50)	17% (13/75)
KOIDE et al. 1992	RT PCR (enterovirus specific)	32% (8/25)	ND
ZOLL et al. 1992	RT PCR (enterovirus specific)	20% (1/5)	0% (0/8)
LILJEQVIST et al. 1993	RT PCR (enterovirus specific)	0% (0/35)	0% (0/23)
SCHWAIGER et al. 1993	RT PCR/hybridization (enterovirus specific)	31% (6/19)	0% (0/21)
	CVB3	5% (1/19)	
WERNER et al. 1993	In situ hybridization (enterovirus specific)	35% (6/17)	ND
GIACCA et al. 1994	RT nested PCR (enterovirus specific)	7% (4/53)	3% (1/28)
KHAN et al. 1994	RT nested PCR (enterovirus/CVB)	53% (8/15)	11% (1/9)
ANDREOLETTI et al. 1995	RT PCR/slot blot (enterovirus specific)	66% (10/15)	15% (8/54)
FIGULLA et al. 1995	In situ hybridization (enterovirus specific)	26% (20/77)	ND
UENO et al. 1995	RT PCR (enterovirus specific)	17% (7/42)	11% (3/27)
SEKO et al. 1995	RT PCR/hybridization (enterovirus specific)	0% (0/4)	ND
Summary		25% (124/496)	9% (36/387)

ND, not done

studies, drawing attention to the enormous variation in the detection rates of viral genome both in DCM and the control samples. Some of these differences arise as a result of the selection criteria used. WEISS et al. (1991) for example, could not find viral sequences in DCM patients when they used an assay that detected CVB3 sequences at the exclusion of other enteroviruses. This low positivity rate agreed with the findings of SCHWAIGER et al. (1993) who also showed only one of 19 DCM myocardial biopsy samples to be positive for CVB3 while 31% of the same patients had enteroviral RNA. Others have expressed concern that some of the positive results may be contaminants. In one of these studies a nested PCR was used to amplify a conserved region at the 5' end of the viral RNA. All four DCM patients who showed a positive signal in PCR had identical sequences to the positive control reference strain used in the laboratory (GIACCA et al. 1994). Another group experienced similar difficulties when they showed enteroviral RNA in six of 63 DCM and 14 of 87 control hearts and demonstrated, by sequencing of the amplified viral genomes, that 13 of the control samples and one of the DCM patients were identical (KEELING et al. 1994). Although this group were also amplifying a conserved region of the viral genome, the authors nevertheless raised doubts about the validity of the positive findings in the control population. It is possible therefore that the differences in the positivity rates between the DCM patients and the controls may be higher than represented by these data. KHAN et al. (1994), however, sequenced eight positive samples obtained from 15 DCM patients, using a less conserved region of the viral RNA, and showed sequence differences in the amplified material. Surprisingly they also reported sequence variations in viral RNA from multiple biopsies obtained from the same patient.

The significance of the presence of enteroviral RNA in patients with DCM is not known. There is a 2.5-fold difference between the positivity rates in DCM and other cardiac disorders. WHY et al. (1994) suggested that the detection of enterovirus RNA in the myocardium correlated with adverse prognosis and acted as an independent predictor of clinical outcome. By contrast, KEELING et al. (1992) found no distinguishing clinical characteristics that could segregate the patients with virus from those without. It is possible that active viral replication or expression of some viral proteins may be responsible for the development of disease but this is by no means proven. The almost universal use of a conserved region of enteroviruses for detection of nucleic acids has not helped matters. Although these sequences make useful probes for identification of positive samples, if a particular strain of virus is cardiovirulent using a generic probe will make it impossible to differentiate at risk patients from those carrying harmless viruses as passengers in the heart tissue. Numerous CVBs have now been sequenced and these data show that there are areas of the RNA that are specific for single serotypes. A specific PCR can now be set up to evaluate the role of individual serotypes of CVB in myocardial disease.

Finally on a positive note, the finding of SEKO et al. (1995) must be mentioned; they analyzed the expression of T cell receptor Vα and Vβ genes in infiltrating cells in the hearts of patients with DCM and found repertoires of gene transcripts were restricted suggesting that a specific antigen in the heart is targeted. This target may be a virus or indeed any host protein present in the myocardium.

5 Chronic Relapsing Pericarditis

The clinical features of acute pericarditis include retrosternal chest pain often following a history of viral illness and accompanied by an increase in pericardial fluid. Pericarditis can be present during myocarditis and some studies refer to myopericarditis as a single disease entity. One in five patients with acute pericarditis experiences recurrence of symptoms (BRANDENBURG et al. 1987). MUIR et al. (1989) studied enteroviral IgM response by a capture assay and showed that nine of 14 patients with chronic relapsing myocarditis of at least 1 year duration had persistent IgM class antibodies against enteroviruses. These antibodies remained for 1–10 years from the onset of symptoms. The presence of IgM antibodies is normally suggestive of an acute infection. These antibodies do not remain for prolonged periods in the absence of infectious virus. MUIR et al. (1989) compared the chronic pericarditis group with a group of acute pericarditis patients and found that levels of CVB antibodies were higher in the acute disease patients but were transient. It was therefore suggested that enteroviruses may persist in patients with chronic disease either in the cardiac tissue or elsewhere in the body and give rise to specific IgM antibodies. Studies to show viral proteins or RNA in the pericardium or myocardium have not been carried out as the clinical severity of the disease is not usually such as to warrant the invasive obtaining of biopsies.

6 Mechanisms of Pathogenesis

The evidence provided above suggests that some members of the enterovirus group can potentially cause human myocardial disease in humans. The precise prevalence and the exact nature of the disease caused is still disputed. It is yet unclear which if any enteroviral serotypes are more prominent as pathogens affecting the heart despite there having been attempts dating back to 1979 to discriminate between cardiotropic and avirulent phenotypes of CVB3 (GAUNT et al. 1979). In recent studies the nucleotide sequence of two cardiovirulent (KLUMP et al. 1990; TRACY et al. 1992) and one noncardiovirulent strains (CHAPMAN et al. 1994) were compared and ten sites were identified that differed in these strains. When avirulent strains were genetically reconstructed to contain mutations to the virulent strains (CHAPMAN et al. 1994), tests on mice revealed that most significant difference occurred at position 234, with C being present in the noncardiovirulent strain and U in the virulent phenotype (TU et al. 1995). Cell culture studies showed that strains with C rather than U at position 234 demonstrated decreased viral RNA synthesis. With the possibility that cardiovirulence in wild-type CVBs may also be mapped to the same region, TRACY et al. (1996) sequenced 13 CVB3 strains with varying degrees of cardiovirulence, isolated over four decades, and found that the nucleotide in position 234 was indeed conserved as U. But analyzing the enterovirus sequence databases revealed that the two nucleotides on either side of position 234 were also conserved (CGUUA) and therefore suggested that nucleotide 234 may not be involved in determining cardiovirulence in naturally occurring strains and the attenuated strain with C in position 234 may be a mutant. Similar experiments in mice using chimeric CVB4, containing threonine at amino acid position 129 of the capsid protein instead of methionine was associated with both pancreatic and cardiac virulence (CAGGANA et al. 1993). A more detailed review of the genetics of virulence appears elsewhere in this volume.

Enteroviruses cultured in fibroblasts of human or monkey origin cause cell damage by lysis. Experiments carried out on mice with severe combined immunodeficiency have shown that CVB3 is capable of lytic damage to the myocytes (CHOW et al. 1992). Mechanism of disease however, may be more complex than cell death as a consequence of viral lysis alone. Cardiovirulent strains of CVB3 do not cause rapid lysis of neonatal mouse skin fibroblasts (LUTTON and GAUNTT 1986) or foetal murine heart fibroblasts (TU et al. 1995) yet cause disease in these animals. Experiments carried out on normal susceptible mice have shown that the highest titres of virus are seen in the myocardium 1–7 days after inoculation. Microscopic examination of the heart shows areas of necrosis and infiltration of mononuclear cells. Cellular damage occurs concurrently with high titres of virus with CVB1 and CVB4 (KHATIB et al. 1980), and with virus titres declining in the case of CVB3 (WOODRUFF and WOODRUFF 1974). This suggests that an immunologically mediated killing of myocytes may be an additional factor in cell damage. Indeed it would be unusual if such a mechanism were not at least partly responsible for myocyte death. In common with other viral infections, replication of enteroviruses in the heart and the presentation of viral antigens on the myocardial cells in the context of MHC can initiate an immune response directed towards eliminating the infectious agents (ESTRIN and HUBER 1987). The host, in an attempt to rid the tissue of infectious agents, may be causing sufficient cellular damage to cause symptoms of disease. To determine the relative roles of direct virus lysis of cells and immune mediated damage HENKE et al. (1995) compared CVB3 infections of normal and immunocompromised CD4 or β_2-microglobulin knockout mice. The immunocompetent mice were most susceptible to early death in a virus dose dependent manner. Histological examination of cardiac tissue from normal and immunocompromised mice that survived showed that animals lacking CD4 positive cells were susceptible to severe myocarditis, but the β_2-microglobulin knockout mice that lacked CD8 positive T lymphocytes had increased survival times and reduced incidence of myocarditis. Moreover, other studies have shown that different immunological mechanisms may be significant in different strains of mice. BALB/c mice are susceptible to cytotoxic T cell mediated damage to the myocardium but the DBA/2 strain of mice are more prone to T helper cell and humoral immune damage (HUBER and LODGE 1986). The picture is therefore complex. Removal of T cell subsets may reduce immune mediated damage in some strains of mice but result in disseminated infection. As such differences exist between strains of mice the validity of specific comparisons between the mouse model and disease in humans may be questioned.

Immunological damage to myocytes can also arise as a consequence of an exaggerated or inappropriate response to infection. Virus induced damage to myocytes may trigger an immune response not only against the pathogen but the host cells also. Evidence that such a mechanism may be operating is obtained from experiments in mice in which cytotoxic T cells directed against CVB3 as well as uninfected myocytes are observed following infection with this virus (ESTRIN and HUBER 1987). It has also been shown that T cytotoxic cells taken from patients with myocarditis and inoculated into mice with severe combined immunodeficiency produce cardiac lesions in these animals suggesting an autoimmune phenomenon (Schwimmbeck et al. 1994). The difference between cardiovirulent and avirulent strains of CVB in terms of initiation of autoimmune response has been investigated with respect to the ability of the virulent strains to recruit a different population of cells to the heart tissue. Van Houten et al. (1991) and Huber et al. (1994), who isolated an attenuated mutant of CVB3 which had a single amino acid change in the E-F loop of the capsid protein 1B, showed that while the cardiovirulent parental strain induced a T_h1 cytokine expression profile, the attenuated virus induced a T_h2 profile. Pathogenesis mediated by an autoimmune antibody response has also been considered. Antibodies develop against α and β myosin in murine CVB infection and in patients with DCM (Caforio 1994); there is however, no evidence to suggest that these antibodies cause disease.

In summary, pathogenesis of disease may be multifactorial and depend on the strain of virus involved as well as the type of host cell infected and immune response to the virus. Evidence for viral involvement in neonates and the young is strong. In adults however, interpretation of information from both traditional and molecular tissue based techniques is confounded by the difficulty in studying appropriate control populations. Epidemiological studies are unlikely to resolve this problem. Further studies are necessary to investigate whether the presence of viral genome in replicative or nonreplicative form is capable of causing damage to the myocytes.

References

Andreoletti L, Wattre P, Decoene C, Lobert PE, Dewilde A, Hober D (1995) Detection of enterovirus-specific RNA sequences in explanted myocardium biopsy specimens from patients with dilated or ischaemic cardiomyopathy. Clin Infect Dis 21:1315–1317

Aretz HT, Billingham ME, Edwards WD, Factor SM, Fallon JT, Fenoglio JJ Jr, Olsen EG, Schoen FJ (1987) Myocarditis: A histopathologic definition and classification. Am J Cardiovas Pathol 1:3–14

Basso NG, Fonseca ME, Garcia AG, Zuardi JA, Silva MR, Outani H (1990) Enterovirus isolation from foetal and placental tissues. Acta Virol 34:49–57

Bell EJ, McCartney RA (1984) A study of coxsackie B virus infections, 1972–1983. J Hyg 93:197–203

Bell EF, Meis A (1959) Pericarditis in infection due to coxsackie virus group B type 3. N Engl J Med 261:126–128

Bing HI (1933) Epidemical pericarditis. Acta Med Scand 80:29

Bowles NE, Richardson PJ, Olsen EGJ, Archard LC (1986) Detection of coxsackie B virus specific RNA sequences in myocardial biopsy samples from patients with myocarditis and dilated cardiomyopathy. Lancet I:1120–1123

Bowles NE, Rose ML, Taylor P, Banner NR, Morgan-Capner P, Cunningham L, Archard LC, Yacoub MH (1989) End-stage dilated cardiomyopathy persistence of enterovirus RNA in myocardium at cardiac transplantation and lack of immune response. Circulation 80:1128–1136

Brandenburg RO, McGoon DC (1987) The pericardium. In: Brandenburg RO, Fuster V, Guilliami ER, McGoon DC (eds), Cardiology Fundamentals and Practice. Year Book, Chicago

Brown GC, Karunas RS (1972) Relationship of congenital anomalies and maternal infection with selected enteroviruses. Am J Epidemiol 95:207–217

Burch GE, Sun SC, Chu K, Sohal R, Colcolough H (1968) Interstitial and coxsackie virus B myocarditis in infants and children. A comparative histologic and immunofluorescent study of 50 autopsied hearts. JAMA 203:1–8

Caforio ALP (1994) Role of autoimmunity in dilated cardiomyopathy. Br Heart J 72:S30-S34

Caggana M, Chan P, Ramsingh A (1993) Identification of a single amino acid residue in the capsid protein VP1 of coxsackievirus B4 that determines the virulent phenotype. J Virol 67:4797–4803

Cambridge G, MacArthur CG, Waterson AP, Goodwin JF, Oakley CM (1979) Antibodies to coxsackie B viruses in congestive cardiomyopathy. Br Heart J 41:692–696

Chapman NM, Tu Z, Tracy S, Gauntt CJ (1994) An infectious cDNA copy of the genome of a non-cardiovirulent coxsackie B3 strain: its complete sequence analysis and comparison to the genomes of cardiovirulent coxsackieviruses. ArchVirol 135:115–130

Chiang FT, Lin Li, Tseng YZ, Tseng CD, Hsu KL, Wu TL, Ho SW (1992) Detection of enterovirus RNA in patients with idiopathic dilated cardiomyopathy by polymerase chain reaction. Taiwan I Hsueh Hui Tsu Chih 91:569–574

Chow LH, Beisel KW, McManus BM (1992) Enteroviral infections of mice with severe combined immunodeficiency. Evidence for direct viral pathogenesis of myocardial injury. Lab Invest 66:24–31

Clements GB (1993) Characteristics of viruses inducing cardiac disease. In: Banatvala JE (eds), Viral infections of the heart. Edward Arnold, London

Dommergues M, Petitjean J, Aubry MC, Delezoide AL, Narcy F, Fallet-Bianco C, Freymuth F, Dumez Y, Lebon P (1994) Fatal enteroviral infection with cerebral ventriculomegaly and cardiomyopathy. Fetal Diag Ther 9:77–78

El-Hagrassy MMO, Banatvala JE, Coltart DJ (1980) Coxsackie-B-virus-specific IgM responses in patients with cardiac and other diseases. Lancet II:1160–1162

Estrin M, Huber SA (1987) coxsackievirus B3 induced myocarditis. Autoimmunity is L3T4+ T helper cell and IL-2 independent in BALB/c mice. Am J Pathol 127:335–341

Figulla HR, Stille-Siegener M, Mall G, Heim A, Kreuzer H (1995) Myocardial enterovirus infection with left ventricular dysfunction: a benign disease compared with idiopathic dilated cardiomyopathy. J Am Coll Cardiol 25:1170–1175

Fletcher E, Brenan CF (1957) Cardiac complications of Coxsackie-virus infection. Lancet I:913–915

Fletcher E, Brenan CF (1958) Cardiac complications of Coxsackie virus infection. Lancet II:585

Frisk G, Torfason EG, Diderholm H (1984) Reverse radioimmunoassay of IgM and IgG antibodies to coxsackie B viruses in patients with acute myopericarditis. J Med Virol 14:191–200

Garel D, Wood C, Bader B, Dussaix E, Fournier E (1986) Infarctus du myocarde neonatal associe a une infection materno-foetale a virus coxsackie B4. Arch Fr Pediatr 43:495–496

Gauntt CJ, Trousdale MD, LaBadie DR, Paque RE, Nealon T (1979) Properties of coxsackievirus B3 varients which are amyocarditic or myocarditic for mice. J Med Virol 3:207–220

Giacca M, Severini GM, Mestroni L, Salvi A, Lardieri G, Falaschi A, Camerini F (1994) Low frequency of detection by nested polymerase chain reaction of enterovirus ribonucleic acid in endomyocardial tissue of patients with idiopathic dilated cardiomyopathy. J Am Coll Cardiol 24:1033–1040

Gillet R (1959) Acute benign pericarditis and the Coxsackie viruses. N Engl J Med 261:838–843

Gordon RB, Lennette EH, Sandrock RS (1959) The varied clinical manifestations of Coxsackie virus infections. Arch Int Med 103:63–75.

Gowrishanker K, Rajajee S (1994) Varied manifestations of viral myocarditis. Indian J Pediatr 61:75–80

Grasso M, Arbustini E, Silini E, Diegoli M, Percivalle E, Ratti G, Bramerio M, Gavazzi A, Vigano M, Milanesi G (1992) Search for coxsackie virus B3 RNA in idiopathic dilated cardiomyopathy using gene amplification by polymerase chain reaction. Am J Cardiol 69:658–664

Haarmann CM, Schwimmbeck PL, Mertens T, Schultheiss HP, Strauer BE (1994) Identification of serotype-specific and nonserotype-specific B-cell epitopes of coxsackie B virus using synthetic peptides. Virology 200:381–389

Haddad J, Gut JP, Wendling MJ, Astruc D, Jernite M, Obert G, Messer J (1993) Enterovirus infections in neonates. A retrospective study of 21 cases. Eur J Med 2:209–214

Helin M, Savola J, Lapinleimuk K (1968) Cardiac manifestations during a coxsackie B5 epidemic. Br Med J 3:97–99

Henke A, Huber S, Stelzner A, Whitton L (1995) The role of CD8+ T lymphocytes in coxsackievirus B3-induced myocarditis. J Virol 69:6720–6728

Herdy GV, al Odeh CS, Cunha J de O, Amorim AM, Ono I de C (1991) Virus coxsackie B em criancas com miocardite aguda. Arq Bras Cardiol 56:201–205

Hilton DA, Variend S, Pringle JH (1993) Demonstration of coxsackie virus RNA in formalin-fixed tissue sections from childhood myocarditis cases by in situ hybridization and polymerase chain reaction. J Pathol 170:45–51

Hodgson J, Bendig J, Keeling P, Booth JC (1995) Comparison of two immunoassay procedures for detecting enterovirus IgM. J Med Virol 47:29–34

Huber SA, Lodge PA (1986) Coxsackievirus B3 myocarditis. Identification of different pathogenic mechanisms in DBA/2 and Balb/c mice. Am J Pathol 122:284–291

Huber S Polgar J, Schultheiss P, Schwimmbeck P (1994) Augmentation of pathogenesis of coxsackivirus B3 infections in mice by exogenous administration of interleukin-1 and interleukin-2. J Virol 68:195–206

Iwasaki T, Monma N, Satodate R, Segawa I, Oyama K, Kawana R, Kurata T (1985) Myocardial lesions by Coxsackie virus B3 and cytomegalovirus infection in infants. Heart Vessels [Suppl] 1:167–172

Javett SN, Heyman S, Mundel B, Pepler WJ, Lurie HI, Gear J, Maesroch V, Kirsch Z (1956) Myocarditis in the newborn infant. J Pediatr 48:1–22

Jenista JA, Powell KR, Menegus MA (1984) Epidemiology of neonatal enterovirus infection. J Pediatr 104:685–690

Jiang J (1992) Possible relationship between acute myocarditis and dilated cardiomyopathy. Chun Hua Hsin Hsueh Kuan Ping Tsa Chih 20:4–6

Jin O, Sole MJ, Butany JW, Chia WK, McLaughlin PR, Liu P, Liew CC (1990) Detection of Enterovirus RNA in myocardial biopsies from patients with myocarditis and cardiomyopathy using gene amplification by polymerase chain reaction. Circulation 82:8–16

Johnson RT, Portnoy B, Rogers NG, Buescher EL (1961) Acute benign pericarditis: Virologic study of 34 patients. Ann Intern Med 108:823–832

Joy J, Rao YY, Raveenderanath M, Shanmugam J, Balakrishnan KG (1990) Coxsackie viral myocarditis: a clinical and echocardiographic study. Indian Heart J 42:441–444

Kanda T, Kobayashi I, Suzuki T, Murata K, Radio SJ, McManus BM (1995) Elevation of ALT to AST ratio in patients with enteroviral myocarditis. J Med 26:153–162

Kandolf R, Hofschneider PH (1985) Molecular cloning of the genome of a cardiotropic coxsackie B3 virus: full-length reverse-transcribed recombinant cDNA generates infectious virus in mammalian cells. Proc Natl Acad Sci USA 82:4818–4822

Keeling PJ, Jeffrey S, Caforio ALP, Taylor R, Bottazzo GF, Davies MJ, McKenna WJ (1992) Similar prevalence of enteroviral genome within the myocardium from patients with idiopathic dilated cardiomyopathy and controls by the polymerase chain reaction. Br Heart J 68:554–559

Keeling PJ, Kukaszyk A, Poloniecki J, Caforio ALP, Davies MJ, Booth JC, McKenna WJ (1994) A prospective case-control study of antibodies to coxsackie B virus in idiopathic dilated cardiomyopathy. J Am Coll Cardiol 23:593–598

Khan M, Why H, Richardson P, Archard L (1994) Nucleotide sequencing of PCR products shows the presence of coxsackie-B3 virus in endomyocardial biopsies from patients with myocarditis or dilated cardiomyopathy. Biochem Soc Trans 22:176S

Khatib R, Chason JL, Silberberg BK, Lerner AM (1980) Age-dependent pathogenicity of group B Coxsackieviruses in Swiss-Webster mice: infectivity for myocardium and pancreas. J Infect Dis 141:394–403

Kibrick S, Benirschke K (1958) Severe generalized disease (encephalohepatomyocarditis) occurring in the newborn period and due to infection with coxsackievirus group B. Pediatrics 22:857–875

Klump WM, Bergmann I, Muller BC, Ameis D, Kandolf R (1990) Complete nucleotide sequence of infectious coxsackievirus B3 cDNA: two initial 5'uridine residues are regained during plus strand RNA synthesis. J Virol 64:1573–1583

Koide H, Kitaura Y, Deguchi H, Ukimura A, Kawamura K, Hirai K (1992) Genomic detection of enteroviruses in the myocardium studies on animal hearts with coxsackie B3 myocarditis and endomyocardial biopsies from patients with myocarditis and dilated cardiomyopathy. Jpn Circ J 56:1081–1093

Koro'lkova EL, Lozovskaia LS, Tadtaeva LI, Khellenov EA (1989) Znachenie antenata'lnoi koksakivirusnoi infektsii v etiologii vrozhdennykh porokov serdtsu u detei. Kardiologiia 29:68–71

Kuhl U, Noutsian M, Seeberg H, Schultheiss HP (1996) Immunohistochemical evidence for a chronic intramyocardial inflammatory process in dilated cardiomyopathy. Heart 75:295–300

Lau G (1994) Acute fulminant, fatal coxsackie B virus infection: a report of two cases. Ann Acad Med Singapore 23:917–920

Lewes D, Lane WF (1961) Acute benign pericarditis due to coxsackie virus group B, type 3. Lancet II:1385–1386

Liljeqvist JA, Bergstrom T, Holmstrom S, Samuelson A, Yousef GE, Waagstein F, Jeansson S (1993) Failure to demonstrate enterovirus aetiology in Swedish patients with dilated cardiomyopathy. J Med Virol 39:6–10

Longson M, Cole FM, Davies D (1969) Isolation of a coxsackie virus group B, type 5, from the heart of a fatal case of myocarditis in an adult. J Clin Pathol 22:654–658

Lutton CW, Gauntt CJ (1986) Coxsackievirus B3 infection alters plasma membrane of neonatal skin fibroblasts. J Virol 60:294–296

Martin AB, Webber S, Fricker J, Jaffe R, Demmler G, Kearney D, Zhang YH, Bodurtha J, Gelb B, Ni J, Bricker T, Towbin J (1994) Acute myocarditis. Rapid diagnosis by PCR in children. Circulation 90:330–339

McCartney RA, Banatvala JE, Bell EJ (1986) Routine use of u-antibody capture ELISA for serological diagnosis of coxsackie B virus infections. J Med Virol 19:205–212

Melnick JL (1996) Enteroviruses: polioviruses, coxsackieviruses, echoviruses, and newer enteroviruses. In: Fields BN, Knipe DM, Howley PM, Chanock RM, Joseph LM, Monath TP, Roizman B, Straus SE (eds), Fields Virology, Third Edition. Lippincott-Raven, Philadelphia

Melnick JL, Wimberly IL (1985) Lyophilised combination pools of enterovirus equine antisera:new LBM pools prepared from reserves of antesera stored frozen for two decades. Bull WHO 63:543–550

Minor PD, Morgan-Capner P, Schild GC (1990) The Enteroviruse. In: Zuckerman AJ, Banatvala JE, Pattison JR (eds), Principles and Practice of Clinical Virology. John Wiley, Chichester

Monaldi TD, Benedetto A, Montalto TT (1963) Pericarditis infection due to coxsackie virus group B, type 2. Br Med J 2:1451–1452

Montgomery J, Gear J, Prinsloo FR, Kahn M, Kirsh ZG (1955) Myocarditis of newborn; outbreak in maternity home in Southern Rhodesia associated with coxsackie group B virus infection. S Afr Med J 29:608–618

Muir P, Nicholson F, Tilzey AJ (1989) Chronic relapsing pericarditis and dilated cardiomyopathy:serological evidence of persistent enterovirus infection. Lancet I:804–807

Okada I, Matsumori, A, Kawai C, Yodi J, Tracy S (1990) The viral genome in experimental murine coxsackievirus B3 myocarditis: A northern blotting analysis. J Molec and Cell Cardiol 22:999–1008

Parks WP, Queiroga LT, Melnick JL (1967) Studies of infantile diarrhoea in Karachi, Pakistan. II. Multiple virus isolations from rectal swabs. Am J Epidemiol 85:469–478

Petitjean J, Kopecka H, Freymuth F, Langlard JM, Scanu P, Galateau F, Bouhour JB, Ferriere M, Charbonneau P, Komajda M (1992) Detection of enteroviruses in endomyocardial biopsy by molecular approach. J Med Virol 37:76–82

Pollen RH (1963) Myocarditis due to Coxsackie virus group B type 5. Am J Cardiol 12:736–740

Ramamurthy S, Talwar KK, Goswami KC, Shrivaastava S, Chopra P, Broor S, Malhotra A (1993) Clinical profile of biopsy proven idiopathic myocarditis. Int J Cardiol 41:225–232

Robart H, Levin M, Villareal L, Tracy S, Semler B, Wimmer E (1985) Factors affecting the detection of enteroviruses in cerebrospinal fluid with coxsackievirus B3 and poliovirus type 1 cDNA probes. J Clin Microbiol 22:220–224

Rueckert RR (1996) Picornaviridae: The viruses and their replication. In: Fields BN, Knipe DM, Howley PM, Chanock RM, Joseph LM, Monath TP, Roizman B, Straus SE (eds), Fields Virology, Third Edition. Lippincott-Raven, Philadelphia

Sainani GS, Dekate MP, Rao CP (1975) Heart disease caused by coxsackievirus B infection. Br Heart J 37:319–323

Sainani GS, Krompotic E, Slodkis J (1968) Adult heart disease due to the coxsackievirus B infection. Medicine 47:133–147

Satoh M, Tamura G, Segawa I (1994) Enteroviral RNA in endomyocardial biopsy tissues of myocarditis and dilated cardiomyopathy. Pathol Int 44:345–351

Schwaiger A, Umlauft F, Weyrer K, Larcher C, Lyons J, Muhlberger V, Dietze O, Grunewald K (1993) Detection of enteroviral ribonucleic acid in myocardial biopsies from patients with idiopathic dilated cardiomyopathy by polymerase chain reaction. Am Heart J 126:406–410

Schwimmbeck PL, Badorff C, Schultheiss HP, Strauer BE (1994) Transfer of human myocarditis into severe combined immunodeficiency mice. Circulation Res 75:156–164

Seko Y, Ishiyama S, Nishikawa T, Kasajima T, Hiroe M, Kagawa N, Osada S, Suzuki S, Yagita H, Okumura K, Yazaki Y (1995) Restricted usage of T cell receptor Va-Vb genes in infiltrating cells in the hearts of patients with acute myocarditis and dilated cardiomyopathy. J Clin Invest 96: 1035–1041

Soutar CA (1971) Unusual case of viral pericarditis. Lancet I:498

Swan NH (1961) Epidemic pleurodynia, orchitis, and myocarditis in an adult due to Coxsackie virus group B, type 4. Ann Int Med 54:1008–1013

Tanimura A, Nakashima N, Komatsu Y, Arakawa M, Ookita Y, Senoo T, Toshima H, Hoike S (1980) Clinicopathological study of coxsackie B viral myocarditis. Acta Pathol Jpn 30:161–170

Towbin JA (1996) Diagnosis and management of myocarditis in children. Presented at the Second International Meeting on Advances in Understanding and Management of the Cardiomyopathies.

Tracy S (1984) A comparison of genomic homologies among the coxsackievirus B group: Use of fragments of the cloned coxsackievirus B3 genome as probes. J Gen Virol 65:2167–2172

Tracy S (1985) Comparison of genomic homologies in the coxsackievirus B group by use of RNA:cDNA dot blot hybridization. J Clin Microbiol 21:371–374

Tracy S, Chapman N, Liu H (1985a) Molecular cloning and partial characterization of the coxsackievirus B3 genome. Archives Virol 85:157–163

Tracy S, Liu H, Chapman N (1985b) Coxsackievirus B3: Primary structure of the 5' non-coding and the capsid protein coding regions of the genome. Virus Research 3:263–270

Tracy S, Chapman NM, McManus BM, Pallansch MA, Beck MA, Carstens J (1990) A molecular and serologic evaluation of enteroviral involvement in human myocarditis. J Mol Cell Cardiol 22:403–414

Tracy S, Chapman NM, Tu Z (1992) Coxsackievirus B3 from an infectious cDNA copy of the genome is cardiovirulent in mice. Arch Virol 122:398–409

Tracy S, Chapman NM, Romero J, Ramsingh AI (1996) Genetics of Coxsackievirus B cardiovirulence and inflammatory heart muscle disease. Trends Microbiol 4:175–179

Tu Z, Chapman NM, Hufnagel G, Tracy S, Romero JR, Barry WH, Zha OL, Curry K, Shapiro B (1995) The cardiovirulent phenotype of coxsackievirus B3 is determined at single site in the genomic 5' non translated region. J Virol 69:4607–4618

Ueno H, Yokota Y, Shiotani H, Yokoyama M, Itoh H, Ishido S, Maeda S, Katayama Y, Hotta H (1995) Significance of detection of enterovirus RNA in myocardial tissues by reverse transcription-polymerase chain reaction. Int J Cardiol 51:157–164

Van Creveld S, DeJager H (1956) Myocarditis in new-borns caused by Coxsackie virus. Clinical and pathological data. Ann Pediatr 187:100–112

Van-Houten N, Bouchard PE, Moraska A, Huber SA (1991) Selection of an attenuated coxsackievirus B3 variant, using a monoclonal antibody reactive to myocyte antigen. J Virology 65:1286–1290

Weiss L, Movahed L, Billingham ME, Cleary ML (1991) Detection of coxsackie B3 RNA in myocardial tissues by polymerase chain reaction. Am J Pathol 138:497–503

Wells RG, Ruskin JA, Sty JR (1986) Myocardial imaging. Coxsackie myocarditis. Clin Nucl Med 11:661–662

Werner GS, Figulla HR, Munz DL, Klingel K, Kandolf R, Emrich D, Kreuzer H (1993) Myocardial indium-111 antimyosin uptake in patients with idiopathic dilated cardiomyopathy: its relation to haemodynamics, histomorphometry, myocardial enteroviral infection, and clinical course. Eur Heart J 14:175–184

Wesselen L, Waldenstrom A, Lindblom B, Hoyer S, Friman G, Fohlman J (1993) Genotypic and serotypic profile in dilated cardiomyopathy. Scand J Infect Dis 88(Suppl):87–91

Why HJF, Meany BT, Richardson PJ, Olsen EGJ, Bowles NE, Cunningham L, Freeke CA, Archard LC (1994) Clinical and prognostic significance of detection of Enteroviral RNA in the myocardium of patients with myocarditis or dilated cardiomyopathy. Circulation 89:2582–2589

Wiegand V, Tracy S, Chapman N, Wucherpfenning C (1990) Enterviral infection in end-stage dilated cardiomyocpathy . Klin Wochensche 68:914–920

Woodruff JF, Woodruff JJ (1974) Involvement of T lymphocytes in the pathogenesis of coxsackievirus B3 heart disease. J Immunol 113:1726–1734

Zoll GJ, Melchers WJG, Kopecka H, Jambroes G, van der Poel HJA, Galama MD (1992) General primer-mediated polymerase chain reaction for detection of enteroviruses: Application for diagnostic routine and persistent infections. J Clin Microbiol 30:160–165

Zvizdic S, Seric K, Radovic S, Selak I (1990) Izolacija virusa iz obdukcionog materijala za vrijeme epidemije coxsackie viroza, u Sarajevu 1985. Med Arh 44:39–42

Group B Coxsackie Disease in Children

J. F. Modlin[1] and H. A. Rotbart[2]

[1]Dartmouth Medical School, Infection Disease Section, Dartmouth-Hitchcock Medical Center, 1 Medical Center Drive, Lebanon, NH, 03756, USA

[2] Pediatric Infectious Disease, University of Colorado Health Sciences Center, 4200 East 9th Avenue, Denver, CO 80262, USA

1 Introduction

The group B coxsackieviruses (CVB) cause a wide variety of clinical manifestations in children. Some disease syndromes are identical to, or share many common features with, diseases caused by other enteroviruses. Indeed, most clinical virology laboratories no longer identify isolated enteroviruses by group or by serotype, since this information is rarely of immediate importance to physicians and their patients. Nonetheless, there are several clinical syndromes which are uniquely associated with CVB infection, such as myopericarditis and pleurodynia (Bornholm disease). This chapter covers several important CVB-related diseases in childhood including CNS infection, respiratory tract infections, pleurodynia, and vertically transmitted infections. A complete discussion of myopericarditis, which is also an important pediatric disease, can be found in Chap. 3.

2 Vertical Transmission and Neonatal Disease

2.1 Background

Neonates are uniquely susceptible to severe CVB infection. Infection of the newborn is acquired most commonly via vertical transmission from the infected mother and less frequently from postnatal transmission from a nosocomial or other source. Both types of transmission occur sporadically and in the setting of seasonal community outbreaks of CVB disease. This section reviews current knowledge regarding the pathophysiology, clinical manifestations, and management of CVB infection in the newborn.

2.2 Incidence

The true incidence of neonatal CVB infections is not known. Non-polio enterovirus infections are not nationally reportable, and there are no population based, prospective studies of the incidence of neonatal infection. One study, based on laboratory records, suggested a minimum rate of CVB infections of 50 per 100 000 among infants born in Nassau County, NY from 1970 to 1979 (Kaplan, et al. 1989). The CDC Enteroviral Surveillance Program, which is largely based on passive reporting from public health laboratories, provides some data regarding the occurrence of neonatal disease for each of the enterovirus classes relative to one another. Over the 4 year period 1972–1975, 45% of reported 338 infants under 2 months of age had CVB infections, compared with 51% and 4%, respectively, for the echoviruses and the group A coxsackieviruses (Morems 1979). Of these cases, 74% were classified as "severe" based on a diagnosis of CNS disease, systemic illness, hepatitis, myocarditis, or death.

2.3 Experimental Studies

2.3.1 Host Factors

Like humans, mice exhibit a marked age-related susceptibility to CVB infection (CURNEN et al. 1949; KAPLAN and MELNICK 1951). The reasons for this unique vulnerability of the newborn to severe infection are not well understood. One major hypothesis holds that newborn tissues possess an innate susceptibility to CVB infection (Holland 1961). The knowledge that CVB tropism is dependent on the presence of specific receptors on the cell membrane (HSU et al. 1988; SHAFREN et al. 1995) now offers the opportunity to determine whether the presence or density of these receptor proteins may be related to neonatal disease; i.e., it is possible that the expression of certain cell surface proteins that double as enterovirus receptors is up-regulated early in life, thus enhancing the susceptibility of specific tissues or organs to CVB infection.

Another important host factor is the relative immaturity of the neonatal immune response to to CVB infection. The relative inability of human neonatal macrophages to function as well as those of older individuals (MILLS 1983) may contribute substantially to the susceptibility of newborn infants to overwhelming enterovirus infection. It has been shown experimentally that newborn mice challenged with CVB type 3, which causes death within 7 days, are protected when adult murine macrophages are transferred prior to infection (RAGER-ZISMAN and ALLISON 1973).

In contrast, human neonates are capable of responding to enterovirus infection with high titers of humoral antibody (EICHENWALD and KOTSEVALOV 1960). Furthermore, while neonatal enterovirus infections are often severe, they are not unusually prolonged, compared to infections in older persons. Cytotoxic T lymphocytes generated in response to CVB infection in the mouse actually contributes to necrosis of both infected and uninfected cells (HUBER et al. 1984), while inhibition of the T cell response has little effect on animal survival (WOODRUFF and WOODRUFF 1974). These data suggest that the relative immaturity of neonatal T cell function is not directly responsible for the severity of neonatal enterovirus infection.

2.3.2 Animal Models of Vertical Transmission

Pregnancy enhances infection by CVB (DALLDORF and GIFFORD 1954; MODLIN and CRUMPACKER 1982) and by other enteroviruses (KNOX 1950; FARBER and GLASGOW 1970) in laboratory mice. CVB-infected pregnant mice develop higher titers of virus in the blood and tissues and remain viremic longer than nonpregnant mice (MODLIN and CRUMPACKER 1982). Furthermore, susceptibility to enteroviruses increases with advancing gestation and then rapidly reverts to that of nonpregnant animals within days of delivery (KNOX 1950; FARBER and GLASGOW 1970). The mechanism of diminished resistance of the late gestational animal to enterovirus infection may be mediated by hormonal influences. FARBER and GLASGOW(1970) demonstrated that resistance of nonpregnant female mice to encephalomyocarditis virus infection could be reduced by administration of corticosterone or estrogen but was not altered by exogenous progesterone.

The CVB-infected pregnant mouse transmits virus to the fetus prior to delivery or to the newborn mouse intrapartum (SOIKE 1967; SELZER 1969; MODLIN and BOWMAN 1987), depending on the gestational stage at which the mouse is infected. Infection early in gestation results in a high rate of fetal loss due to resorption and abortion (DALLDORF and GIFFORD 1954; SELZER 1969; MODLIN and CRUMPACKER 1982; [CE1]Modlin 1987). In contrast, infection later than 12 days of gestation (which typically spans 19–21 days) uncommonly results in fetal infection in utero, instead resulting in a high rate of stillbirth and postnatal deaths among newborn mice born at term (DALLDORF and SICKLES 1948; SOIKE 1967; MODLIN et al. 1981; MODLIN and BOWMAN 1987). In these late gestational infections the placenta has proven to be an effective, but not impenetrable, barrier to fetal infection by CVB (MODLIN et al. 1981; MODLIN and BOWMAN 1987) and by other enteroviruses (ABZUG et al. 1989). The mechanisms by which the placenta protects the fetus during late gestational enterovirus infection are unknown; however, both human and murine placental tissues contain a high density of macrophages (Hofbauer cells) which may contribute to the effectiveness of the placental barrier (MODLIN and BOWMAN 1988).

Experimental studies in mice (MODLIN and BOWMAN 1987) have confirmed observations in humans (see below), that neutralizing antibody passively acquired from the mother reduces the risk of serious disease and death from perinatally acquired CVB infection.

2.4 Infections in Pregnant Women and Vertical Transmission

2.4.1 CVB Infections in Pregnant Women

Group B coxsackievirus infections occur frequently during pregnancy, although vertical transmission to the fetus or neonate is relatively uncommon. The Collaborative Perinatal Research Project found diagnostic complement fixation antibody titer rises to CVB types 1 and 5 occurred in 2%–8% of 198 women in Boston and Philadelphia during the last 4 months of pregnancy (SEVER et al. 1963). Similar rates of gestational CVB infections have been confirmed by others. For example, Cherry and colleagues detected seroconversion to CVB types 2 or 3 in five of 55 pregnant women followed prospectively during the last 2–7 weeks of pregnancy (CHERRY et al. 1968).

It is not certain how many maternal CVB infections cause symptomatic illness. Point prevalence surveys of term pregnant women suggest that asymptomatic enterovirus infections are rarely identified (JENISTA et al. 1984). However surveillance conducted over time indicate that maternal enterovirus infections may or may not result in illness (CHERRY et al. 1968). Data on non-polio enterovirus infections in pregnant women culled from case reports of women giving birth to infected infants indicate that about two thirds have had symptomatic illnesses during the perinatal period. Reports of CVB infections in pregnancy include nonspecific febrile illness, upper respiratory tract symptoms, pleurodynia, myocarditis, and aseptic meningitis (DELANEY and FUKUNGA 1958; KIBRICK and BENIRSCHKE 1958; RAPMUND et al. 1959; MCLEAN et al. 1961; ROBINO et al. 1962; ARTENSTEIN et al. 1964; NOGAN and LEPOW 1967; CHERRY et al. 1968; HALL and MILLER 1969; KAPLAN et al. 1983).

Although it is not known whether CVB infections in pregnant women are more likely to result in symptomatic illness or more severe illness than in nonpregnant women, there are reasons to suspect that pregnancy is a risk factor for more serious enterovirus illness. Paralytic poliomyelitis occurred at about three times the expected incidence among pregnant women, and poliomyelitis developing in the last trimester was associated with a higher risk of bulbar paralysis and death (WEINSTEIN et al. 1951; Siegel, Greenberg 1956). Furthermore, experimental murine CVB infections produce prolonged viremia, higher titers of virus in blood and tissues, and enhanced mortality in late gestational animals compared with nonpregnant animals (DALLDORF and GIFFORD 1954; SOIKE 1967; MODLIN and CRUMPACKER 1982). Therefore, greater viral replication and prolonged maternal enterovirus excretion occurring in late pregnancy may well enhance the risk of infection of the newborn infant in the perinatal period.

2.4.2 Mechanism of Vertical Transmission

Group B coxsackievirus transmission from mother to infant may occur in utero, during parturition, or during the postpartum period, depending on the timing of the acute maternal infection.

The evidence for prenatal CVB infection includes recovery of virus recovery from amniotic fluid late in gestation (KAPLAN et al. 1983) and documented intrauterine infections in late pregnancy associated with stillbirth (BURCH et al. 1968; BATES 1970) and with serious neonatal disease presenting at birth or shortly after birth (KAPLAN et al. 1983). However prenatal CVB infections are rare. Studies in mice with CVB types 1 and 3 have consistently shown the placenta to be an effective, though not impenetrable, barrier to intrauterine transmission (SOIKE 1967; SELZER 1969; MODLIN and BOWMAN 1987). The extremely low risk of human fetal infection suggests that the human placenta serves a similar protective function.

Most neonates with serious CVB infection have onset of symptoms between 3 and 10 days of age (KAPLAN et al. 1983; KRAJDEN 1983) strongly suggesting that many infants are exposed and infected during the peripartum period. Multiple routes of transmission are possible. Although the experience with CVB is limited, other enteroviruses have been recovered from vaginal (JONES et al. 1980) and cervical (REYES et al. 1983) secretions of mothers of infected newborns, in addition to the upper respiratory tract and lower gastrointestinal tract.

The potential for nosocomial and other types of postpartum transmission is amply documented by numerous reports of outbreaks of CVB infection in newborn nurseries (reviewed in reference MODLIN 1994). Both full-term and intensive care nurseries have experienced outbreaks coinciding with seasonal peaks of CVB activity. Spread of infection via the hands of nursery personnel has been strongly implicated. Nursing activities such as gavage feeding and routine mouth care may increase the risk of neonatal enterovirus infection (KINNEY et al. 1986).

2.4.3 Role of Passively Transferred Maternal Antibody

Clinical studies conducted during community outbreaks of echovirus infection noted that infants who developed fatal, vertically transmitted infections had no detectable neutralizing antibody to the specific echovirus serotype, while similarly exposed infants who had acquired antibody transplacentally by the time of birth either remained uninfected or developed only asymptomatic infections (NAGINGTON et al. 1978; MODLIN 1980; MODLIN et al. 1981; BERRY and NAGINGTON 1982). Earlier studies had established the efficacy of passively administered immune serum globulin (ISG) for the prevention of poliomyelitis (STEVENS 1959), and the known interference of oral poliovirus vaccine immunization of the newborn by maternal antibody (WARREN et al. 1964).

Even though there are no corresponding data for CVB infections in newborns, these data, in addition to the animal data cited above, strongly suggest that the timing of the mother's infection in the perinatal period is likely to be a critical determinant of the outcome of neonatal CVB infection. Maternal infections beginning more than 5–7 days prior to delivery allow for the development and transplacental transport of specific IgG antibody which protects the newborn infant from severe disease (but not necessarily from infection). Conversely, maternal infections occurring closer to delivery, or in the immediate peripartum period, place the infant at much higher risk of infection with a poor prognosis.

2.5 Clinical Manifestations of Group B Coxsackievirus Disease in the Neonate

The clinical expression of neonatal CVB disease depends on several factors, including the timing of maternal infection, the age of the infant at the time of exposure, and the presence or absence of passively acquired maternal antibody. CVB cause a considerable range of clinical manifestations in infected neonates, including exanthems, aseptic meningitis, meningoencephalitis, pneumonia, myocarditis, gastroenteritis, necrotizing enterocolitis, hepatitis, pancreatitis, generalized hemorrhage, and sudden death (CHERRY 1990).

The most commonly encountered syndrome caused by CVB in the neonatal period is probably simple, undifferentiated fever with or without aseptic meningitis. Because the clinical features of this syndrome for infants in the first days of life are not much different than for infants up to 3 months of age (Berlin et al. 1993), the reader is referred to the section below on "Undifferentiated Fever and Aseptic Meningitis" for a more complete discussion. In contrast, the syndrome of acute myocarditis, which occurs with or without meningoencephalitis, is characteristic of more serious neonatal CVB disease with onset of symptoms in the first 7–10 days of life.

2.5.1 Myocarditis

Although CVB myocarditis is extensively reviewed in Chap. 3, there are features of this disease that are unique when it occurs within the first 10–14 days of life, which will be briefly covered in this section.

Neonatal myocarditis is most frequently caused by CVB types 2, 3, 4, and 5; type 1 has been reported less frequently (Kaplan et al. 1983; Cherry 1990). Onset of symptoms beyond the tenth day of life is rare. Approximately one third of cases present with a biphasic illness with nonspecific signs and symptoms of lethargy, poor feeding, or mild respiratory distress preceding the onset of cardiac manifestations by 2–5 days (Kibrick 1961). The onset of frank myocarditis is often abrupt with respiratory distress, tachycardia, cyanosis, jaundice, and diarrhea. Initial examination will often reveal temperature instability, tachycardia, arrhythmias, hepatomegaly, and signs of poor peripheral circulation. The ECG may show low voltage and other electrophysiologic abnormalities, and echocardiographic studies often indicate poor left ventricular or biventricular function.

Unlike older children and adults, infants with CVB myocarditis often develop concomitant meningoencephalitis, pneumonia, hepatitis, pancreatitis, or adrenalitis. Although a limited number of organs may be involved pathologically, the degree of involvement is often severe, giving rise to the use of such adjectives in the literature as "disseminated,", "systemic," or "overwhelming" CVB disease in the neonatal period. Mortality among infants with myocarditis alone is generally reported to be 30%–50%, but is inevitably higher when other organs are involved.

2.5.2 Other Neonatal Disease

Meningoencephalitis presents with lethargy, seizures, hemiparesis, flaccid paralysis, and coma. The signs and symptoms differentiate the more serious forms of CNS involvement that often accompany myocarditis or hepatitis from the far more benign disease observed with aseptic meningitis. Inflammation of the brain or spinal cord can be found in approximately two thirds of infants dying with myocarditis (Kaplan et al. 1983). For these patients, diffuse or scattered lesions consisting of perivascular infiltration with mononuclear cells and polymorphonuclear leukocytes are described in the cerebrum, cerebellum, pons, medulla, and spinal cord. Meningoencephalitis in the absence of other organ involvement is quite rare.

Hepatitis may occur along with myocarditis and other manifestations of "generalized" neonatal CVB infection; however, unlike the echoviruses, CVB rarely causes severe hepatitis alone.

Pneumonitis is often found at postmortem examination of infants dying of CVB myocarditis, along with pulmonary vascular congestion and extravasation of edema fluid and red cells into alveoli (Gear and Measroch 1973). Pathologic evidence of viral pneumonitis is generally limited to focal areas of interstitial inflammation (Kibrick and Benirschke 1958; Robino et al. 1962; Case Records of the Massachusetts General Hospital 1965; Kaplan et al. 1983; Wong et al. 1989).

2.6 Management

The management of CVB disease in the newborn is mainly supportive. However, the severity of disease and the poor prognosis have generated interest in the use of immune globulin for treatment of neonatal enterovirus infections. Anecdotal reports describe successful outcomes following the administration of intravenous immune globulin (IVIG; 1 g/kg) (Johnston and Overall 1989) and maternal plasma (Jantausch et al. 1995) to echovirus type 11-infected infants with hepatitis, but similar reports relating to CVB-infected neonates have not been made.

Recently, Abzug and colleagues have reported a randomized trial of IVIG (750 ml/kg) among enterovirus infected newborns (Abzug et al. 1995). Treatment was associated with an antiviral effect when the IVIG neutralizing antibody titer was $\geq$ 1:800 against the infant's enterovirus isolate; however the study was too small to assess clinical benefit.

Although several antipicornavirus agents are undergoing early clinical trials in humans (see Sect. 11), none have been given to infants with CVB disease.

2.7 Prevention

Virtually all serious neonatal CVB infections are acquired in the peripartum period either directly from the mother or from nosocomial sources. Thus, while many maternal enterovirus infections may be unavoidable because of the pervasive presence of enteroviruses during the summer and fall, the risk of maternal infection late in gestation can probably be reduced by special attention to hand washing, particularly after changing diapers and other close contact with objects contaminated by feces, urine, or respiratory secretions. Since transmission via the hands of nursing personnel have been implicated in the nosocomial spread of enteroviruses among infants in newborn nurseries, strict enforcement of recommended infection control practices should diminish enterovirus transmission in nurseries for both preterm and term newborns.

Finally, when women present with signs and symptoms of enterovirus infection in late gestation, i.e., fever, rash, headache, pharyngitis, or lower abdominal pain, the temptation to terminate pregnancy by induction of labor or cesarian delivery is best avoided. Since passively acquired antibody appears critical to the development and outcome of neonatal enterovirus infection, it would be prudent for the obstetrician to delay delivery long enough to allow the development and transplacental transfer of specific maternal IgG antibody, typically 5–10 days after the onset of symptoms. Unfortunately, the signs and symptoms of acute maternal enterovirus infection may suggest other diagnoses such as chorioamnionitis or abruptio placentae, which often dictate immediate delivery. In the future, rapid diagnostic methods such as the PCR may assist the obstetrician in making these difficult management decisions.

3 Undifferentiated Fever and Aseptic Meningitis in Infants Up to 3 Months of Age

3.1 Background

Like other non-polio enteroviruses, infection due to CVB is a commonly identified cause of fever without an apparent focus among infants under 3 months of age (LEGGIADRO and DARRAS 1983; DAGAN et al. 1985, 1989; KROBER et al. 1985). Overall, enteroviruses account for at least 53%–63% of these cases during the summer and fall months (LEGGIADRO and DARRAS; DAGAN et al. 1989). Fever is often the sole finding although some infants will present with other nonspecific findings such as irritability, lethargy, poor feeding, vomiting, diarrhea, exanthems, and signs of upper respiratory tract infection (RORABAUGH et al. 1993). Clinical management for infants of this age with persistent or moderate to high fever usually includes a full "septic workup" to rule out serious bacterial disease. This evaluation often includes examination of the cerebrospinal fluid (CSF), which demonstrates the presence of aseptic meningitis in approximately half of infants ultimately proven to have enterovirus infection (KROBER et al. 1985; DAGAN et al. 1988; RORABAUGH et al. 1993), even though there are no clinical features that distinguish those with meningitis prior to performing the lumbar puncture (DAGAN et al. 1988).

3.2 Etiology

Among the different CVB serotypes, types 2, 4 and 5 appear to be the most commonly associated with undifferentiated fever and aseptic meningitis in young infants (BERLIN et al. 1993). Dagan and colleagues found that the infants infected with CVB (and echovirus serotypes 11 and 30) were more likely to have meningitis than infants infected with other enterovirus serotypes (DAGAN et al. 1988).

3.3 Course

Most infants hospitalized with enterovirus induced fever or aseptic meningitis recover within 2–10 days without complications. In fact, one study (DEANGELIS et al. 1983) has shown greater morbidity from complications of therapy and from nosocomial infections, than from the underlying viral infection itself. Nonetheless, approximately 10% of infants and children under 2 years of age hospitalized with aseptic meningitis experience acute CNS complications such as seizures, obtundation, or increased intracranial pressure (RORABAUGH et al. 1993). Although some early reports suggested a moderate risk of long-term neurological and cognitive sequelae for infants experiencing viral meningitis at an early age (FARMER et al. 1975; SELLS et al. 1975; JENISTA et al. 1984), recent prospective studies with larger numbers of subjects have indicated virtually no measurable effects (BERGMAN et al. 1987; GONZALEZ-DEL-RAY et al. 1991; RORABAUGH et al. 1992), including those patients with neurological findings during the acute illness (RORABAUGH et al. 1993). Aseptic meningitis appears to have few short-term complications, and virtually no long-term sequelae for even the youngest infants with CVB meningitis.

4 Central Nervous System Infections

4.1 Aseptic Meningitis and Encephalitis

The clinical disease observed during CVB meningitis is indistinguishable from that due to the other enteroviruses and varies with the host's age and immune status. Neonates are at risk for severe systemic illness (see below) of which meningitis or meningoencephalitis is commonly a part (KAPLAN et al. 1983; ABZUG et al. 1993; MODLIN 1994; ABZUG 1995). CVB were associated with aseptic meningitis in 62% of infants less than 3 months of age in one study (KAPLAN et al. 1983). In a recent prospective study of neonates (<2 weeks of age) with proven enterovirus infection (of which a number were due to CVB), 75% had clinical or laboratory evidence of meningitis (ABZUG et al. 1993).

CVB meningitis outside of the immediate (<2 weeks of age) neonatal period is rarely associated with severe disease or poor outcome. Onset is usually sudden, and fever of 38°–40°C is the most consistent clinical finding, occurring in 76%–100% of patients (LEPOW et al. 1962; SINGER et al. 1980; WILFERT et al. 1983). The fever pattern may be biphasic (HORSTMANN and YAMADA 1968), appearing first with nonspecific constitutional symptoms, followed by resolution and reappearance with the onset of meningeal signs. Nuchal rigidity is found in more than half of the patients, particularly in children older than 1–2 years of age (LEPOW et al. 1962; SINGER et al. 1980; WILFERT et al. 1983). Headache is nearly always present in adults and children old enough to report it, and photophobia is also common.

Nonspecific and constitutional signs and symptoms of viral infection, in decreasing order of occurrence, include vomiting, anorexia, rash, diarrhea, cough and upper respiratory findings (particularly pharyngitis), diarrhea, and myalgias (LEPOW et al. 1962 HORSTMANN and YAMADA 1968; SINGER et al. 1980; WILFERT et al. 1983). Symptoms other than fever may also be biphasic, a presentation observed more often in adults than children. Neurologic abnormalities are unusual; the literature on enterovirus meningitis rarely includes more than 5% of patients with abnormal neurologic examinations. Febrile seizures are among those neurologic "abnormalities" which may complicate aseptic meningitis in children without implicating parenchymal brain involvement. A history of concomitant family illness is often obtained, including rashes and upper respiratory or gastrointestinal symptoms, and occasionally meningitis (FAULKNER et al. 1957). The duration of illness due to CVB meningitis is usually less than 1 week, with many patients feeling better immediately after the lumbar puncture (JAFFE et al. 1989), presumably due to the transient reduction of intracranial pressure with fluid removal. Occasional adult patients may have symptoms which persist for several weeks (LEPOW et al. 1962), an observation not frequently made in children. Comparison of case series of adult enterovirus meningitis (MOSES et al. 1977; KINNUNEN et al. 1987; RICE et al. 1995) with similar series in children (RAY et al. 1966; SUMAYA and CORMAN 1982; WILFERT et al. 1983; BERLIN et al. 1993; DAI-MING et al. 1993) suggest that the disease is more severe in adults. Difficulties in eliciting symptoms from young infants may contribute to this impression, in contrast to adults who readily attest to prolonged headaches, dizziness, photophobia, etc.

Table 1. Controlled long-term outcome studies of enteroviral meningitis occurring in the first year of life

Investigator	n	Controls	Abnormal findings: Neurologic	Cognitive	Developmental/ Language
SELLS et al. (1975)	13	Age-match	+	+	+
FARMER et al. (1975)	15	Matched	+	+	+
WILFERT et al. (1981)	9	Matched	–	–	+
JENISTA et al. (1984)	71	[a]Matched	–	ns	+
BERGMAN et al. (1987)	42	Siblings	–	–	–
GONZALEZ-DEL-RAY et al. (1991)	16	Matched	–	+	+
RORABAUGH et al. (1991)	140	[b]Matched	–	–	–

+, abnormalities found; –, abnormalities not found; ns, not stated.
[a] Neonatal patients with enterovirus infection, not necessarily all meningitis.
[b] Patients with aseptic meningitis, not necessarily all enteroviral.

The short-term prognosis of young children with enterovirus meningitis early in life appears to be good; however, there has been some controversy as to possible later sequellae. Uncontrolled studies found numerous, subtle, long-term behavioral and neurologic abnormalities (KAWANA et al. 1965). Neurologic, cognitive, and developmental/language abnormalities have been reported in controlled studies of long-term outcome in children with enterovirus meningitis during infancy (Table 1) (SELLS et al. 1975; WILFERT et al. 1981; JENISTA et al. 1984; BERGMAN et al. 1987; GONZALEZ-DEL-REY et al. 1991;). In the largest and most carefully controlled study, however, no differences between patients and controls could be demonstrated in any of the neurodevelopmental parameters studied (Table 1) (RORABAUGH et al. 1992). Less well studied are the ultimate outcomes of aseptic meningitis cases in older children and adolescents; preliminary data suggest possible school and learning difficulties, but control patients were not studied (ETTER et al. 1991). Examination of the CSF of patients with CVB meningitis usually reveals a modest monocytic pleocytosis (100–1000 cells/mm^3) with normal or slightly depressed glucose levels and normal to slightly increased protein. Importantly, however, wide variations in these findings have been reported (reviewed in ROTBART 1995), with cell counts ranging from 0 to several thousand, marked hypoglycorrhachia, and markedly elevated CSF protein amounts. Neutrophils may be the predominant cell type in early CVB meningitis with a shift to monocyte predominance within 24–48 h (1995).

Encephalitis due to the CVB is well known but thought to be an unusual complication (WHITLEY et al. 1989; ROTBART 1995b). Improved diagnostic methods such as the PCR (see below) may result in increased recognition of CVB in cases of encephalitis (ROTBART et al. 1990a). Unlike aseptic meningitis, encephalitis due to enteroviruses may have more profound acute disease and long-term sequellae. In contrast to the typical focal disease seen with herpes simplex virus encephalitis, enteroviruses have been more commonly associated with global encephalitis and generalized neurologic depression (LEPOW et al. 1962; HOWDEN et al. 1966; THIVIERGE and DELAGE 1982; DAI-MING et al. 1993). The illness usually begins like aseptic meningitis with a prodrome of fever, myalgias, and upper respiratory symptoms. Onset of CNS signs and symptoms is often abrupt, with confusion, weakness, lethargy, drowsiness, or irritability. Progression to coma or generalized seizures may occur. When meningeal signs and CSF pleocytosis accompany these findings, meningoencephalitis is the appropriate term. Unusual but occasional findings in CVB encephalitis include blurred optic discs and other signs of increased intracranial pressure, multifocal encephalomyelopathy, apnea, truncal ataxia, abnormalites of cranial nerves, and paralysis; the latter sign is usually a manifestation of spinal cord involvement and, when accompanying central signs and symptoms, is appropriately termed encephalomyelitis.

Focal CVB encephalitis is less commonly reported than global disease, but may be underappreciated. Enteroviruses are demonstrable by brain biopsy in 13% of patients suspected of having herpes simplex encephalitis, the most commonly identified cause of focal encephalitis (WHITLEY et al. 1981; 1989); several of those patients had CVB serotypes. Modlin et al. reported four cases of focal enterovirus encephalitis and reviewed an additional three from the literature (MODLIN et al.

1991). A variety of focal neurologic findings were seen as well as abnormalities by imaging studies; one of the seven patients had CVB2 documented infection, but most were coxsackievirus A serotypes.

The unique situation of a child or adult with absent or deficient humoral immunity illustrates an important "experiment of nature" with regard to enterovirus infections of the CNS. Unlike other viruses which are largely controlled by cellular immune mechanisms, the enteroviruses are cleared from the host primarily by antibody-mediated mechanisms. Agammaglobulinemic individuals infected with the enteroviruses may develop chronic meningitis or meningoencephalitis lasting many years, often with fatal outcome (McKinney et al. 1987); although this syndrome can occur in patients with mixed humoral and cellular immune deficiencies, normal cell-mediated immune function can be documented in most affected patients. The echovirus serotypes have been most commonly associated with this form of chronic meningoencephalitis, but cases of CVB infection have been reported and are clinically indistinguishable from those due to other enteroviruses (McKinney et al. 1987). Although CSF culture-negative periods occur, more consistent evidence of persistent virus has been obtained using the PCR (Rotbart et al. 1990a). Alteration of the viral genome over time in the same individual has also been documented in a patient with chronic coxsackievirus A infection (O'Neil et al. 1988). Approximately 50% of these infected patients also develop a rheumatologic syndrome, most often dermatomyositis, which is also felt to be a direct result of enterovirus infection of the affected tissues (McKinney et al. 1987). Treatment with antibody preparations intravenously and intrathecally or intraventricularly has resulted in stabilization of some of these patients; however, virus persistence has been documented during therapy (McKinney et al. 1987; Rotbart et al. 1990). With the availability of intravenous preparations of immune globulin and the early recognition of this illness, fewer patients appear to be progressing to the classic description of this disease, and atypical neurologic presentations have appeared, including emotional lability, dementia, ataxia, paresthesias, deafness, memory loss, dysarthria (Webster et al. 1993). The expanded clinical spectrum of chronic enterovirus meningoencephalitis has been revealed by PCR studies of CSF from patients with these atypical manifestations (Webster et al. 1993; Rotbart 1995).

4.2 Other Neurologic Syndromes Associated with Group B Coxsackievirus Infections

Numerous non-polio enteroviruses have been associated with outbreaks of paralytic myelitis, including CVB serotypes (Cherry 1992). Some of these associations are weakened by identification of the enterovirus only in stool specimens, a situation in which causality is difficult to establish due to the known prolonged fecal shedding period of the enteroviruses (see below). Cerebellar ataxia has occasionally been associated with CVB infections, as have Guillain-Barre syndrome and transverse myelitis (CHERRY 1992). All such associations suffer from the same difficulty in distinguishing pathogenicity of a throat or stool isolate from coincidental shedding.

5 Respiratory Tract Diseases

Many CVB infections, like those due to other enterovirus serotypes, are accompanied by nonspecific respiratory signs and symptoms, usually mild in nature. The common cold (summer cold), pharyngitis, tonsillitis, and laryngotracheobronchitis (croup) have been frequently reported (CHONMAITREE and MANN 1995). Bronchiolitis and pneumonia are less commonly seen. Most CVB respiratory illness is benign, but symptoms may persist for many days and the resultant disruption in school and work days may be substantial. The enteroviruses are responsible for approximately 15% of upper respiratory infections (URIs) for which an etiology is identified (CHONMAITREE and MANN 1995); conversely, respiratory illness is the major manifestation reported in 15–20% of cases of enterovirus infections (GRIST et al. 1978). In a recent 10 year review of enterovirus-associated respiratory illnesses, 46% of cases presented with URIs, 13% with respiratory distress/apnea, 13% with pneumonia, 12% with otitis media, and fewer cases with bronchiolitis, wheezing, croup, and pharyngotonsillitis (CHONMAITREE and MANN 1995). Many enterovirus serotypes are identified in respiratory infections, approximately equally divided among the major subgroups (GRIST et al. 1978; MOORE 1982). The clinical manifestations of CVB-associated URIs, otitis media, and pharyngotonsillitis, are indistiguishable from those due to other respiratory viruses. Pneumonia due to the enteroviruses has been associated with numerous serotypes in infants and children (CHERRY 1992), including the CVB. The clinical manifestations caused by these agents include fever, hyperpnea, and cyanosis. The laboratory findings usually include a normal leukocyte count, although extreme leukocytosis is occasionally encountered. Chest X-ray may reveal perihilar infiltrates. Fatalities have occurred in infants and young children. Histopathologic study of the lungs reveals thickening and infiltration of the alveolar septa but no necrosis or giant cells. In adults, bronchopneumonia has been associated with coxsackievirus B3 and echovirus 9 (CHERRY 1992; CHONMAITREE and MANN 1995). Several distinctive syndromes of enterovirus respiratory illness have been described, one of which is discussed in the following paragraph.

5.1 Herpangina

Group A coxsackieviruses are the most common causes of herpangina, but the syndrome has been reported with the CVB and the echoviruses as well (CHERRY and

JAHN 1965). The highest incidence is among children 1–7 years old, but infection has also been described in neonates and adults. There is usually an abrupt onset of fever associated with a sore throat, dysphagia and malaise. One fourth of the patients may have vomiting and abdominal pain. Early in the illness, grayish-white vesicles measuring 1 to 4 mm in diameter appear over the posterior portion of the palate, uvula, tonsillar pillars, and occasionally on the oropharynx. These vesicles are discrete, surrounded by erythema, and usually number fewer than 20. The vesicles usually rupture, leaving punched-out ulcers that may enlarge slightly, while new vesicles may appear. There may also be mild cervical adenopathy, headache, myalgia, arthralgia, and rarely parotitis or aseptic meningitis. Coryza and other respiratory symptoms are lacking and clinical laboratory studies are usually normal. The fever lasts 1–4 days; local and systemic symptoms begin to improve in 4–5 days, and recovery is usually complete within a week of onset (PARROTT et al. 1951; CHERRY and JAHN 1965).

6 Diseases of Skeletal Muscle

6.1 Pleurodynia

Pleurodynia (epidemic pleurodynia, Bornholm disease) is an acute febrile illness caused by CVB which affects the intercostal and abdominal muscles, leading to sharp, spasmodic pain in the chest wall or abdomen. Pleurodynia is typified by local outbreaks with high attack rates within affected households, and major epidemics which may occur as infrequently as every 10–20 years. Many reported outbreaks have been in sparsely populated geographical areas. Older children and adults are recognized to have disease more commonly than infants and young children.

The etiologic role of the CVB in pleurodynia was established shortly after the discovery of the coxsackieviruses (CURNEN et al. 1949; WELLER et al. 1950). Although both group A coxsackieviruses and echoviruses have been reported, the CVB are the dominant cause of pleurodynia.

The initial description of pleurodynia (SYLVEST 1934) was so complete that few details regarding the clinical features have since been added. Complaints of abdominal muscle involvement may be more common in younger children than in older children and adults who more often present with thoracic pain. Although the location may vary, it is the spasmodic and paroxysmal nature of the pain that is its hallmark. Sore throat and headache may accompany muscle pain, but cough and other symptoms of respiratory infection are notably absent. Involvement of the intercostal muscles and other muscles of respiration results in enhanced pain with deep breathing and a sensation of "pleuritic" pain which may result in splinting of the chest and rapid, shallow breathing. Muscle inflammation can be detected by eliciting pain with direct palpation, or less commonly by careful observation of localized swelling (SYLVEST 1934). Frank pleural involvement is documented in fewer than 10% of patients by detection of a pleural friction rub on auscultation or pleural fluid by chest X-ray (BAIN et al. 1961). Other systemic manifestations of CVB disease may accompany pleurodynia, including aseptic meningitis and orchitis (SYLVEST 1934; WARIN et al. 1953; GORDON et al. 1959).

Pain in the chest may mimic pneumonia, pulmonary embolus, myocardial ischemia, or the pre-eruptive phase of herpes zoster. Although pleurodynia is often misdiagnosed clinically as pneumonia, the chest X-ray inevitably reveals clear lung fields. Abdominal pain brought on by pleurodynia may suggest a variety of causes of acute abdomen. In fact, during a 1951 outbreak in Birmingham, England, nine of 49 children admitted to hospital with pleurodynia underwent laparotomy (DISNEY et al. 1953).

Analgesics and limitation of physical activity are helpful in reducing pain, which is generally most severe on presentation and then progressively diminishes during the course of illness which is typically 4–6 days, but may last as long as 3 weeks.

6.2 Inflammatory Myositis

In much the same way that diabetes is felt to be an autoimmune response to a triggering CVB infection, there is also evidence for such a mechanism in polymyositis and dermatomyositis. The evidence for associating these rheumatologic diseases with CVB, as with diabetes, includes serologic studies (TRAVERS et al. 1977). In addition, enterovirus-like particles have been seen by electron microscopy in muscle biopsy specimens from these patients (CHOU and GUTMANN 1970). Certain in situ and dot-blot hybridization and PCR studies of muscle tissue suggest the presence of enterovirus RNA (BOWLES et al. 1987; ROSENBERG et al. 1989). Other investigators, using the same molecular techniques, failed to find enterovirus traces in affected muscles from myositis patients (LEFF et al. 1992; LEON-MONZON and DALAKAS 1992). It is known that enteroviruses can cause inflammatory muscle disease because about 50% of agammaglobulinemia patients with chronic CNS enterovirus infections also develop myositis (MCKINNEY et al. 1987). In these patients, cultivable virus is recovered from muscle tissue. An occasional patient with myositis and normal immunoglobulin concentrations has responded to gammaglobulin therapy, with improvement or resolution of the disease.

7 Gastrointestinal Illnesses

Enteroviruses derive their name from their site of replication and shedding in the gastrointestinal tract; however, enteric illness (vomiting and diarrhea) is usually a minor manifestation of CVB infections (CHERRY 1992) and CVB infection has not been epidemiologically implicated as an important primary cause of acute gastroenteritis. Isolation of a CVB from the feces of a patient with gastroenteritis must be interpreted with caution because it may represent asymptomatic carriage in a patient made ill by a noncultivable agent (see below). The CVB have been rarely associated with acute abdominal pain and mesenteric adenitis syndromes, which may mimic acute appendicitis. In addition to the occurrence of hepatitis as one aspect of general disease in the newborn, coxsackieviruses B2, B3, and A9 have been associated with hepatitis in older children (CHERRY 1992). In adults, coxsackieviruses B3 and B5 have been associated with hepatitis on rare occasions.

8 Exanthems

8.1 Hand-Foot-and-Mouth Disease

Although hand-foot-and-mouth disease is one of the more common and unique syndromes associated with coxsackievirus A16, other serotypes, including the CVB, are occasionally isolated (CHERRY 1992). In outbreaks, the highest attack rates are among children younger than 4 years of age but adults are also frequently affected. The disease is usually mild and the onset is associated with a sore throat with or without a low-grade fever. Scattered vesicular lesions occur randomly on the oral structures, the pharynx, and the lips; these ulcerate readily, leaving shallow lesions with red areolae. About 85% of patients also develop sparse grayish vesicles (3–5 mm in diameter, surrounded by erythematous areolae) on the dorsum of the fingers, particularly in periungal areas, and on the margins of heels. Occasionally, palmar, plantar, and groin lesions may appear particularly in young children (CHERRY 1992). Resolution is usually complete within 1 week.

8.2 Other Exanthems

Infection with any of the CVB may be associated with a nonspecific exanthem; however CVB5 is the most commonly exanthematous of these agents, observed in as many as 10% of cases during outbreak situations (CHERRY 1992).

9 Laboratory Diagnosis of Group B Coxsackievirus Infections

9.1 Virus Isolation

Dr. John Enders and his colleagues received the Nobel Prize in 1954 for successfully propagating poliovirus in a continuous cell culture system, paving the way for vaccine development (ENDERS et al. 1949). Isolation of CVB in cell culture remains the gold standard for diagnosis. Among the many commercially available continuous cell lines, CVB grow best in cells of simian kidney (Buffalo green monkey, rhesus, cynomolgous) and human (HeLa, HEp-2) origin (ROTBART 1995).

Isolation of CVB in cell culture and recognition of cytopathic effect require a high level of expertise and may be quite labor intensive. Specimens from as many as 25%–35% of patients with characteristic CVB infections do not produce cytopathic effect in cell culture (CHONMAITREE et al. 1982), because of antibody neutralization in situ, inadequate collection, handling, and processing of the samples, or because of intrinsic insensitivity to the cell lines used. CVB which do grow in cell culture may do so slowly. Reported mean isolation times from CSF range from 3.7 to 8.2 days (JARVIS and TUCKER 1981; CHONMAITREE et al. 1982; WILFERT and ZELLER 1985); CVB from other sites, where viral titers are higher, often grow more rapidly (MINTZ and DREW 1980; JARVIS and TUCKER 1981; CHONMAITREE et al. 1982; WILFERT and ZELLER 1985).

9.2 Antigen Detection

The CVB share a common antigen, making antigen detection a possibility for this group of six enterovirus serotypes (YOLKEN and TORSCH 1980); however, it is seldom relevant to test a specimen for CVB serotypes only because of the clinical similarity of all enterovirus serotypes with the CVB. Hence, a useful antigen detection system must be "pan-enteroviral". Recent reports of polyclonal (ROMERO et al. 1986) and monoclonal antibodies (YOUSEF et al. 1987) that cross-react with multiple enterovirus serotypes are promising, but further testing is required to determine the clinical relevance of those observations.

9.3 Nucleic Acid Detection

Because of the significant overlap between the clinical manifestations of the CVB with those of other enterovirus serotypes, a clinically relevant nucleic acid detection test should be pan-enteroviral. In 1984, two laboratories reported the successful detection of multiple enterovirus serotypes of different subclassifications with single molecular probes derived from a single serotype (HYYPIA et al. 1984; ROTBART et al. 1985). This accomplishment demonstrated that genetic homology among the enteroviruses extended well beyond the few serotypes for which genomic sequence information was known at that time. In the ensuing 6 years, numerous investigators extended those observations using complementary DNA probes, RNA probes, and oligomeric probes (reviewed in [CE6]ROTBART 1991). Although enteroviruses were readily detected in body fluids during reconstruction experiments, the sensitivity in actual clinical specimens was only 33% or less. The limiting variable in all of these hybridization-based assays was the low titer of enteroviruses in many specimens, particularly CSF from patients with aseptic meningitis, which may contain as few as 1–10 titratable virions per ml (WILFERT and ZELLER 1985).

Several sets of pan-enteroviral PCR primers and probes have been studied, each with broad reactivity among many enterovirus serotypes and high specificity for the enteroviruses (HYYPIA et al. 1969; CHAPMAN et al. 1990; ROTBART 1990). All are directed at highly conserved regions of the 5' noncoding region of the viral genome and designed for reverse transcription combined with PCR (RT-PCR). Enterovirus RT-PCR using these primer/probe sets has been found to be consistently more sensitive than culture and virtually 100% specific (OLIVE et al. 1990; ROTBART 1990; MUIR et al. 1993; ROTBART et al. 1994; SAWYER et al. 1994). Development of a a one-step, microwell-based, colorimetric RT-PCR kit which is user-friendly sufficiently to be performed in 5 h by personnel in the routine diagnostic virology (ROTBART et al. 1994) has recently been accomplished. An accurate diagnosis of enterovirus infections available in less than 1 day has the potential of significantly impacting the quality and cost of patient management (CHONMAITREE et al. 1982; RICE et al. 1995).

9.4 Serotyping and Serologic Assays

The determination of the specific serotype of infecting CVB is often unnecessary because the diseases caused by the CVB and, for that matter, the enteroviruses in general, are not serotype-specific. In most circumstances, therefore, it is adequate and useful for the diagnostic laboratory to report the presence of "an enterovirus" without further detail. The most common exception to this principle is in pediatrics, where distinguishing between vaccine strain polioviruses and non-polioviruses is critical to interpretation of viral culture results. Further specific identification of a non-polio enterovirus by serotype is useful under certain circumstances. The gold standard for enterovirus serotype determination continues to be the use of intersecting pools of lyophillized antisera, as established by Lim and Benyesh-Melnick (LBM pools) (MELNICK et al. 1973, 1977). Broadly reactive and serotype-specific enterovirus monoclonal antibodies have now been developed (YAGI et al. 1992; ABZUG et al. 1993) and are commercially available for tissue culture confirmation by immunofluorescence. Preliminary studies have shown that these reagents, used singly and in pools, may find an important role in serotype identification. Immunofluorescence is a more rapid and simpler procedure than traditional neutralization-based serotyping and the supply of monoclonal antibodies is unlimited.

Serologic testing, like immunoassays, has had only a limited role in CVB diagnosis. If the specific serotype of an infecting CVB is known or suspected, e.g., in community-wide outbreaks, confirmatory immunoglobulin G serology can be performed on individual patients to document a rise in antibody titer from the acute to the convalescent phase of infection, thus providing useful epidemiologic information; however, little actual benefit accrues to the individual patient. In a situation in which CVB are recovered from feces or throat of a patient with unusual clinical manifestions, the etiologic role of the CVB may be more firmly established by documenting a four-fold rise in antibody titer to that serotype in paired acute and convalescent sera. Serosurveys of populations for past exposure to specific CVB have been performed by testing for antibody in single serum specimens; such studies may be useful for retrospective disease associations, such as with CVB and diabetes mellitus or inflammatory myositis (see above). In the usual scenario, when a patient presents with meningitis or other acute manifestations of illness and an CVB is suspected, serology is not a practical option.

Three traditional types of antibody determinations have been applied to CVB: neutralization, complement fixation, and hemagglutination inhibition. Neutralizing antibodies rise early in infection with the CVB and persist for many years or for life. They may be assessed in standard tube dilutions or by microtiter plate. In either format a fixed titer of a single serotype of CVB is inoculated onto cells in the presence of serial dilutions of the patient's serum (reviewed in ROTBART 1995); ideally, acute and convalescent sera are tested in parallel simultaneously. The rapid appearance of serum neutralizing antibodies following CVB infection may make it difficult to document a fourfold rise in titer with convalescence. Complement fixation antibody assays are of much less utility, as these antibodies are transient (weeks to months) and broadly cross-reactive among enterovirus serotypes, i.e., nonspecific. Historically, this assay was successfully applied only in testing for infection with the polioviruses. Only about a quarter to a third of all enterovirus serotypes agglutinate red blood cells, making hemagglutination inhibition of minimal usefulness; this type of antibody is also cross-reactive among enterovirus serotypes, hence nonspecific.

More recently, CVB immunoglobulin M assays, which take advantage of the shared antigen among the six CVB serotypes and the early appearance of the immunoglobulin M class of antibodies, have appeared promising (DORRIES and TER MEULEN 1983; BELL et al. 1986; POZZETTO et al. 1989). Although the IgM response measured in these assays may be nonspecific, i.e. a response to a non-CVB enterovirus may result in a positive assay, the nonspecificity may actually make this test more broadly reflective of recent exposure to any of numerous enterovirus serotypes; cross-reactivity with non-enterovirus pathogens causing infection has not been thoroughly studied. Many patients in the reported studies had positive IgM assays but were negative for CVB infection by culture, making the assay either more sensitive or less specific than traditional culture diagnosis. These tests are not yet widely available, and clinical experience with them remains limited.

9.5 Interpretation of Results

The body site at which CVB are detected is critical to the interpretation of enterovirus assays and to the differentiation between CVB "colonization" and actual CVB-associated disease. The nasopharynx and GI tract are permissive sites of infection, i.e., CVB have ready access to these sites and may remain as "colonizers" for weeks to months. Detection of CVB by virus isolation or PCR at these sites must be interpreted cautiously because their presence alone does not establish causality of the illness in question (JOHNSON et al. 1992). Indeed, virtually 100% of patients with CVB aseptic meningitis will have detectable CVB in feces (MINTZ and DREW 1980), but most persons shedding CVB in the feces at any particular time are asymptomatic – feces are thus the most sensitive and least specific site for detecting true CVB-associated illness. Since the shedding period in the nasopharynx after CVB infection is shorter than in the feces, the specificity of an enterovirus isolate from the nasopharynx for true causation of current symptoms is better than with feces, but far short of a definitive association.

In contrast, the central nervous system, bloodstream, and genitourinary tract are not usually colonized with CVB, i.e. detection of virus in specimens from these sites implies true invasive infection and a high liklihood of association with current illness. Reports of co-infections of the cerebrospinal fluid by bacteria and enteroviruses have appeared (WRIGHT et al. 1962; EGLIN et al. 1984). In these patients, the bacteria-associated clinical sequellae usually dominate. In the much more common situation, where the clinical presentation is typical of viral meningitis, co-infection with clinically "silent" bacteria would be extraordinarily unlikely. Hence, identification of a CVB from a site not ordinarily colonized in a patient with a clinically compatible illness, is usually sufficient evidence for establishing causality.

10 Prevention

Vaccines are available only for the polioviruses and provide no protection against CVB.

11 Treatment

Specific antiviral therapy for the CVB is not currently available for clinical use, although a number of drugs have been developed which have activity in vitro and in animal studies (McKinlay and Steinberg 1986; O'Connell et al. 1995). This class of drugs, of which there are now several in development and early trials (McKinlay and Steinberg 1986; O'Connell et al. 1995), slip through a pore at the base of the receptor canyon (Chap. 9) and bind within a hydrophobic pocket beneath the canyon floor, preventing either the steric change required by the CVB for uncoating or virus binding to a cellular receptor.

As noted above, clearance of CVB by the host is primarily antibody-mediated and, as might be expected, exogenously administered antibody has been useful in certain CVB infections. Some patients with agammaglobulinemia and chronic enterovirus meningitis or meningoencephalitis have shown stabilization and improvement during therapy with gammaglobulin, often administered by multiple routes including directly into the CNS (McKinney et al. 1987); other patients are less responsive to this therapy. Neonates with overwhelming enterovirus sepsis, including meningitis, have received IV gammaglobulin, maternal plasma, and exchange transfusions in attempts to reverse their otherwise bleak course, with occasional success (Kaplan et al. 1983; Modlin 1986). A single randomized trial of IV gammaglobulin plus standard therapy vs standard therapy alone in neonates suspected of having enterovirus infection in the first 2 weeks of life was undertaken; however, too few patients were enrolled for a definitive conclusion regarding clinical efficacy. A reduction in viral shedding was seen in the treatment group (Abzug et al. 1995).

Supportive care for the patient with CVB meningitis is usually adequate to assure complete recovery. Attention to fluid balance is necessary to avoid or ameliorate the syndrome of inappropriate antidiuretic hormone or brain edema. Electrolytes and, on occasion, urine and serum osmolality may require monitoring. Brain edema is a rare complication of CVB meningitis, but is readily managed with mannitol. Seizures may result from fever alone, or may reflect direct viral or indirect inflammatory damage of brain parenchyma (in which case "encephalitis" is the more apt term). Phenytoin or phenobarbital are the preferred agents for managing this complication. Rapidly progressive deterioration requiring more intensive support speaks strongly against a CVB etiology, and other potentially treatable causes must be immediately considered. Adequate hydration is the only indicated therapy in children with herpangina and hand-foot-mouth syndrome due to the enteroviruses. Other respiratory manifestations of enterovirus infections are managed symptomatically.

References

Abzug MJ (1995) Perinatal enterovirus infections. In: Rotbart HA (ed), Human enterovirus infections. American Society of Microbiology, Washington, DC pp 221–238

Abzug MJ, Keyserling HL, Lee ML, Levin MJ, Rotbart HA (1995) Neonatal enterovirus infection: virology, serology, and effects of intravenous immune globulin. Clin Infect Dis 20:1201–1206

Abzug MJ, Levin MJ, Rotbart HA (1993) Profile of enterovirus disease in the first two weeks of life. Pediatr Infect Dis J 12:820–824

Abzug MJ, Rotbart H, Levin M (1989) Demonstration of a barrier to transplacental passage of murine enteroviruses in late gestation. J Infect Dis 159:761–765

Artenstein MS, Cadigan FC, Buescher EL (1964) Epidemic coxsackie virus infection with mixed clinical manifestations. Ann Int Med 60:196–203

Bain HW, McLean DM, Walker SJ (1961) Epidemic pleurodynia (Bornholm disease) due to coxsackie B-5 virus: The interrelationship of pleurodynia, benign pericarditis, and aseptic meningitis. Pediatr 27:889–903

Bates HR (1970) Coxsackie virus B3 calcific pancarditis and hydrops fetalis. Am J Obstet Gyn 106:629–630

Bell EJ, McCartney RA, Basquill D, Chaudhuri AKR (1986) Antibody capture ELIZA for the rapid diagnosis of enterovirus infections in patients with aseptic meningitis. J Med Virol 19:213–217

Bergman I, Painter MJ, Wald ER, Chiponis D, Holland AL, Tayor HG (1987) Outcome in children with enteroviral meningitis during the first year of life. J Pediatr 110:705–709

Berlin LE, Rorabaugh ML, Heldrich F, Roberts K, Doran T, Modlin JF (1993) Aseptic meningitis in infants less than two years of age: diagnosis and etiology. J Infect Dis 168:888–892

Berry PJ, Nagington J (1982) Fatal infection with echovirus 11. Arch Dis Child 57:22–29

Bowles NE, Dubowitz V, Sewry CA, Archard LC (1987) Dermatomyositis, polymyositis, and coxsackie-B-virus infection. Lancet i:1004–1007

Burch GE, Sun SC, Chu KC, Sohal RS, Colcolough HL (1968) Interstitial and coxsackie B myocarditis in infants and children. J Am Med Assn 203:1–8

Case Records of the Massachusetts General Hospital (1965) Case 20–1965. N Eng J Med 272:907–914

Chapman NM, Tracy S, Gauntt CJ, Fortmueller U (1990) Molecular detection and identification of enteroviruses using enzymatic amplification and nucleic acid hybridization. J Clin Microbiol 28:843–850

Cherry JD (1990) Enteroviruses. In: Remington JS, Klein JO (eds) Infectious diseases of the fetus and newborn infant, 3rd edn. W B Saunders, Philadelphia, pp 325–366

Cherry JD (1992) Enteroviruses: Polioviruses (poliomyelitis), coxsackieviruses, echoviruses, and enteroviruses. In: Feigin RD, Cherry JD (eds), Textbook of Pediatric Infectious Diseases, 3rd edn. WB Saunders, Philadelphia pp 1705–1753

Cherry JD, Jahn CL (1965) Herpangina: The etiologic spectrum. Pediatr 36:632–634

Cherry JL, Soriano F, Jahn CL (1968) Search for perinatal enterovirus infection. Am J Dis Child 116:245–250

Chonmaitree T, Mann L (1995) Respiratory infections. In: Rotbart HA (ed) Human enterovirus infections. American Society of Microbiology, Washington, DC pp 255–270

Chonmaitree T, Menegus MA, Powell KR (1982) The clinical relevance of CSF viral culture. A two-year experience with aseptic meningitis in Rochester, New York. J Am Med Assn 247:1843–1847

Chou SM, Gutmann L (1970) Picornavirus-like crystals in subacute polymyositis. Neurol 20:205–213

Curnen EC, Shaw EW, Melnick JL (1949) Disease resembling nonparalytic poliomyelitis associated with virus pathogenic for infant mice. J Am Med Assn 141:894–901

Dagan R, Hall CB, Powell KR, Menegus MA (1989) Epidemiology and laboratory diagnosis of infection with viral and bacterial pathogens in infants hospitalized with suspected sepsis. J Pediatr 115:351–356

Dagan R, Jenista J, Menegus MA (1988) Association of clinical presentation, laboratory findings, and virus serotypes with the presence of meningitis in hospitalized infants with enterovirus infection . J Pediatr 113:975–978

Dagan R, Powell KR, Hall CB, Menegus MA (1985) Identification of infants unlikely to have serious bacterial infection although hospitalized for suspected sepsis. J Pediatr 107:855–860

Dai-ming W, Guo-chang Z, Shi-mei Z, Yu-cai Z (1993) An epidemic of encephalitis and meningoencephalitis in children caused by echovirus type 30 in Shanghai. Chin Med J 106:767–769

Dalldorf G, Gifford R (1954) Susceptibility of gravid mice to coxsackie virus infection. J Exp Med 99:21–27

Dalldorf G, Sickles G (1948) An unidentified, filtrable agent isolated from the feces of children with paralysis. Science 108:61–63

DeAngelis C, Joffe A, Wilson M, Willis E (1983) Iatrogenic risks and financial costs of hospitalizing febrile infants. Am J Dis Child 137:1146–1149

Delaney TB, Fukunga FH (1958) Myocarditis in a newborn infant with encephalomeningitis due to coxsackie virus group B, type 3. N Eng J Med 259:234–236

Disney ME, Howard EM, Wood BSB (1953) Bornholm disease in children. Br Med J 1:1351–1354

Dorries R, Ter Meulen V (1983) Specificity of IgM antibodies in acute human coxsackievirus B infec-

tions, analyzed by direct solid phase enzyme immunoassay and immunoblot technique. J Gen Virol 64:159–167
Eglin RP, Swann RA, Isaacs D, Moxon ER (1984) Simultaneous bacterial and viral meningitis. Lancet 2:984
Eichenwald HF, Kotsevalov O (1960) Immunologic responses of premature and full-term infants to infection with certain viruses. Pediatr 25:829–839
Enders JF, Weller TH, Robbins FC (1949) Cultivation of the Lansing strain of poliomyelitis virus in culture of various human embryonic tissues. Science 109:85–99
Etter CG, Wedgwood J, Schaad UB (1991) Aseptische meningitiden in der padiatrie. Schweizerische Medizinische Wochenschrift. Journal Suisse de Medecine (Basel) 121:1120–1126
Farber PA, Glasgow LA (1970) Viral myocarditis during pregnancy: encephalomyocarditis virus infection in mice. Am Heart J 80:96–102
Farmer K, MacArthur BA, Clay MM (1975) A follow-up study of 15 cases of neonatal meningoencephalitis due to coxsackie virus B5. J Pediatr 87:568–571
Faulkner RS, MacLeod AJ, Van Rooyen CE (1957) Virus meningitis-seven cases in one family. Can Med Assn J 77:439–444
Gear JHS, Measroch V (1973) Coxsackievirus infection of the newborn. Prog Med Virol 15:42–62
Gonzalez-del-Rey JA, Baker RC, Kummer AW, Schultz JR, Ho M (1991) Neurodevelopmental outcome of infants with enteroviral meningitis in the first three months of life. Pediatr Res 29(2):172A
Gordon RB, Lennette EH, Sandrock RS (1959) The varied clinical manifestations of coxsackie viral infections. Arch Int Med 103:63–73
Grist NR, Bell EJ, Assad F (1978) Enteroviruses in human disease. Prog Med Virol 24:114–157
Hall CB, Miller DG (1969) The detection of silent Coxsackie B-5 virus perinatal infection. J Pediatr 75:124–127
Holland J (1961) Receptor affinities as major determinants of enterovirus tissue tropisms in humans. Virol 15:312–326
Horstmann DM, Yamada N (1968) Enterovirus infections of the central nervous system. Research Publications-Association for Research in Nervous and Mental Disease 44:236–253
Howden CW, Lang WR, Laws J (1966) An epidemic of ECHO 9 virus meningitis in Auckland. N Z Med J 65:763–684
Hsu KHL, Lonberg-Holm K, Alstein B, Crowell RL (1988) A monoclonal antibody specific for the cellular receptor for the group B coxsackieviruses. J Virol 62:1647–1652
Huber SA, Job LP, Woodruff JF (1984) In vitro culture of coxsackievirus group B, type 3 immune spleen cells on infected endothelial cells and biological activity of the cultured cells in vivo. Infect Immun 43:567–573
Hyypia T, Auvinen P, Maaronen M (1969) Polymerase chain reaction for human picornaviruses. J Gen Virol 70:3261–3268
Hyypia T, Stalhardsk R, Vainionpaa R, Pettersson U (1984) Detection of enteroviruses by spot hybridization. J Clin Microbiol 19:436–438
Jaffe M, Srugo I, Tirosh E, Colin AA, Tal Y (1989) The ameliorating effect of lumbar puncture in viral meningitis. Am J Dis Child 143:682–685
Jantausch BA, Luban NLC, Duffy L, Rodriguez WJ (1995) Maternal plasma transfusion in the treatment of disseminated neonatal echovirus 11 infection. Ped Infect Dis J 14:154–155
Jarvis WR, Tucker G (1981) Echovirus type 7 meningitis in young children. Am J Dis Child 135:1009–1012
Jenista JA, Dalzell LE, Davidson PW, Menegus M (1984) Outcome studies of neonatal enterovirus infection. Pediatr Res 18(2):230A
Jenista JA, Powell KA, Menegus MA (1984) Epidemiology of neonatal enterovirus infection. J Pediatr 104:685–690
Johnson GM, McAbee GA, Seaton ED, Lipson SM (1992) Suspect value of non-CSF viral cultures in the diagnosis of enteroviral CNS infection in young infants. Dev Med Child Neurol 34:876–884
Johnston JM, Overall JC (1989) Intravenous immunoglobulin in disseminated neonatal echovirus 11 infection. Ped Infect Dis J 8:254–256
Jones MJU, Kolb M, Votava HJ, Johnson RL, Smith TF (1980) Intrauterine echovirus type 11 infection. Mayo Clin Proc 55:509–512
Kaplan AS, Melnick JL (1951) Oral administration of coxsacie viruses to newborn and adult mice. Proc Soc Exp Biol Med 76:312–315

Kaplan MH, Klein SW, McPhee J, Harper RG (1983) Group B coxsackievirus infections in infants younger than three months of age: A serious childhood illness. Rev Infec Dis 5:1019–1032
Kibrick S (1961) Viral infections of the fetus and newborn. Perspect Virol 2:140–159
King DL, Karzon DT (1975) An epidemic of aseptic meningitis syndrome due to EHCO virus type 6. Pediatr 29:432–437
Kinney JS, McCray E, Kaplan JE, Low DE, Hammond GW, Harding G, Pinsky PF, Davi MJ, Kovnats SF, Riben P, Nartone WJ, Schonberger LB, Anderson LJ (1986) Risk factors associated with echovirus 11 infection in a newborn nursery. Ped Infect Dis J 5:192–197
Kinnunen E, Hovi T, Stenvik M, Hellström O, Porras J, Kleemola M, Kantanen ML (1987) Localized outbreak of enteroviral meningitis in adults. Acta Neurol Scand 75:346–351
Knox AW (1950) Influence of pregnancy in mice on the course of infection with murine poliomyelitis virus. Proc Soc Exp Biol Med 73:520–523
Krajden S, Middleton PJ (1983) Enterovirus infections in the neonate. Clin Pediatr 22:87–92
Krober MS, Bass JW, Powell JM, Smith FR, Seto DSY (1985) Bacterial and viral pathogens causing fever in infants less than 3 months old. Am J Dis Child 139:889–892
Leff RL, Love LA, Miller FW, Greenberg SJ, Klein EA, Dalakas MC, Plotz PH (1992) Viruses in idiopathic inflammatory myopathies: absence of candidate viral genomes in muscle. Lancet 339:1192–1195.
Leggiadro RJ, Darras BT (1983) Viral and bacterial pathogens of suspected sepsis in young infants. Ped Infect Dis J 2:287–289
Leon-Monzon M, Dalakas MC (1992) Absence of persistent infection with enterovirus in muscle of patients with inflammatory myopathies. Ann Neurol 32:219–222
Lepow ML, Carver DH, Wright HT, Woods WA, Robbins RC (1962) A clinical, epidemiologic and laboratory investigation of aseptic meningitis during the four-year period, 1955–1958. I. Observations concerning etiology and epidemiology. N Eng J Med 266:1181–1187
McKinlay MA, Steinberg BA (1986) Oral efficacy of WIN 51711 in mice infected with human poliovirus. Antimicrob Agents Chemother 29:30–32
McKinney RE, Katz SL, Wilfert CM (1987) Chronic enteroviral meningoencephalitis in agammaglobulinemic patients. Rev Infect Dis 9:334–356
McLean DM, Donohue WL, Snelling CE (1961) Coxsackie B5 virus as a cause of neonatal encephalitis and myocarditis. Can Med Assn J 85:1046–1048
Melnick JL, Rennick V, Hampil B, Schmidt NJ, Ho HH (1973) Lyophilized combination pools of enterovirus equine antisera: preparation and test procedures for the identification of field strains of 42 enteroviruses. Bull World Health Organ 48:263–268
Melnick JL, Schmidt NJ, Hampil B, Ho HH (1977) Lyophilized combination pools of enterovirus equine antisera: preparation and test procedures for the identification of field strains of 19 group A coxsackievirus serotypes. Intervirol 8:172–181
Mills EL (1983) Mononuclear phagocytes in the newborn: their relation to the state of relative immunodeficiency. Am J Pediatr Hematol Oncol 5:189–198
Mintz L, Drew WL (1980) Relation of culture site to the recovery of nonpolio enteroviruses. Amer Soc Microbiol 74:324–326
Modlin JF (1980) Fatal echovirus 11 disease in premature neonates. Pediatr 66:775–780
Modlin JF (1986) Perinatal echovirus infection: Insights from a literature of 61 cases of serious infection and 16 outbreaks in nurseries. Rev Infect Dis 8:918–926
Modlin JF (1994) Neonatal enterovirus infections. In: Baker C, Feigin R (eds) Neonatal infections. Seminars in pediatric infectious diseases, vol 5. WB Saunders, Philadelphia, pp 70–77
Modlin JF, Bowman M (1987) Perinatal transmission of coxsackie B3 virus in a murine model. J Infect Dis 156:21–25
Modlin JF, Bowman M (1988) Localization of coxsackievirus B3 (CB3) antigen in the mouse placenta. Presented to the Society for Pediatric Research. Washington, DC
Modlin JF, Crumpacker CS (1982) Coxsackie B1 virus infection in the pregnant mouse and transmission to the fetus. Infect Immun 37:222–226
Modlin JF, Dagan R, Berlin LE, Virshup DM, Yolken RH, Menegus M (1991) Focal encephalitis with enterovirus infections. Pediatr 88:841–845
Modlin JF, Polk BF, Horton P, Etkind P, Crane E, Spiliotes A (1981) Perinatal echovirus 11 infection: risk of transmission during a community outbreak. N Eng J Med 305:368–371
Moore M (1982) Enteroviral disease in the United States. J Infect Dis 146:103–108

Moses EB, Dean AG, Hatch MH, Burron AL (1977) An outbreak of aseptic meningitis in the area of Fort Smith, Arkansas, 1975, due to echovirus type 4. J Arkansas Med Soc 74:121–125

Muir P, Nicholson F, Jhetam M, Neogi S, Banatvala JE (1993) Rapid diagnosis of enterovirus infection by magnetic bead extraction and polymerase chain reaction detection of enterovirus RNA in clinical specimens. J Clin Microbiol 31:31–38

Nagington J, Wreghitt TG, Gandy G (1978) Fatal echovirus 11 infections in outbreak in special-care baby unit. Lancet ii:725–728.

Nogan AG, Lepow ML (1967) Enteroviral meningitis in very young infants. Pediatr 40:617–626

O'Connell J, Albin R, Blum D, Grint P, Schwartz J (1995) Development of antiviral agents for picornavirus infections. In: Rotbart HA (ed) Human enterovirus infections. American Society for Microbiology, Washington DC pp 419–434

O'Neil KM, Pallansch MA, Winkelstein JA, Lock TM, Modlin JF (1988) Chronic group A coxsackievirus infection in agammaglobulinemia: Demonstration of genomic variation of serotypically identical isolates persistently excreted from the same patient. J Infect Dis 157:183–186

Olive DM, Al-Mufti S, Al-Mulla W, Khan MA, Pasca A, Stanway G, Al-NakibW (1990) Detection and differentiation of picornaviruses in clinical samples following genomic amplification. J Gen Virol 71:2141–2147

Parrott RH, Ross S, Burke FG, Rice EC (1951) Herpangina: clinical studies of a specific infectious disease. N Eng J Med 245:275–280

Pozzetto B, Gaudin OG, Aouni M, Ros A (1989) Comparative evaluation of immunoglobulin M neutralizing antibody response in acute-phase sera and virus isolation for the routine diagnosis of enterovirus infection. J Clin Microbiol 27:705–708

Rager-Zisman B, Allison AC (1973) The role of antibody and host cells in the resistance of mice against infection by coxsackie B-3 virus. J Gen Virol 19:329–338

Rapmund G, Gauld JR, Rogers NG, Holmes GE (1959) Neonatal myocarditis and meningoencephalitis due to coxsackie virus group B, type 4. N Eng J Med 260:819–821

Ray CG, McCollough RH, Doto IL, Todd JC, Glezen WP, Chin TDY (1966) Echo 4 ilnness: epidemiological, clinical and laboratory studies of an outbreak in a rural community. Am J Epidemiol 84:253–267

Ray CG, McCollough RH, Doto IL, Todd JC, Glexen WP, Chin TDY (1987) Echo 4 illness Epidemiological, clinical and laboratory studies of an outbreak in a rural community. Am J Epidemiol 84:253–267

Reyes MP, Ostrea EM, Roskamp J, Lerner AM (1983) Disseminated neonatal echovirus 11 disease following antenatal maternal infection with a virus-positive cervix and virus negative gastrointestinal tract. J Med Virol 12:155–159

Rice SK, Heinl RE, Thornton LL, Opal SM (1995) Clinical characteristics, management strategies, and cost implications of a statewide outbreak of enterovirus meningitis. Clin Infect Dis 20:931–937

Robino G, Perlman A, Togo Y, Reback J (1962) Fatal neonatal infection due to coxsackie B_2 virus. J Pediatr 61:911–918

Romero J, Putnak JR, Wimmer E (1986) Use of poliovirus proteins VP3 and 2C as group antigens for the detection of enterovirus infections by indirect immunofluorescence. Abstract number 967. Pediatr Res 20:319

Rorabaugh ML, Berlin LE, Heldrich F, Roberts K, Rosenberg LA, Doran T, Modlin JF (1993) Aseptic meningitis among infants less than two years of age: acute illness and neurologic complications. Pediatr 92:206–211

Rorabaugh ML, Berlin LE, Rosenberg L, Rossman M, Allen M, Modlin JF (1992) Absence of neurodevelopmental sequelae from aseptic meningitis. Abstract presented at APS/SPR. Pediatric Research 31(4), Part 2:1047, p 177A Baltimore, MD

Rosenberg NL, Rotbart HA, Abzug MJ, Ringel SP, Levin MJ (1989) Evidence for a novel picornavirus in human dermatomyositis. Ann Neurol 26:204–209

Rotbart HA (1990) Diagnosis of enteroviral meningitis with the polymerase chain reaction. J Pediatr 117:85–89

Rotbart HA (1990) Enzymatic RNA amplification of the enteroviruses. J Clin Microbiol 28:438–442

Rotbart HA (1991) Nucleic acid detection systems for enteroviruses. Clin Microbiol Rev 4:156–168

Rotbart HA (1995) Enteroviruses. In: Murray PR, Baron EJ, Pfaller MA, Tenover FC, Yolken RH (eds), Manual of clinical microbiology, 6th edn. American Society of Microbiology, Washington DC, pp 1004–1011

Rotbart HA (1995) Enteroviral infections of the central nervous system. Clin Infect Dis 20:971–981
Rotbart HA (ed) (1995) Meningitis and encephalitis. In: Human enterovirus infections. American Society of Microbiology, Washington, DC pp 271–289
Rotbart HA (1991) Viral meningitis and the aseptic meningitis syndrome. In: Scheld WM, Whitley RJ, Durack DT (eds) Infections of the central nervous system. Raven, New York pp 19–40
Rotbart HA, Kinsella JP, Wasserman RL (1990) Persistent enterovirus infection in culture-negative meningoencephalitis: demonstration by enzymatic amplification. J Infect Dis 161:787–791
Rotbart HA, Levin MJ, Villarreal LP (1985) Use of subgenomic poliovirus DNA hybridization probes to detect the major subgroups of enteroviruses. J Clin Microbiol 20:1105–1108
Rotbart HA, Sawyer MH, Fast S, Lewinski C, Murphy N, Keyser EF, Spadoro J., Kao SY, Loeffelholz M (1994) Diagnosis of enteroviral meningitis by using PCR with a colorimetric microwell detection assay. J Clin Microbiol 32:2590–2592
Shafren DR, Bates RC, Agrez MV, Herd RL, Burns GF, Barry RD (1995) Coxsackieviruses B1, B3, and B5 use decay accelerating factor as a receptor for cell attachment. J Virol 69:3873–3877
Sawyer M, Holland D, Aintablian N, Connor J, Keyser EF, Waecker NJ (1994) Diagnosis of enteroviral central nervous system infection by polymerase chain reaction during a large community outbreak. Ped Infect Dis J 13:177–182
Sells CJ, Carpenter RL, Ray CG (1975) Sequelae of central-nervous-system enterovirus infections. N Eng J Med 293:1–4
Selzer G (1969) Transplacental infection of the mouse fetus by coxsackie virus. Israel J Med Sci 5:125–127
Sever JL, Huebner RJ, Castellano GA, Bell JA (1963) Serologic diagnosis "en masse" with multiple antigens. Am Rev Resp Dis 88(2):342–349
Siegel M, Greenberg M (1956) Poliomyelitis in pregnancy: effect on fetus and newborn infant. J Pediatr 49:280–288
Singer JI, Mauer PR, Riley JP, Smith PB (1980) Management of central nervous system infections during an epidemic of enteroviral aseptic meningitis. J Pediatr 96:559–563
Soike K (1967) Coxsackie B-3 infection in the pregnant mouse. J Infect Dis 117:203–208
Stevens KM (1959) Estimate of molecular equivalent of antibody required for prophylaxis and therapy of poliomyelitis. J Hyg (Lond) 57:198–201
Sumaya CV, Corman LI (1982) Enteroviral meningitis in early infancy: significance in community outbreaks. Pediatr Infect Dis 3:151–154
Sylvest E (1934) Epidemic myalgia: Bornholm disease. Oxford University, London
Thivierge B, Delage G (1982) Infections du systeme nerveux central a enterovirus: 223 cas vus a un hopital pediatrique entre 1973 et 1981. Can Med Assn J 127:1097–1102
Travers RL, Hughes GRV, Cambridge G, Sewell G Jr (1977) Coxsackie B neutralisation titres in polymyositis/dermatomyositis. Lancet 1:1268
Warin JF, Davies JBM, Sanders FK, Vizoso AD (1953) Oxford epidemic of Bornholm disease, 1951. Brit Med J i:1345–1354
Warren RJ, Lepow ML, Bartsch GE, Robbins FC (1964) The relationship of maternal antibody, breast feeding, and age to the susceptibility of newborn infants to infection with attenuated poliovirus. Pediatr 34:4–13
Webster ADB, Rotbart HA, Warner T, Rudge P, Hyman N (1993) Diagnosis of enterovirus brain disease in hypogammaglobulinemic patients by polymerase chain reaction. Clin Infect Dis 17:657–661
Weinstein L, Aycock WL, Feemster RF (1951) Relation of sex, pregnancy, and menstruation to susceptibility in poliomyelitis. N Eng J Med 245:54–58
Weller TH, Enders JF, Buckingham M, Finn JJ Jr (1950) Etiology of epidemic pleurodynia: Study of two viruses isolated from typical outbreak. J Immunol 65:337–346
Whitley RJ, Cobbs CG, Alford CA, Soong SJ, Hirsch MS, Connor JD, Corey L, Hanley DF, Levin M, Powell DA (1989) Diseases that mimic herpes simplex encephalitis. J Am Med Assn 262:234–239
Whitley RJ, Soong S-J, Hirsch MS, Karchmer AW, Dolin R, Galasso G, Dumick JK, Alford CA (1981) Herpes simplex encephalitis: vidarabine therapy and diagnostic problems. N Eng J Med 304:313–318
Wilfert CM, Lehrman SN, Katz SL (1983) Enteroviruses and meningitis. Pediatr Infect Dis 2:333–341
Wilfert CM, Thompson RJ, Sunder TR, O'Quinn AO, Zellee J, Blacharsh J (1981) Longitudinal assessment of children with enteroviral meningitis during the first three months of life. Pediatr 67:811–815
Wilfert CM, Zeller J (1985) Enterovirus diagnosis. In: de la Maza LM, Peterson EM (eds), Medical virology IV. Lawrence Erlbaum Associates, Hillsdale, NJ, pp 85–107

Wong SN, Tam AYC, Ng THK, Ng WF, Tong CY, Tang TS (1989) Fatal coxsackie B1 virus infection in neonates. Ped Infect Dis J 8:638–641
Woodruff JF, Woodruff JJ (1974) Involvement of T lymphocytes in the pathogenesis of coxsackievirus B3 heart disease. J Immunol 113:1726–1734
Wright HT, McAllister RM, Ward R (1962) Mixed " meningitis: reports of a case with isolation of Haemophilus influenzae type b and echovirus type 9 from the cerebrospinal fluid. N Eng J Med 267:142–144
Yagi S, Schnurr D, Lin J (1992) Spectrum of monoclonal antibodies to coxsackievirus B-3 includes type- and group-specific antibodies. J Clin Microbiol 30:2498–2501
Yolken RH, Torsch VM (1980) Enzyme-linked immunosorbent assay for detection and identification of coxsackie B antigen in tissue cultures and clinical specimens. J Med Virol 6:45–52
Yousef GE, Brown IN, Mowbray JF (1987) Derivation and biochemical characterization of an enterovirus group-specific monoclonal antibody. Intervirol 28:163–170

Effects of Nutritional Antioxidants and Other Dietary Constituents on Coxsackievirus-Induced Myocarditis

M. A. BECK[1] AND O. A. LEVANDER[2]

1 Introduction

Nutrition of the host has been largely neglected in studies of viral pathogenesis. Much more emphasis has been placed on the effect of nutrition on the host's immune system. A functioning immune system is required for both initial protection from viral illness as well as subsequent viral clearance once the pathogen has gained entrance to the host. Any perturbation of the immune system induced by a change in host nutritional status would be expected to affect the course of a viral infection. For example, severely malnourished children are more susceptible to infection and severe disease with rotavirus, which is thought to be due in part to a dysfunctional immune response (DAGAN et al. 1990). Therefore, the vast majority of nutrition research related to viral infection has emphasized the effect of nutrition on the host immune system, which in turn leads to a change in host susceptibility to infectious disease. Indeed, the current nutrition-infectious disease paradigm can be diagrammed as follows:

[1]University of North Carolina at Chapel Hill, Frank Porter Graham Child Development Center and Department of Pediatrics, 105 Smith Level Rd., CB #8180 Chapel Hill, NC 27599-8180; USA

[2]United States Department of Agriculture, ARS, Beltsville Human Nutrition Research Center, Beltsville, MD 20705-2350, USA

inadequate host nutrition
↓
dysfunctional host immunity
↓
enhanced host susceptibility to virus

This chapter will discuss issues related to nutrition and coxsackievirus. Although this interaction has not been studied as well as other areas of coxsackie virology, it nonetheless represents an exciting new field of collaboration among virologists, nutritionists and immunologists. It is hoped that this chapter will engage the interest of the reader and encourage more interdisciplinary research in this area. Finally, we hope to convince the reader to accept a new paradigm of nutrition-infectious disease interactions based on our work with coxsackievirus.

2 Malnutrition and Coxsackievirus Infection

WOODRUFF and KILBOURNE (1970) found that graded underfeeding (marasmus) of weanling mice led to an increase in the severity of infection with coxsackievirus B3 (Nancy strain; CVB3). The severity of the CVB3-induced damage was in general proportional to the severity of the malnutrition. Elevated virus titers in organs (heart, spleen, pancreas and liver) were also associated with the severest malnutrition. In addition, mononuclear cellular infiltrates were absent from the CVB3-infected hearts, prompting the authors to suggest that the tissue damage was due directly to the virus and local tissue metabolic changes. A subsequent paper (WOODRUFF 1970) demonstrated that lymphoid tissue weight was decreased in direct proportion to the degree of malnutrition. In the most severely nutritionally deprived mice, neutralizing antibody titers were decreased at 7 days postinfection, but this difference disappeared at 12 and 16 days postinfection. Clearly, severe malnutrition can affect the lymphoid tissues and subsequently influence CVB3 pathogenesis.

LORIA et al. (1976) fed mice either an atherogenic and hypercholesterolemic diet (high in animal fat, sucrose, cholesterol and cholic acid) or a normal diet for 3 months and then inoculated the mice with CVB5. A mortality rate of 97% occurred in infected mice fed the hypercholesterolemic diet. In contrast, no mortality occurred in mice fed the normal diet. A decrease in the mortality rate could be achieved by using a lower viral inoculum. The hearts of the CVB5-infected hypercholesterolemic animals at 10–12 days postinfection had discrete focal areas of myocytolysis without cellular infiltrate, whereas mice fed the control diet developed inflammatory lesions typical of CVB infection. Virus titers were found to be higher in mice fed the control than in mice fed the test diet, demonstrating that the increased mortality could not be explained by increased virus replication. These authors suggested that the hypercholesterolemic diet induced metabolic changes in the animals which were increased with viral infection. The lack of a heart inflammatory infiltrate in these animals suggested a decreased immune response, and indeed, leukopenia occurred following viral infection in the animals fed the test diet.

However, the Loria et al. paper contains a problem that can occur in the design of nutritional studies. The test diet was a purified casein-based diet, whereas the diet fed the control animals was not a purified diet. Therefore, it is impossible to compare the two diets nutritionally. An additional problem with the diets used in this study is that the selenium (Se) content of the purified diet was unknown: it was not reported how much Se the casein contained and the salt mix used contained no added Se. Moreover, the casein content of the diet was low (15%), particularly for a 20% fat diet. Because of the unknown nutritional differences between the control and test diets, the difference in the mortality rates between the two groups is impossible to interpret.

A study by ILBÄCK et al. (1976) used more carefully controlled dietary conditions. These investigators fed mice either a control diet or the same diet supplemented with 1% cholesterol for 7 weeks prior to infection with CVB3. In contrast to LORIA et al. (1976), their hypercholesterolemic mice infected with CVB3 did not differ from mice fed control diets based on intensity of the inflammation, mortality or immune function. However, ILBÄCK et al. (1976) found an increase in cholesterol accumulation by 7 days after CVB3 infection: cholesterol uptake increased 75% in the heart and 92% in the aorta. This increase in cholesterol accumulation was independent of diet. The study of Ilbäck et al. suggests that CVB3 may play a role in the generation or initiation of atherosclerotic plaques.

The difference in outcomes between the LORIA et al. (1976) and ILBÄCK et al. (1976) studies may have been due to several factors, including viral strain (CVB5 vs CVB3), cholesterol content of diet (2% cholesterol vs 1% cholesterol) and the fact that the Loria et al. diet was not controlled for other nutritional variables.

Single nutrient deficiencies uncomplicated by other dietary deficits are relatively uncommon in humans. However, in the 1970s, scientists from China postulated a novel etiology for Keshan disease, which presents as a cardiomyopathy, in the form of a specific trace element deficiency. Of particular interest, Keshan disease is also thought to require an infectious cofactor, possibly a coxsackievirus.

3 Keshan Disease

Keshan disease is an endemic cardiomyopathy that was first described in Keshan County, Heilongjiang Province, Northeast China in 1935 (LI et al. 1985). The heart pathology is characterized by foci of necrosis throughout the myocardium (GU 1983). The lesions can exhibit varying degrees of cellular infiltration and calcification. The clinical course of the disease can be divided into four types: acute, subacute, chronic and latent (JIN 1984). The acute form of the disease is characterized by an abrupt onset and severe degeneration and necrosis of the myocardium which is generally fatal. The subacute form is noticed approximately 1 week after onset, characterized clinically as congestive heart failure and cardiogenic shock. Pathologically, cardiac lesions can be found, consisting of fibrosis and calcification. The chronic form has a slower onset, with the myocardium demonstrating both fresh and old tissue lesions. Clinically, the patient presents with moderate to severe congestive heart failure, which may have a course of months or years. The latent form is characterized by compensated cardiac function in a dilated heart.

During the course of Keshan disease, all the forms may be manifested at varying times in an individual patient. Keshan disease was predominantly found in women of child-bearing age and in post-weaned children. The epidemiology of Keshan disease suggested an environmental factor because the distribution of the disease occurred in a belt-like fashion from northeast to southwest China and in very specific hilly and mountainous areas (YU et al. 1981). Further investigations led to the finding that the soils in Keshan disease endemic areas were poor in the trace element Se (GE et al. 1983). Subsequent testing of individuals in Keshan disease areas demonstrated low Se levels in both hair and blood samples. In 1974, a placebo-controlled randomized treatment trial with sodium selenite was begun with children 1–9 years of age (Keshan Disease Research Group 1979). The incidence of Keshan disease dropped significantly in the groups treated with Se. Therefore, children at risk of Keshan disease were supplemented with sodium selenite.

Although improved economi conditions and the introduction of selenium supplementation have essentially eradicated Keshan disease, there are aspects of the epidemiological pattern which suggest that other factors, in addition to Se deficiency, may be involved. Keshan disease has a seasonal and annual incidence: it usually occurred in winter and spring in the northeast and in summer in the southwest region. In some years, Keshan disease had a high prevalence rate – in 1962 Keshan disease was responsible for 42% of all cardiac diseases in Beian County of Heilongjiang Province. In other years, the prevalence rate was only 11%. In addition, not every individual with low Se status developed Keshan disease. Because the epidemiological data suggested involvement of an infectious cofactor, scientists in China examined blood and tissue of Keshan disease patients for the presence of viruses. A number of enteroviruses, including coxsackievirus, were isolated from Keshan disease tissues (SU et al. 1979), suggesting a possible role for coxsackieviruses in the etiology of Keshan disease.

BAI et al. (1980), using a strain of CVB4 isolated from the blood of a Keshan disease patient, infected suckling mice born to Se-deficient dams and Se-adequate dams. Myocardial damage was found to be more extensive in the suckling mice born to Se-deficient dams vs suckling mice born to Se-adequate dams. The incidence of heart lesions was found to correlate inversely with blood Se concentrations. These results suggested that CVB may be involved in Keshan disease.

Utilizing the polymerase chain reaction (PCR), LI et al. (1995) found 85.7% of myocardial specimens from Keshan disease patients (18/21) were positive for enteroviral RNA, whereas only 3% of controls were positive. ROATH et al. (1996)

found that 30% of blood samples from Keshan disease victims were positive by PCR for enteroviral RNA. A large number of controls (30%) living in Keshan endemic areas were also positive for enteroviral RNA suggesting that enteroviruses are ubiquitous in some areas of China.

In order to investigate further the role of coxsackieviruses and Se deficiency in Keshan disease, we utilized a well-characterized model of virus-induced myocarditis and imposed a specific nutritional deficiency of Se.

4 Selenium and Coxsackievirus B3-Induced Myocarditis

Selenium was first reported to have a nutritional function as an essential trace element in 1957 (SCHWARZ and FOLTZ 1957). In 1973, Se was found to be an essential cofactor for glutathione peroxidase, a peroxide destroying enzyme (ROTRUCK et al. 1973). Through this enzyme, Se plays an essential role in the protection of host tissues from oxidative stress by mediating the removal of hydrogen peroxide and organic hydroperoxides. The importance of Se in human nutrition was clearly shown by its role in the etiology of the Se-responsive Keshan disease.

In 1994, BECK et al. (1994a) fed mice a diet adequate (0.2 μg/g) or deficient in Se for 4 weeks, at which time the mice were inoculated with CVB3/20, a myocarditic strain of CVB3 (TRACY et al. 1992). Mice fed diets adequate in Se developed mild to moderate myocarditis 10 days postinfection. However, mice fed the diets deficient in Se developed much more severe myocarditis: lesions were larger and increased calcification occurred. Virus titers in the heart and liver were elevated in the Se-deficient mice, although both Se-adequate and Se-deficient mice cleared the virus from the heart at an identical time point. Neutralizing antibody titers at 14 days postinfection were identical between Se-adequate and Se-deficient mice. However, T cell proliferation against both mitogen and specific viral antigens was depressed in the Se-deficient mice, suggesting that the immune system was compromised.

Although deficits in the immune response were noted, the fact that the Se-deficient mice could clear virus as readily as the Se-adequate mice suggested that an impairment in immune function alone was not sufficiently severe to account for the great difference in heart pathology observed between the two dietary groups. This conclusion led us to test the hypothesis that the virus itself may have changed phenotype as a result of replication in a Se-deficient host. Therefore, the following experiments were performed.

Beck and coworkers isolated virus from Se-deficient mice, which was renamed CVB3/20se-, and passed it into Se-adequate mice (BECK et al. 1994a). Hearts from Se-adequate mice infected with CVB3/20se- virus developed much more severe disease than Se-adequate mice infected with virus obtained from Se-adequate mice (CVB3/20se^{+}). The increased heart damage due to CVB3/20se- was accompanied by a cardiac virus titer 10^3-fold higher than that observed in the hearts of mice infected with CVB3/20se+. Thus, the viral phenotype had changed as a consequence of replicating in a Se-deficient host. These results demonstrate the importance of examining not only the host response to a nutritional deficiency, but the response of the pathogen as well.

In a second series of experiments, BECK et al. (1994b) inoculated Se-adequate and Se-deficient mice with CVB3/0 (GAUNTT et al. 1979; CHAPMAN et al. 1994), a strain of CVB3 which is normally avirulent in mice, although virus is recoverable from heart tissue. Hearts from CVB3/0-infected Se-adequate mice did not show any myocarditic lesions, whereas hearts from CVB3/0-infected Se-deficient mice developed moderate to severe myocarditis. Similar to findings in the CVB3/20 infected mice, virus titers were elevated in the Se-deficient mice and T cell proliferative responses were impaired. Viral passage experiments again demonstrated that the viral phenotype had changed: CVB3/0 virus obtained from Se-deficient mice (CVB3/0se--) caused disease in Se-adequate mice, whereas CVB3/0 virus obtained from Se-adequate mice (CVB3/0se+) did not.

To determine whether the phenotype change was due to a change in the viral genome, both CVB3/0se- and CVB3/0se+ viruses were sequenced (BECK et al. 1995). The genomic sequence of CVB3/0se+ was found to be identical to CVB3/0. However, CVB3/0se- differed from CVB3/0 at six nucleotide positions: nucleotide numbers (5′-3′) 234, 788, 2271, 2438, 3324 and 7334. Each point mutation found in CVB3/0Se- was identical to the nucleotide found in the myocarditic strains (CHAPMAN et al. 1994). Nucleotide 2690, which differs between myocarditic strains and the amyocarditic CVB3/0 strain, was not altered in the CVB3/0se- strain. Thus, the myocarditic CVB3/0se- strain was a hybrid between known myocarditic strains and the original amyocarditic CVB3/0 strain.

These results are important in that they demonstrate for the first time that a specific nutritional deficiency in the host alters the genotype of an infecting virus, thereby resulting in a more virulent pathogen. More work in this area is required to understand how host diet influences the genetic mutations in the CVB3 virus. One possibility is that the increased extent of replication of the virus in the Se-deficient host leads to an increase in replication errors, which are not repaired due to a lack of proofreading by the enterovirus polymerase. Other mutations may be occurring, but either are lethal, or are out-competed by the CVB3/0se- mutant strain, which is able to achieve the highest titer at the time of virus isolation. Clearly, further work must be done in the area of CVB3 infection and host nutrition.

If a deficiency in Se can increase the pathology of CVB3-induced myocarditis, can supplementation with Se prevent or limit the pathology? ILBÄCK et al. (1989a) injected normal mice intraperitoneally with 2 μg/kg body weight Se three times prior to infection with CVB3. CVB3-infected mice treated with Se had statistically fewer myocardial lesions than did mice not treated with Se. In addition, Se prophylaxis increased class II expressing cells. A dose-dependent increase in survival of infected mice was also noted, although high levels of Se (2000 μg/kg body weight) increased mortality, presumably due to Se toxicity. The mice in this study were fed a diet containing 0.14 ppm selenium, but a marginal level of vitamin E. At that concentration of Se in the diet, one would expect tissue glutathione peroxidase activities to be maximized. Therefore, it seems likely that the protective effect of the additional Se given intraperitoneally is due to factor(s) other than glutathione peroxidase activity. ILBÄCK et al. (1989a) suggested that Se may have an immunostimulating effect by increasing cytotoxic activity directed to virus-infected cells and increasing B cell activity, thus providing protection with antibodies. In a later report, 5 ppm Se in the drinking water, generally regarded as a toxic dose (National Research Council 1976), improved survival of CVB3-infected mice fed an unspecified diet, but had no effect on myocardial inflammatory or necrotic lesions (ILBÄCK et al. 1995a). Thus, it appears that luxus levels of Se may partially protect mice from CVB3-induced myocarditis by an unknown mechanism.

Heavy metal toxins, such as methyl mercury, cadmium and nickel, have been shown to interact with coxsackievirus. Mice inoculated with CVB3 have marked changes in distribution of nickel and cadmium (ILBÄCK et al. 1992a, b). Nickel accumulation occurred in the pancreas and walls of the ventricles of CVB3-inoculated mice, whereas uninfected control mice had almost no accumulation in these tissues. Increased myocardial damage occurred in nickel-treated CVB3-inculated mice as compared with CVB3-inculated mice that were not exposed to nickel. Cadmium accumulation in CVB3-inculated mice increased 47% in the kidneys when compared with uninfected mice. Mice inculated with CVB3 and exposed to methyl mercury had increased myocardial lesions (ILBÄCK et al. 1995b).

These studies point out that infection with CVB3 can cause alterations in the distribution of certain metallic environmental pollutants known in turn to interact metabolically with Se. For example, both Se and vitamin E protect against the neurotoxicity of methylmercury (GANTHER 1980). Since the only common known metabolic action of these two nutrients is their antioxidant properties, it was hypothesized that they both counteract methylmercury poisoning by preventing lipid peroxidation induced by free radicals generated from the homolytic cleavage of the carbon:metal bond in the alkyl metal. Se also prevents the testicular damage caused by cadmium (WHANGER et al. 1980). In that case, Se is thought to divert the binding of cadmium from low molecular weight testicular target proteins to proteins of higher molecular weight. More research is needed to clarify any three-way virus-nutrient-toxicant interactions that may be occurring.

5 Vitamin E and Coxsackievirus B3-Induced Myocarditis

Vitamin E is a lipid-soluble antioxidant which scavenges lipophilic radicals within membranes. Although both vitamin E and Se are antioxidant nutrients, they act by two very different antioxidant mechanisms and therefore the nature and degree of the oxidative stress caused by deficiencies of either nutrient would be expected to be different. None the less, these two nutrients can spare one another's requirements and many of the patients with Keshan disease, in addition to being Se-deficient, were also of marginal vitamin E status.

Because Se and vitamin E are closely related metabolically and because vitamin E deficiency may have played a role in Keshan disease, BECK et al. (1994c) inoculated both vitamin E-deficient and vitamin E-adequate mice with either CVB3/20 or CVB3/0 to determine whether or not the previously described results with Se were unique to Se. Similar to findings with the Se-deficient mice, vitamin E-deficient mice also developed more severe myocarditis when infected with CVB3/20 than did vitamin E-adequate mice. Likewise, vitamin E-adequate mice did not develop myocarditis when infected with CVB3/0, whereas vitamin E-deficient mice did develop heart inflammation. If fish oil were added to the vitamin E-deficient diet, the intensity of the lesions was greater. Dietary unsaturated fat increases the nutritional requirement for vitamin E possibly by increasing the tendency of tissue lipids to peroxidation. GAUNTT et al. (1989) reported a decrease in natural killer activity in mice fed fish oil, which led to an increase in severity of CVB3-induced myocarditis. However, under our conditions, addition of fish oil to the diet inhibited natural killer activity in both normal and vitamin E-deficient mice to the same extent so that there was no correlation between decreased natural killer activity and increased heart pathology.

As with mice fed Se-deficient diets, neutralizing antibody titers were not affected by the vitamin E-deficient diet, although T cell proliferation to either mitogen or antigen was again decreased. Mice fed a vitamin E-deficient diet that contained N,N′-diphenyl-p-phenylenediamine (DPPD), which mimics the antioxidant effects of vitamin E, did not develop increased inflammation when infected with CVB3/20, suggesting that the antioxidant effect of vitamin E, rather than nonantioxidant mechanisms, e.g., membrane stabilization, is the most important in conferring protection from enhanced myocarditis.

Viral passage experiments again demonstrated that the phenotype of the virus had changed in the vitamin E-deficient mice, such that mice of normal vitamin E status would develop myocarditis when infected with CVB3/0 virus obtained from a vitamin E-deficient animal. Recently, the RNA from CVB3/0 virus obtained from vitamin E-deficient mice has been sequenced (BECK et al. 1996) and was found to be identical to the CVB3/0se- virus. These results suggest that a mechanism common to vitamin E and Se depletion induces identical nucleotide changes in the CVB3/0 virus.

Because of the observations that both Se and vitamin E deficiency increase the pathogenicity of CVB3 infection, that fish oil intensifies the pathogenicity of CVB3 infection in vitamin E-deficient mice, and that DPPD can inhibit the effects of a vitamin E-deficient diet, we believe that the common mechanism responsible for the viral genetic changes is oxidative stress.

6 Oxidative Stress and Coxsackievirus Infection

Our work described in the previous section suggests that the effect of diet on CVB-induced pathology is due to changes in the antioxidant/pro-oxidant balance in the host. CVB-induced myocarditis is characterized by an inflammatory infiltrate that consists of macrophages and neutrophils early in the infection which are later replaced by T cells. Both macrophages and neutrophils produce hydrogen peroxide and superoxide anion free radicals which are part of the host's defense system against pathogens. Thus, these cells could contribute to a localized state of oxidative stress which could then contribute to myocardial damage.

What do dietary deficiencies of Se and vitamin E have in common so that a deficiency of either might be expected to have similar effects in intensifying the cardiotoxicity of the coxsackievirus? Both vitamin E and Se are antioxidant nutrients and a deficiency in either would be expected to result in varied increases in oxidative stress in the host, in addition to the localized state of oxidative stress induced by the viral infection itself. Oxidative stress could affect the virulence of the virus in a number of different ways. For example, the integrity of the heart cells (fibroblasts, myocytes, endothelial cells) could be compromised due to oxidative stress and thus become more vulnerable to viral attack. Alternatively, oxidative stress is known to affect the immune response (MEEKER et al. 1985; HALLIWELL et al. 1992). Thus, an impaired immune response could lead to heightened viral replication, resulting in increased heart pathology. In addition, the CVB3 RNA polymerase lacks proofreading ability. Therefore, any increase in extent of viral replication might be expected to favor the emergence of viral mutants with altered cardiovirulent properties. It is also possible that the oxidative stress status of the host directly affects the virus by stimulating specific nucleotides in the genome to mutate, leading to the evolution of more cardiovirulent progeny. Although the exact biochemical and/or genetic mechanism(s) whereby dietary Se and/or vitamin E deficiency alter the genetic expression of coxsackievirus-induced pathology is not understood, these experiments show the potent influence that host nutritional status can have on CVB3 pathogenicity and point out the need for additional research in this area.

HIRAOKA et al. (1992, 1993) performed a series of experiments demonstrating the effect of oxidative stress on CVB3-infected mice. Mice were treated at the time of CVB3 infection with either polyethylene glycol conjugated with superoxide dismutase (PEG-SOD) or with PEG-SOD combined with PEG-catalase. SOD is a potent scavenger of superoxide anion free radicals. Catalase breaks down hydrogen peroxide. The goal of the Hiraoka et al. study was to determine whether oxygen-derived free radicals are involved in the development of CVB3-induced myocarditis. These studies found that mice treated with PEG-SOD developed fewer inflammatory infiltrates, less necrosis, and less calcification in a dose-dependent manner than untreated control mice. No additional benefit was seen when mice were treated with PEG-SOD combined with PEG-catalase, suggesting that H_2O_2 does not play a major role in myocyte injury. No differences in viral titers were noted between treated and untreated groups. These studies suggest the superoxide anion is an important feature of myocardial injury in CVB3-infected mice.

Several workers have examined the effect of exercise on CVB3-induced myocarditis (GATMAITAN 1970; REYES et al. 1982, ILBÄCK 1989b). These studies used swimming or running on a treadmill as the exercise stressor. All of these studies found more extensive heart lesions in exercised CVB3-infected mice vs nonexercised CVB3-infected mice. Altered immune function(s) were found to be associated with the exercised group: increased heart-infiltrating cytotoxic T cells, decreased antibody titers, decreased interferon production and decreased class II expressing cells in the myocardium.

In addition to altered immune function, another possible explanation for the increase in heart damage in infected exercised animals is oxidative stress. Exercise increases the metabolic rate, which in turn leads to an increase in free radical production (DAVIES et al. 1982). Previous studies have found an increase in superoxide dismutase (SOD) and a decrease in glutathione peroxidase and catalase in exercised rats (KUMAR et al. 1992). Heart tissue in exercised animals was found to have increased free radical production, measured by the malondialdehyde (MDA) reaction. Supplementation of exercised rats with vitamin E completely prevented the accumulation of MDA, suggesting that exercised-induced oxidative stress could be ameliorated with an antioxidant supplement.

Because a deficiency in Se or vitamin E, as well as exercise, increased the pathology of CVB3-induced myocarditis, a likely explanation for the commonality among all three conditions is an increase in oxidative stress in the host. The fact that fish oil, a pro-oxidant stimulator, increased the CVB3 pathology and that DPPD, a synthetic antioxidant, protected from the increased damage in vitamin E-deficient animals, also supports oxidative stress as a common mechanism.

What about other nutrients with antioxidant properties? For example, copper and zinc are cofactors for the activity of SOD. The work of HIRAOKA et al. (1992, 1993) demonstrated that mice treated with PEG-SOD were partially protected from CVB3-induced myocarditis, suggesting that copper and zinc may play a role in CVB3 infection. Further, copper-deficient rats develop heart damage, which has been related to the weak antioxidant defenses of the heart (CHEN et al. 1994).

Other possible nutrients which may interact with CVB3-induced pathology include manganese, because of its role in the functioning of mitochondrial SOD, and iron, because it is a constituent of catalase, and conversely, because it can also function as a pro-oxidant. Other possible nutrient candidates include riboflavin, a component of glutathione reductase and vitamin C and β-carotene. Clearly, we are just beginning to appreciate the tremendous influence that nutrition may have on CVB3 infection.

7 Coxsackievirus B3, Selenium and Potential New Genes

A new and interesting approach to examining the effect of Se on CVB3 has been offered by TAYLOR et al. (1996). Early work suggested that, because of size limitations, viruses are under a powerful constraint to maximize the information contained in their genome and that the placing of overlapping genes in different reading frames becomes a viable method for increasing their coding density. TAYLOR et al. (1996) used the following computer method for identifying potential new genes: (1) translate the genome in all three reading frames; (2) search for novel open reading frames (ORFs) by assuming that a UGA codon may be read as sense rather than as stop; (3) if the 5′ region of the ORF overlaps with a known coding region, search for a frameshift sequence in the overlap; (4) if frameshifts are found, look for RNA stems or pseudoknots no more than 15 bases past the start of the frameshift; (5) do sequence analysis and search databases to determine if the potential hypothetical protein has significant similarity to known proteins.

If a frameshift and an appropriate pseudoknot are present, then this is a potential new gene. The UGA codon, normally seen as a stop codon, can under specific conditions encode for a selenocysteine, thus allowing read-through to occur. In eukaryotes, selenocysteine incorporation at UGA depends on the presence of a SECIS element (a structural signal in the mRNA 3′-untranslated region). In bacteria, a stem-loop structure is required immediately following the UGA codon.

Using the technique described above, TAYLOR et al. (1994) analyzed CVB3/20 for novel genes and found two potential selenoprotein genes. In the beginning of the VPG coding region, a potential -1 frameshift was found following the first two NH_2-terminal amino acids. This potential selenoprotein would have 122 amino acids and overlap the VPG gene and much of the p3C protease. This gene could possibly be cleaved at the same protease site as VPG.

A second potential selenoprotein was found between bases 3442 and 3749 which has a strong similarity to a family of proteins that includes epidermal growth factor. This ORF has a start codon and eight in-frame UGA codons.

In order for selenocysteine to be inserted at UGA, a SECIS element must be present, which is found in CVB3 at 300 nucleotides following the 3′ end of the ORF containing the eight in-frame UGA codons. This is equivalent to a 3′-UTR region in mammalian genes, where SECIS elements are always located.

When TAYLOR et al. (1994) looked at CVB4 for potential new genes, they did not find any: the potential new genes in CVB3 are inactivated in CVB4 due to base changes. These authors theorize that perhaps CVB4 lost the selenoprotein genes because of prolonged exposure to low-Se conditions, which provides no selective pressure to maintain the genes.

Although the results by TAYLOR et al. (1994) are intriguing – they have described similar novel genes in HIV – actual expression of these putative genes remains to be tested experimentally.

8 Concluding Remarks

In this chapter, the reader has been introduced to a new paradigm for the study of nutrition-infection interactions. The old paradigm was essentially a unidirectional linear model that considered only host effects:

inadequate host nutrition
↓
dysfunctional host immunity
↓
increased host susceptibility to virus

On the basis of our work with coxsackievirus-induced murine myocarditis, we now propose a new paradigm which represents more accurately the observed interactions between host nutritional status and the virus:

inadequate host nutrition
↓ ↘
dysfunctional host immunity ↔ virus
↓
enhanced susceptibility to virus

As before, this new paradigm describes the effect of host nutritional status on the host immune system. However, the new paradigm now also describes the direct influence that host nutritional status has on the virus itself, as demonstrated experimentally by our work with CVB3/0 (Chap. 11). Just as host nutritional status impacts the virus directly, so can a dysfunctional immune system affect the virus and vice versa. Thus, nutritional status of the host not only can bring about alterations in the virus genotype, but also can cause changes in host immunity. Both of these in turn can then increase the host's susceptibility to the coxsackievirus.

The biggest challenge ahead is to define the scope of this phenomenon. Up to now, all of our experiments have dealt only with deficiencies of Se or vitamin E. This was a reasonable approach, since the driving hypothesis was to clarify the role of these deficiencies and infection with coxsackievirus in the etiology of the Se-responsive cardiomyopathy known as Keshan disease. The next logical step would be to investigate the effect of deficiencies of a variety of other nutrients that are thought to have antioxidant functions in vivo, such as zinc, copper, manganese, β-carotene, and vitamin C, on the virus.

Once having accepted the fact that antioxidant nutrients can have direct effects on the virus, there is a priori no reason to exclude the possibility that nutrients with no known antioxidant function might also have some effect on the virus. Indeed, the classical work of Woodruff and Kilbourne (1970) showed that the relatively simple manipulation of withholding food had pronounced effects on the pathogenesis of coxsackievirus B3 infection in mice. Was this due to alterations in the virus itself? It is perhaps not too far-fetched to visualize a systematic survey of nutritional experiments being carried out to determine which deficiencies affect the virus and how they affect it. And if they don't affect the virus, then why not?

The studies conducted with metal pollutants (e.g., mercury and cadmium) are provocative because these metals have strong metabolic interactions with Se. It is possible that the modifying effect of these metals reflects an altered distribution or function of Se. The role of other environmental contaminants, many of which have pro-oxidant properties (ozone, halogenated hydrocarbons, etc.), needs to be investigated. Clearly, virologists should be talking not only with nutritionists, but with toxicologists as well.

Based on the results showing that exercise stress worsens the effects of coxsackievirus infection in mice, virologists may wish to initiate a dialogue with physiologists also. Because exercise releases free radicals, such studies should be carried out in the future under controlled dietary conditions. Is the fact that athletes are sometimes targeted by coxsackieviruses merely a chance event with often tragic consequences? Would antioxidant supplements help such people?

The other side of the nutrition-virus coin (i.e., the virus itself) would also seem to offer exciting possibilities for experimentation. Are coxsackieviruses other than B3 affected by host malnutrition? Diabetes induced chemically with alloxan depends upon free radical production and the β-cells of the pancreas are thought to be the cells in the body most sensitive to oxidative stress. What would a deficiency of antioxidant nutrients do in coxsackievirus-induced diabetes? Would nutritional deficiencies in the host alter tropism of the virus?

On a mechanistic level, our current understanding of the host nutrition-viral infection interaction is very poor. Conceivably, host nutrition could be exerting its effects on the host heart cells, the host immune system, or the virus per se (without the intervention of any of the aforementioned host systems). In any case, research generated along these lines is likely to produce information that is useful both practically and theoretically.

From the epidemiological point of view, it is now possible to track the mutational history of poliovirus as it progresses through an epidemic. Would the same be possible with coxsackievirus in Keshan disease areas? Would one expect to find different strains of the coxsackievirus in Keshan disease victims and in healthy individuals living in the same area? Would one expect to find more virulent strains of the coxsackievirus in areas with populations of poor Se status than in areas with populations of better Se status?

Just as we do not know now how many nutritional deficiencies can impact on coxsackievirus infection, we also do not know how many viral infections can be impacted by malnutrition. That is, we do not know how widely applicable our observations are either from the nutritional or the virological point of view. If we limit ourselves to the Picornaviridae, we would have to concern ourselves about polio, foot and mouth disease, and the common cold. That would seem to be enough, but if we venture that our results are applicable to RNA viruses in general, the list of pathogens becomes impressive: influenza, measles, retroviruses, etc. One could speculate, for example, on whether it would be possible to alter the population of HIV quasispecies in an AIDS patient through nutritional manipulation. Such research might also prove helpful in clarifying certain poorly understood problems in contemporary virology, such as the emerging "new" viral diseases and the mechanisms of cross-species transfer.

Whatever the ultimate applicability of our work, its true impact can only be judged through the formation of successful collaborations between two groups that have not interacted to any significant degree in the past, namely virologists and nutritionists. The authors personally believe that the research described in this chapter is just the tip of the iceberg regarding host nutrition-viral infection interactions. The extent to which this chapter succeeds in stimulating such interdisciplinary interactions is the extent to which this chapter will have succeeded in its purpose.

References

Bai J, Wu S, Ge K, Deng X, Su C (1980) The combined effect of selenium deficiency and viral infection on the myocardium of mice. Acta Acad Med Sinica 2:31–33

Beck MA, Kolbeck PC, Shi A, Rohr LH, Morris VC, Levander OA (1994a) Increased virulence of a human enterovirus (coxsackievirus B3) in selenium-deficient mice. J Infect Dis 170:351–357

Beck MA, Kolbeck PC, Rohr LH, Shi Q, Morris VC, Levander OA (1994b) Benign human enterovirus becomes virulent in selenium-deficient mice. J Med Virol 43:166–170

Beck MA, Kolbeck PC, Rohr LH, Shi Q, Morris VC, Levander OA (1994c) Vitamin E deficiency intensifies the myocardial injury of coxsackievirus B3 infection of mice. J Nutr 124:345–358

Beck MA, Shi Q, Morris VC, Levander OA (1995) Rapid genomic evolution of a non-virulent Coxsackievirus B3 in selenium-deficient mice results in selection of identical virulent isolates. Nat Med 1:433–436

Beck MA, Shi Q, Morris VC, Levander OA (1996) From avirulent to virulent: Vitamin E deficiency in mice drives rapid genomic evolution of a coxsackie B3 virus. FASEB J 10:A191

Chapman, N.M., Tu, Z., Tracy, S., Gauntt, C.J. (1994) An infectious cDNA copy of the genome of a non-cardiovirulent coxsackievirus B3 strain: its complete sequence analysis and comparison to the genomes of cardiovirulent coxsackieviruses. Arch Virol 134:115–130.

Chen Y, Saari JT, Kang YJ (1994) Weak antioxidant defenses make the heart a target for damage in copper-deficient rats. Free Rad Biol Med 17:529–536

Dagan R, Bar-David Y, Sarov B, Katz M, Kassis I, Greenberg D, Glass RI, Margolis CZ, Sarov I (1990) Rotavirus diarrhea in Jewish and Bedouin children in the Negev region of Israel: epidemiology, clinical aspects and possible role of malnutrition in severity of illness. Ped Inf Dis J 9:314–321

Davies KJA, Quintanilha T, Brooks GA, Packer L (1982) Free radicals and tissue damage produced by exercise. Biochem Biophy Res Commun 107:1178–1205

Ganther HE (1980) Interactions of vitamin E and selenium with mercury and silver. In: Levander OA, Cheng L (eds) Micronutrient interactions: vitamins, minerals, and hazardous elements, Annals of the NY Acad Sci, vol. 355, New York Acadademy of Sciences, New York, pp 212–226

Gatmaitan BG, Chason JL, Lerner AM (1970) Augmentation of the virulence of murine coxsackievirus B-3 myocardiopathy by exercise. J Exp Med 131:1121–1136

Gauntt CJ, Godeny EK, Lutton CW, Fernandes G (1989) Role of natural killer cells in experimental murine myocarditis. Springer Semin Immunopathol 11:51–59

Gauntt CJ, Trousdale MD, LaBadie DRL, Paque RE, Nealon T (1979) Properties of coxsackievirus B3 variants which are amyocarditic or myocarditic for mice. J Med Virol 3:207–220

Ge KY, Xue A, Bai J (1983) Keshan disease- an endemic cardiomyopathy in China. Virchows Arch {A} 401:1–14

Gu BQ (1983) Pathology of Keshan disease. A comprehensive review. Chinese Med J 96:251–261

Halliwell B, Gutteridge JMC, Cross CE (1992) Free radicals, antioxidants, and human disease: Where are we now? J Lab Clin Med 119:598–620

Hiraoka Y, Kishimoto C, Masahiko K, Ochiai H, Sasayama S (1992) Effects of polyethylene glycol conjugated superoxide dismutase on coxsackievirus B3 myocarditis in mice. Cardiovas Res 26:956–961

Hiraoka Y, Kishimoto C, Takada H, Durokawa M, Ochiai H, Shiraki K, Sasayama S (1993) Role of oxygen derived free radicals in the pathogenesis of coxsackievirus B3 myocarditis in mice. Cardiovas Res 27:957–961

Ilbäck N-G, Mohammed A, Fohlman J, Friman G (1976) Cardiovascular lipid accumulation with coxsackievirus infection in mice. Am J Pathol 136:159–167

Ilbäck N-G, Fohlman G, Friman G (1989a) Protective effect of selenium on the development of coxsackievirus B3-induced inflammatory lesions in the murine myocardium. J Trace Elements Exp Med 2:257–266

Ilbäck N-G, Fohlman J, Friman G (1989b) Exercise in coxsackie B3 myocarditis: Effects on heart lymphocyte subpopulations and the inflammatory reaction. Am Heart J 117:1298–1302

Ilbäck N-G, Fohlman J, Friman G (1992a) A common viral infection can change nickel target organ distribution. Toxicol Appl Pharmacol 114:166–170

Ilbäck N-G, Fohlman J, Friman G, Glynn AW (1992b) Altered distribution of ^{109}cadmium in mice during viral infection. Toxicol 71:193–202

Ilbäck N-G, Fohlman J, Friman G (1995a) Selenium (Se) supplementation decreases myocardial injury and increases survival in coxsackie B3 (CB3) virus infected mice. J Invest Med 43:269A

Ilbäck N-G, Lindh U, Wesslso L, Fohlman J, Priman G (1995b) Heart trace elements are changed after viral infection (coxsackie B3) in methyl mercury exposed mice. J Invest Med 43:269A

Jin S (1984) III. Keshan disease in children. Chinese Med 2: 348–352

Keshan Disease Research Group of the Chinese Academy of Medical Sciences (1979) Observations on effect of sodium selenite in prevention of Keshan disease. Chin Med J 92:471–476

Kumar CT, Reddy VK, Prasad M, Thyagaraju K, Reddanna P (1992) Dietary supplementation of vitamin E protects heart tissue from exercise-induced oxidant stress. Mol Cell Biochem 111:109–115

Li G, Wang F, Kang D, Li C. (1985) Keshan disease: an endemic cardiomyopathy in China. Hum Path 16: 602–609

Li Y, Zhang H, Yang Y, Chen H, Archard LC (1995) High prevalence of enteroviral genomic sequences in an endemic cardiomyopathy (Keshan disease) detected by nested polymerase chain reaction. Progress in Clinical Virology 1995 Joint Meeting, Prague, Czech Republic, pg 172

Loria RM, Kibrik S, Madge GE (1976) Infection of hypercholesterolemic mice with coxsackievirus B. J Inf Dis 133:655–662

Meeker HC, Eskew ML, Scheuchenzuber W, Scholz RW, Zarkower A (1985) Antioxidant effects on cell-mediated immunity. J Leuk Biol 38:451–458

National Research Council. Committee on Medical and Biologic Effects of Environmental Pollutants (1976) Selenium. Division of Medical Sciences, Assembly of Life Sciences, National Academy of Sciences, National Academy Press, Washington DC

Reyes MP, Thomas JA, Ho KL, Smith FE, Lerner AM (1982) Elevated thymocyte norepinephrine and cyclic guanosine 3',5' monophosphate in sensitized T-lymphocytes from exercised mice with coxsackievirus B-3 myocarditis. Biochem Biophys Res Commun 109:701–708

Roath PD, Beck MA, Chen X, Xue A, Wang S, Levander OA (1996) Detection of enterovirus RNA from heart tissue blocks and blood obtained from a Keshan disease endemic area of China using reverse transcriptase-polymerase chain reaction (RT-PCR). FASEB J 10:A537

Rotruck JT, Pope AL, Ganther HE, Swanson AB, Hafeman DG, Hoekstra WG (1973) Selenium: biochemical role as a component of glutathione peroxidase. Science 179:585–590

Schwarz K, Foltz CM (1957) Selenium as an integral part of factor 3 against dietary necrotic liver degeneration. J Am Chem Soc 79:3292–3293

Su C, Gong C, Li J, Chen L, Zhou D, Jin Q (1979) Preliminary results of viral etiology of Keshan disease. Chinese J Med 59:466–472

Taylor EW, Ramanathan C, Madimpalli RG (1996) A general method for predicting potential new genes in nucleic acid sequences: application to the human immunodeficiency virus. In: Witten M (ed) Computational medicine, public health and technology. World Scientific, London, 1:285–309

Taylor EW, Ramanathan CS, Jalhari RK, Nadimpalli RG (1994) A basis for new approaches to the chemotherapy of AIDS: novel genes in HIV-1 potentially encode selenoproteins expressed by ribosomal frameshifting and termination suppression. J Med Chem 37:2637–2654

Tracy S, Chapman N, Tu Z (1992) Coxsackievirus B3 from an infectious cDNA copy of the genome is cardiovirulent in mice. Arch Virology 122:399–409.

Whanger PD, Ridlington JW, Holcomb CL (1980) Interactions of zinc and selenium on the binding of cadmium to rat tissue proteins. In: Levander OA, Cheng L (eds) Micronutrient interactions: vitamins, minerals, and hazardous elements. Annals of the NY Acad Sci, New York Academy of Sciences, New York vol. 355 pp 333–346

Woodruff JF (1970) The influence of quantitated post-weaning undernutrition on coxsackievirus B3 infection of adult mice. II. Alteration of host defense mechanisms. J Inf Dis 121: 164–181

Woodruff JF, Kilbourne ED (1970) The influence of quantitated post-weaning undernutrition on coxsackievirus B3 infection of adult mice. I. Viral persistence and increased severity of lesions. J Inf Dis 121: 137–163

Yu WH (1981) Analysis of the incidence of Keshan disease in 25 years in Heilongjiang province. Chinese J Epidemiol 2: 26–34

Genetic Divergence Among the Group B Coxsackieviruses

J. R. ROMERO[1,2,4], C. PRICE[3,4], AND J. J. DUNN[4]

1 Introduction

Six serotypes comprise the group B coxsackieviruses (CVB1–6) (MELNICK 1996). Similar to other members of the family Picornaviridae, the CVB genome is a single stranded, messenger sense, polyadenylated RNA molecule (Rueckert 1996). Analysis of the genomes from the fully sequenced CVB (CVB1, 3, 4, and 5) (Table 1) further documents their relationship to the Picornaviridae by demonstrating that

[1]Joint Division of Pediatric Infectious Diseases, Departments of Pediatrics, Creighton University and University of Nebraska Medical Center, Creighton University, 2500 California Plaza, Criss II Rm 409, Omaha, Nebraska 68178, USA

[2]Department of Medical Microbiology and Immunology, Creighton University, 2500 California Plaza, Criss II Rm 409, Omaha, Nebraska 68178; USA

[3]Genetic Sequence Analysis Facility, University of Nebraska Medical Center. 600 South 42nd Street, Omaha, Nebraska 68198, USA

[4]Department of Pathology and Microbiology, University of Nebraska Medical Center. 600 South 42nd Street, Omaha, Nebraska 68198, USA

Table 1. Reported full length sequences of the group B coxsackieviruses (CVB)

Serotype	Strain	Genome length[a]	Accession number	Reference
CVB1	Japan	7389	M16560	Iizuka et al. 1987
CVB3	Nancy	7396	M16572	Lindberg et al. 1987
CVB3	Nancy	7399	M33854	Klump et al. 1990
CVB3	20	7399	—	Tracy et al. 1992
CVB3	0	7399	—	Chapman et al. 1994
CVB4	JVB Benschoten	7395	X05690	Jenkins et al. 1987
CVB4	E2	7396	S76772	Kang et al. 1994
CVB4	P-CB4	7395	—	Titchener et al. 1994
CVB5	1954/UK/85	7402	X67706	Zhang et al. 1993

[a]Nucleotides.

they are organized into a 5' nontranslated region, a protein coding region containing a single open reading frame, a 3' nontranslated region and a terminal polyA tail.

The protein coding region of the CVB genome can be subdivided into three regions: P1, P2, and P3 (RUECKERT and WIMMER 1984). The P1 region encodes the four capsid proteins VP4 (1A), VP2 (1B), VP3 (1C), and VP1 (1D). The P2 and P3 regions code for nonstructural proteins required for the CVB lifecycle: 2A (protease), 2B, 2C, 3A, 3B (VPg), 3C (protease), and 3D (polymerase) (RUECKERT 1996).

Since the publication by Palmenberg (PALMENBERG 1989) of her exhaustive comparative analysis of the picornaviral capsid protein sequences, several complete and multiple, partial, CVB sequences have been made available. This new information can be used to draw stronger conclusions about the CVB genome and its proteins. The complete nucleotide sequences of nine strains of CVB have been published and/or deposited into GenBank (Release 95.0 [6/96]) (Table 1) (IIZUKA et al. 1987; JENKINS et al 1987; LINDBERG 1987; KLUMP et al. 1990; TRACY et al. 1992; ZHANG et al. 1993; Chapman et al. 1994; KANG et al. 1994; TITCHENER et al. 1994). In addition, partial genomic sequences various CVB strains have been determined by several laboratories (LINDBERG et al. 1992; DEBROT et al. 1994, ROMERO et al. 1994; KOPECKA et al. 1995; ROMERO et al. 1995b; GAUNT and PALLANSCH 1996; LEE et al. 1996). Undoubtedly, many more partial or complete CVB sequences have been determined by investigators around the world but have yet to be published or otherwise made available.

For CVB1 and 5 a single complete genomic sequence has been entered into GenBank and no other sequences are currently available (IIZUKA et al. 1987; ZHANG et al. 1993). In the case of CVB3 the complete sequences of four strains have been made available (LINDBERG et al. 1987; KLUMP et al. 1990; TRACY et al. 1992; CHAPMAN et al. 1994). For the purpose of this review, the 20 strain of CVB3 Nancy (CVB3 20), published by Tracy (TRACY et al. 1992), was employed as the representative strain of CVB3 and used for intertypic comparisons. Recently, a consensus CVB3 sequence has been entered into GenBank (accession number

M88483) that lists and reconciles the differences between the four published complete sequences.

Three complete sequences of CVB4 have been published (JENKINS et al. 1987; KANG et al. 1994; TITCHENER et al. 1994). Only the sequence from the prototype JVB Benschoten strain (JENKINS et al. 1987) and a murine diabetogenic strain, E2 (KANG et al. 1994), are currently entered into GenBank. The sequence of the JVB Benschoten strain was used for the CVB intertypic comparisons described below.

In addition to the sequences mentioned previously, the 5' nontranslated region of six strains of CVB2 and nine strains of CVB3 amplified using reverse transcription-polymerase chain reaction (RT-PCR) and directly sequenced by cycle sequencing were used to examine intratypic variation within those CVB serotypes. Complete or partial sequences for CVB6 have not, as of yet, been published or deposited into GenBank.

Sequence alignment and genomic analysis were carried out using the University of Wisconsin Genetics Computer Group Genetic Sequence Analysis package version 8.1-UNIX and the MacVector and Assemblylign software packages (IBI, New Haven, CT). Where necessary, manual corrections of the computer-generated alignments were made to provide a more appropriate fit of the data. Determination of nucleic acid and amino acid identities was performed using the GCG Gap program with the following parameters: For nucleic acid analysis, gap weight: 5.0, average match: 1.0, length weight: 0.300, average mismatch: 0.000. For amino acid analysis, gap weight: 3.0, length weight: 0.1, average match: 0.540, average mismatch: –0.396.2.

2 Intertypic Genetic Divergence Among the Group B Coxsackieviruses

The genomes of the fully sequenced CVB are remarkably similar in length, ranging from 7389 nucleotides in CVB1 to 7402 nucleotides in CVB5 (Table 1). The variations in genomic length observed among the CVB are due principally to differences within the coding region of two of the viral capsid proteins, VP1 and VP2 (discussed below). Minor contributions to the differences in length are also made by the 5' and 3' nontranslated regions of the CVB genome.

2.1 The 5' and 3' Nontranslated Regions

The complete sequences of the 5' nontranslated regions (5' NTRs) of CVB1, 3, 4, and 5 have been determined. In addition, all but the initial 86 nucleotides of the 5' NTR from the prototypic strain of CVB2, Ohio has also been elucidated (Romero JR, Dunn JJ, unpublished data). The overall length of the 5' NTR is remarkably constant among the fully sequenced CVB; ranging from 741 to 743 nucleotides (Fig. 1). The sequenced region of CVB2 Ohio contains no insertions or deletions

```
                              20                  40                  60                  80
     TTAAAACAGCCTGTGGGTTGWWCCCACCCACANGGCCCAHTGGGCGCTAGCACWCTGGTAYYVCGGTACCTTTGTGCGCC

CVB1 TTAAAACAGCCTGTGGGTTGTTCCCACCCACA.GGCCCATTGGGCGCTAGCACTCTGGTATCACGGTACCTTTGTGCGCC
CVB2
CVB3 TTAAAACAGCCTGTGGGTTGATCCCACCCACAGGGCCCATTGGGCGCTAGCACTCTGGTATCACGGTACCTTTGTGCGCC
CVB4 TTAAAACAGCCTGTGGGTTGTACCCACCCACAGGGCCCAATGGGCGCTAGCACACTGGTATTCCGGTACCTTTGTGCGCC
CVB5 TTAAAACAGCCTGTGGGTTGTACCCACCCACAGGGCCCACTGGGCGCTAGCACTCTGGTACCGCGGTACCTTTGTGCGCC

                             100                 120                 140                 160
     TGTTTTAYMNNNCCCYYSCCCNNNWHYGBAAYWTAGAAGNWWWSWMMMAMHGRTCAWYWRBNRRCVYRGYRHVCCARCYR

CVB1 TGTTTTACAT..CCCCTCCCC.AAATTGTAATTTAGAAG.TTTCACACACCGATCATTAGCAAGCGTGGCACACCAGCCA
CVB2        ATCTA.CCCTTGCCC...AATGCAACATAGAAGCTTAGTCCAAAAGGTCAATAGGCAGCCCAGTGCACCAACTG
CVB3 TGTTTTATA..CCCCCTCCCCC.AACTGTAACTTAGAAG.TAACACACACCGATCAACAGTCAGCGTGGCACACCAGCCA
CVB4 TGTTTTATAA..CCCCCCCCC.AGTTCGCAACTTAGAAGCAAAGAAACAATGGTCAATTACTGACGCAGCAACCCAGCTG
CVB5 TGTTTTATATCCCCCCTCCCCTTAACCGGAACTTAGAAG.TTACACACACTGATCAACAGTGGGCATAGCATGCCAGCTA

                             180                 200                 220                 240
     BGTYYWGRYCAAGYACTTCTGTBWCCCCGGACYRAGTATCARTARRCYGCTHRCGCGGYTGAAGGAGAAAVCGTTCGTTA

CVB1 TGTTTTGATCAAGCACTTCTGTTACCCCGGACTGAGTATCAATAGACCGCTAACGCGGTTGAAGGAGAAAACGTTCGTTA
CVB2 GGTCCAGACCAAGCACTTCTGTTACCCCGGACCAAGTATCAGTAGGCTGCTCACGCGGCTGAAGGAGAAAACGTTCGTTA
CVB3 CGTTTTGATCAAGCACTTCTGTTACCCCGGACTGAGTATCAATAGACTGCTCACGCGGTTGAAGGAGAAAGCGTTCGTTA
CVB4 TGTTTTGGCCAAGTACTTCTGTGTCCCCGGACTGAGTATCAATAAGCTGCTTGCGCGGCTGAAGGAGAAACCGTTCGTTA
CVB5 TGTCTTGATCAAGCACTTCTGTCTCCCCGGACCAAGTATCAATAGACTGCTCACGCGGTTGAAGGAGAAAGCGTTCGTTA

                             260                 280                 300                 320
     YCCGRCYAACTACTTCGARAARCCYAGTARCRCYRTGRAVGTTGMRVRGYGTTTCGCTCMRCACWDCCCCMGTGTAGWTC

CVB1 CCCGGCCAACTACTTCGAAAAACCTAGTAACACCATGGAAGTTGCGGAGTGTTTCGCTCAGCACTACCCCAGTGTAGATC
CVB2 CCCGGCCAACTACTTCGAGAAACCTAGTAGCACCATGAAGGTTGCGCAGTGTTTCGCTCCGCACAGCCCCAGTGTAGATC
CVB3 TCCGGCCAACTACTTCGAAAAACCTAGTAACACCGTGGAAGTTGCAGAGTGTTTCGCTCAGCACTACCCCAGTGTAGATC
CVB4 CCCGGCCAACTACTTCGAGAAGCCTAGTAACGCCATGAACGTTGAGGAGTGTTTCGCTCAGCACTTCCCCCGTGTAGTTC
CVB5 CCCGACTAACTACTTCGAGAAACCCAGTAGCACTATGAAAGTTGCGAGGCGTTTCGCTCAACACTTCCCCAGTGTAGATC

                             340                 360                 380                 400
     AGGTCGATGAGYCACCGCRTTCCCCACNGGYGACCGTGGCGGTGGCTGCGTTGGCGGCCTGNCYRYGGGGHAACYCRYRG

CVB1 AGGTCGATGAGTCACCGCGTTCCCCACGGGCGACCGTGGCGGTGGCTGCGTTGGCGGCCTGCCTACGGGGAAACCCGTAG
CVB2 AGGTCGATGAGCCACCGCATTCCCCACGGGTGACCGTGGCGGTGGCTGCGTTGGCGGCCTGCCCATGGGGTAACCCATGG
CVB3 AGGTCGATGAGTCACCGCATTCCCCACGGGCGACCGTGGCGGTGGCTGCGTTGGCGGCCTGCCCATGGGGAAACCCATGG
CVB4 AGGTCGATGAGTCACCGCGTTCCCCACGGGTGACCGTGGCGGTGGCTGCGTTGGCGGCCTGCCTGTGGGGCAACCCGCAG
CVB5 AGGTCGATGAGTCACCGCATTCCCCAC.GGCGACCGTGGCGGTGGCTGCGTTGGCGGCCTG.CTACGGGGCAACTCGTAG
```

Fig. 1. Alignment of the nucleotide sequence of the 5' NTR of the CVB. The consensus sequence for the CVBs is shown underlined above the alignments of sequences from each of CVB serotypes. Nonconserved positions are designated by the single letter genetic code where R=G or A; K=G or T; Y=C or T; S=C or G; M=A or C; W=A or T; B=C, G or T; D=A, G or T; H=A, C or T; V=A, C or G; N=A, C, G or T.

```
                         420                 440                 460                 480
      GACGCTYYRATACWGACATGGTGYGAAGAGYCTATTGAGCTAGTTRGTARTCCTCCGGCCCCTGAATGCGGCTAATCCTA

CVB1  GACGCTCTAATACAGACATGGTGCGAAGAGTCTATTGAGCTAGTTGGTAATCCTCCGGCCCCTGAATGCGGCTAATCCTA
CVB2  GACGCTTCAATACTGACATGGTGTGAAGAGTCTATTGAGCTAGTTGGTAGTCCTCCGGCCCCTGAATGCGGCTAATCCTA
CVB3  GACGCTCTAATACAGACATGGTGCGAAGAGTCTATTGAGCTAGTTGGTAGTCCTCCGGCCCCTGAATGCGGCTAATCCTA
CVB4  GACGCTCTGATACAGACATGGTGTGAAGAGCCTATTGAGCTAGTTGGTAGTCCTCCGGCCCCTGAATGCGGCTAATCCTA
CVB5  GACGCTTCAATACAGACATGGTGCGAAGAGTCTATTGAGCTAGTTAGTAGTCCTCCGGCCCCTGAATGCGGCTAATCCTA

                         500                 520                 540                 560
      ACTGCGGAGCRYRYRYYYKCAAVCCAGNGRGYRGYRYGTCGTAAYGGGYAACTCTGCAGCGGAACCGASTACTTTGGGTG

CVB1  ACTGCGGAGCACATACCCTCAAACCAGGGGGCAGTGTGTCGTAACGGGCAACTCTGCAGCGGAACCGACTACTTTGGGTG
CVB2  ACTGCGGAGCGTGCGCTTGCAACCCAGTGAGTAGCACGTCGTAATGGGTAACTCTGCAGCGGAACCGACTACTTTGGGTG
CVB3  ACTGCGGAGCACACACCCTCAAGCCAGAGGGCAGTGTGTCGTAACGGGCAACTCTGCAGCGGAACCGACTACTTTGGGTG
CVB4  ACTGCGGAGCACACGTTCGCAAGCCAGCGAGTGGTGTGTCGTAACGGGCAACTCTGCAGCGGAACCGAGTACTTTGGGTG
CVB5  ACTGCGGAGCACATACCCTCAACCCAGGGGGCAGTGTGTCGTAACGGGTAACTCTGCAGCGGAACCGACTACTTTGGGTG

                         580                 600                 620                 640
      TCCGTGTTTCMTTTTATYCYTWBHYTGGCTGCTTATGGTGACAATTGAVAGRTTGTTRCCATATAGYTATTGGATTGGCC

CVB1  TCCGTGTTTCATTTTATTCCTATACTGGCTGCTTATGGTGACAATTGACAGGTTGTTACCATATAGTTATTGGATTGGCC
CVB2  TCCGTGTTTCCTTTTATTCCTTGTCTGGCTGCTTATGGTGACAATTGAGAGATTGTTGCCATATAGCTATTGGATTGGCC
CVB3  TCCGTGTTTCATTTTATTCCTATACTGGCTGCTTATGGTGACAATTGAGAGATTGTTACCATATAGCTATTGGATTGGCC
CVB4  TCCGTGTTTCCTTTTATTCTTACCTTGGCTGCTTATGGTGACAATTGAAAGATTGTTACCATATAGCTATTGGATTGGCC
CVB5  TCCGTGTTTCCTTTTATCCTTACATTGGCTGCTTATGGTGACAATTGAAAGATTGTTACCATATAGCTATTGGATTGGCC

                         660                 680                 700                 720
      ATCCRGTRWCHAAYAGRGCDATYATWTAYCTSTTTGTTGGDTTYRTWCCMYTKRRYYWNRRRRWDBWTAADACHYTHHMD

CVB1  ATCCGGTGACTAACAGAGCAATTATATATCTCTTTGTTGGGTTTATACCACTTAGCTTGAAAGAGGTTAAAACACTACAT
CVB2  ATCCGGTATCTAACAGGGCGATCATTTACCTGTTTGTTGGTTTCATACCCTTGAATCTTGAGGTTTATAATACCCTCACT
CVB3  ATCCGGTGACCAATAGAGCTATTATATATCTCTTTGTTGGGTTTATACCACTTAGCTTGAAAGAGGTTAAAACATTACAA
CVB4  ATCCAGTGTCAAATAGAGCAATCATATATCTGTTTGTTGGTTTCGTTCCCTTGGACTACAGAAATCTTAAAACTCTTTAT
CVB5  ATCCAGTGTCTAATAGAGCTATTATATACCTCTTTGTTGGATTTATACCACTTAGCCTAAAAGAAGTTAAGACACTACAG

                         740                 760                 780                 800
      YTYATHDTDMRRYTVAAYACRDYAARATG

CVB1  CTCATCATTAAACTAAATACAACAAAATG
CVB2  CTTATAATAAAACTCAATACATCAAGATG
CVB3  TTCATTGTTAAGTTGAATACAGCAAAATG
CVB4  TTCATATTGAGACTCAATACGATAAAATG
CVB5  CTTATTGTACAATTAAACACAGCAAAATG
```

relative to the fully sequenced CVB, indicating that it also will have a similarly sized 5' NTR.

Alignment of the 5' NTRs of CVB 1–5 reveals absolute conservation at 73% of the nucleotide positions within this region (Fig. 1). Two large regions containing high sequence variability are clearly identifiable within the alignment. The most 3' of these extends from approximately nucleotide 680 to the true initiation codon and corresponds to a previously described "variable region" of the enteroviral 5' NTR (TOYODA et al. 1984). This region is not located within an area of known or predicted higher order structure.

The second region of highly variable sequence begins at a cytosine-rich pyrimidine tract located between nucleotides 90 and 100 and extends downstream for approximately 80 nucleotides. The cytosine-rich region is reminescent of the polycytidylic acid tracts located 150 and 400 bases from the 5' end of the genomes of cardio-and apthoviruses, respectively, but dramatically shorter (CHUMAKOV and AGOL 1976; HARRIS and BROWN 1976; PEREZ-BERCOFF and GANDER 1977; ROWLANDS et al. 1978). This upstream variable reagion is located in between domains I and II (HARBER and WIMMER 1993) of the predicted secondary of the 5' NTR and does not appear to involve either the downstream or upstream halves of stems of these respective structures.

The results of intertypic nucleic acid identities of the CVB 5' NTR are as follows: CVB1:2–81%, CVB1:3–94.2%, CVB1:4–84.7%, CVB1:5–89.2%, CVB3: 2–81.2%, CVB3:4–84%, CVB3:5–89.6%, CVB4:2–81.2%, CVB4:5–83.9%, CVB5:2–82%. As can be seen, the greatest shared identity is found to exist between the Benschoten strain of CVB1 and the Nancy strain of CVB3. These two CVB share identical nucleotides at 698 of 742 (94%) possible positions within the 5' NTR. Of significance is that the high degree of identity is maintained even within the variable region of the 5' NTR; a region known to be dissimilar even among members of the same serotype.

It remains to be determined whether the high degree of nucleotide identity exhibited by these two strains is unique or shared by other members of their respective serotypes. While no other complete CVB1 5' NTR sequences have been made available, partial sequences from clinical isolates of CVB1 have been determined (DREBOT et al. 1994) and can be used to provide some insight into this question. The nucleotide sequence of a 372 base pair fragment of the CVB1 5'

```
                     20                  40                  60                  80                  100                 120
      ATTAGAGRCAATTTSAAHYRATTWDRATTGGCTTAACCCTACYGYRCYAACCGAACYAGAYAACGGTRCRGTRGGGGTAAATTWTCCGCRTTCGGTGCGGG
CVB1  ATTAGAGACAATTTGAACTAATTTTAATTGGCTTAACCCTACCGCACTAACCGAACTAGATAACGGTGCAGTGGGGGTAAATTCTCCGCATTCGGTGCG
CVB3  ATTAGAGACAATTTGAAATAATTTAGATTGGCTTAACCCTACTGTGCTAACCGAACCAGATAACGGTACAGTAGGGGTAAATTCTCCGCATTCGGTGCG
CVB4  ATTAGAGACAATTTGAAACAATTTAAATTGGCTTAACCCTACTGCACTAACCGAACTAGATAACGGTGCAGTAGGGGTAAATTCTCCGCGTTCGGTGCGG
CVB5  ATTAGAGGCAATTTCAATTGATTAGAATTGGCTTAACCCTACCGCACCAACCGAACTAGACAACGGTGCGGTAGGGGTAAATTATCCGCATTCGGTGCGGG
```

Fig. 2. Alignment of the nucleotide sequence of the 3' NTR of the CVB. The consensus sequence for the CVBs is shown underlined above the alignments of sequences from each of CVB serotypes. Nonconserved positions are designated by the single letter genetic code where R=G or A; K=G or T; Y=C or T; S=C or G; M=A or C; W=A or T; B=C, G or T; D=A, G or T; H=A, C or T; V=A, C or G; N=A, C, G or T.

NTR was found to be identical among the nine isolates, but it diverged by 10.5% when compared to the homologous region of CVB1 JVB Benschoten. The 5' NTR sequence for multiple strains of CVB3 have also been determined (ROMERO et al. 1994) and, as will be discussed below, significant nucleotide divergence exists among clinical isolates of CVB3 within this region. Based on these limited data, it is unlikely that the high level of nucleotide identity between the JVB Benschoten and Nancy strains is shared by other members of their respective serotypes.

The reason for the existence of such a high degree of nucleotide identity between the 5' NTR of the JVB Benschoten strain of CVB1 and the Nancy strain of CVB3 is unknown. A possible explanation may be that a recombination event led to the exchange of the 5' NTR between them. Confirmation of this possibility will need to await sequencing of the complete 5' NTR from multiple CVB1 and CVB3 clinical isolates.

The CVB 3' nontranslated region (3' NTR) is considerably shorter than its 5' counterpart (Fig. 2). As is the case for the 5' NTR, the length of this region is also remarkably conserved among the CVB, measuring from 99 to 101 nucleotides. Differences in length are due to minor variations of from one to three guanine residues immediately upstream of the terminal polyA tail.

Of the 101 nucleotide positions of the 3' NTR, 80 are absolutely conserved among the CVB. The CVB intertypic nucleotide identity of the 3' NTR ranges from a low of 82% (CVB3:5) to a high of 94% (CVB1:4). CVB5 appears to be the most divergent of the CVB at the level of the 3' NTR. The overall nucleotide identity of the CVB5 3' NTR with other CVB is the lowest ranging from 82% to 88%. For the other CVB (1, 3, and 4) the intertypic nucleic acid identity ranges from 91% to 94%.

2.2 The Capsid Protein Coding Region (P1 Region)

The P1 region of the CVB genome encodes for the four proteins, VP1–4, that comprise the capsid of the mature virion. Of the four capsid proteins, VP4 (1A), the smallest, maintains the highest intertypic nucleotide and amino acid identity among the CVB. The intertypic nucleotide identity of VP4 ranges from a low of 75.9%, for CVB1 and 2 to a maximum of 83.1% for CVB1 and 3. In general intertypic nucleic acid identities for VP1 cluster around 78%–80%. Genomic differences in the sequence of VP4 are due primarily to variations at the third nucleotide position of the codons encoding the amino acids of this protein.

Alignment of the predicted amino acid sequence of VP4 is shown in Fig. 3. Of the 69 amino acids, 61 (88.4%) are absolutely conserved among the published CVB (Table 1). The CVB intertypic amino acid identity of VP4 is significantly higher than that of its nucleotide identity, ranging from 89.9% to 97.1%. With the exception of the intertypic amino acid identity shared between CVB4 and 5, all other CVB share identities of 92.8% or greater. The intertypic amino acid similarity of VP4 is >95% for all CVB comparisons. The high degree of nucleic and amino acid identity is not unexpected given that VP4 is buried within the capsid and is not

```
              10        20        30        40        50        60
      MG-QVSTQKTGAHET-L-A-G-SIIHYTN-NYYKDAASNSANRQ-F-QDP-KFTEPVKD-MIKS-PAI
CVB1  MGAQVSTQKTGAHETGLNASGNSIIHYTNINYYKDAASNSANRQDFTQDPGKFTEPVKDIMIKSMPAI
CVB2  MGAQVSTQKTGAHETGLSASGGSIIHYTNINYYKDAASNSANRQDFTQDPGKFTEPVKDIMIKSMPAI
CVB3  MGAQVSTQKTGAHETRLNASGNSIIHYTNINYYKDAASNSANRQDFTQDPGKFTEPVKDIMIKSLPAI
CVB4  MGAQVSTQKTGAHETSLSASGNSIIHYTNINYYKDAASNSANRQDFTQDPSKFTEPVKDVMIKSLPAI
CVB5  MGAQVSTQKTGAHETGLRASGNSIIHYTNINYYKDAASNSANRQEFAQDPGKFTEPVKDIMIKSMPAI
```

Fig. 3. Alignment of the amino acid sequence of the VP4 (1A) protein of the CVB. Amino acid positions are given relative to the sequence of CVB3 (Muckelbauer et al. 1995). The consensus amino acid sequence, derived from all published full-length CVB sequences (Table 1), is shown underlined above the alignments of sequences for each of the prototype CVB serotypes. Amino acid positions at which diffferences occur among the sequenced CVB are designated by a dash. See text for discussion.

subjected to mutational pressures exerted by the host immune system as is the case for the major capsid proteins (VP1–3).

The coding region of VP1 (1D), the largest of the capsid proteins, varies in length among the four fully sequenced CVB prototypic strains: CVB1–834 nucleotides (278 amino acids); CVB3–843 nucleotides (281 amino acids); CVB4–843 nucleotides (281 amino acids); and CVB5–849 nucleotides (283 amino acids). This variation in the size of the VP1 coding region contributes to the differences observed in the total length of the CVB genomes.

Of the four CVB capsid proteins (and of all viral encoded proteins), VP1 exhibits the lowest overall conserved identity at both the nucleic and amino acid levels (Fig. 4). The VP1 nucleic acid identity among CVB1, 3–5 ranges from 65% for CVB4 and CVB5 to 68.5% for CVB3 and CVB5. Its intertypic amino acid identity, although somewhat greater (70.7%–79.4%) than that of the nucleotide identity, is also lower than other CVB structural and nonstructural proteins. This low amino acid identity is the result of the diversity of the amino acid sequences that comprise several of the surface loops of this protein (discussed below).

The capsid structure of CVB3 has been solved to a resolution of 3.5 Å (Muckelbauer et al. 1995). By aligning the CVB sequences and comparing them to those of CVB3 it is possible to speculate on the possible relative locations of the regions of major amino acid diversity among the CVB. As depicted in Fig. 4, the VP1 amino acid sequences of the CVB reveal regions of variability at the following positions of the alignment (relative to the numeration of the amino acids comprising the VP1 protein of CVB3): 1003–1012, 1075–1089, 1123–1139, 1152–1158, 1197–1209, 1223–1230 and 1256–1280. These regions would, respectively, be located to the amino-terminal, B-C loop, D-E loop, E-F loop, G-H loop, H-I loop, and the carboxyl-terminal of the VP1 protein. With the exception of the amino-terminal of the VP1 protein, all other areas of amino acid diversity align to regions of CVB3 known to encode for surface exposed residues.

In CVB4 and 5 the VP2 protein coding region, the second largest of the capsid proteins, measures 783 nucleotides, while in CVB1 and 3 it is slightly longer at 789 nucleotides. Intertypic comparisons reveal nucleic acid identities ranging from a low of 69.5% for CVB1 and 5 to a maximum of 74.7% for CVB1 and 4. Amino

```
                 10        20        30        40        50        60        70        80
      GP-------A--RVADT----P-NSE-IPALTA-ETGHTSQV-P-DTMQTRHV-NYHSRSES--ENFLCRSACV----Y-
CVB1  GPVEESVERAMVRVADTVSSKPTNSESIPALTAAETGHTSQVVPSDTMQTRHVKNYHSRSESSIENFLCRSACVYYATYN
CVB3  GPVEDAITAAIGRVADTVGTGPTNSEAIPALTAAETGHTSQVVPGDTMQTRHVKNYHSRSESTIENFLCRSACVYFTEYK
CVB4  GPTEESVERAMGRVADTIARGPSNSEQIPALTAVETGHTSQVDPSDTMQTRHVHNYHSRSESSIENFLCRSACVIYIKYS
CVB5  GPPGEAVERAIARVADTISSGPVNSESIPALTAAETGHTSQVVPADTMQTRHVKNYHSRSESTVENFLCRSACVYYTTYK

                 90       100       110       120       130       140       150       160
      ----------A-WV---RQ-AQL-RRK-E-FTY-R-D-E-TFVITS-Q--ST---------T-QIMYVPPGGPVP-----
CVB1  NNSE    KGYAEWVINTRQVAQLLRRKLE FTYLRFDLELTFVITSAQEPSTATSVDAPVQTQQIMYVPPGGPVPTKVTD
CVB3  NSGA    KRYAEWVLTPRQAAQL RRKLEFFTYVRFDLELTFVITSTQQPSTTQNQDAQILTHQIMYVPPGGPVPDKVDS
CVB4  SAESNNLKRYAEWVINTRQVAQL RRKMEMFTYIRCDMELTFVITSHQEMSTATNSDVPVQTHQIMYVPPGGPVPTSVND
CVB5  NHGTDGDNF AYWVINTRQVAQL RRKLEMFTYARFDLELTFVITSTQEQSTIQGQDSPVLTHQIMYVPPGGPVPTKINS

                170       180       190       200       210       220       230       240
      Y-WQTSTNPS-FWTEG-APPR-S-PF-SIGNAY--FYDGW--F---G-YG-N-LNNMGT-Y-RHVN----G---ST-RIY
CVB1  YAWQTSTNPSVFWTEGNAPPRMSIPFISIGNAYSCFYDGWTQFSRNGVYGINTLNNMGTLYMRHVNEAGQGPIKSTVRIY
CVB3  YVWQTSTNPSVFWTEGNAPPRMSIPFLSIGNAYSNFYDGWSEFSRNGVYGINTLNNMGTLYARHVNAGSTGPIKSTIRIY
CVB4  YVWQTSTNPSIFWTEGNAPPRMSIPFMSIGNAYTMFYDGWSNFSRDGIYGYNSLNNMGTIYARHVNDSSPGGLTSTIRIY
CVB5  YSWQTSTNPSVFWTEGSAPPRISIPFISIGNAYSMFYDGWAKFDKQGTYGINTLNNMGTLYMRHVNDGSPGPIVSTVRIY

                250       260       270       280
      FKPKHVK---PRPPRLCQY-K---VNF----VT--R--------T
CVB1  FKPKHVKAWVPRPPRLCQYEKQKNVNFNPTGVTTTRSNIT    TT
CVB3  FKPKHVKAWIPRPPRLCQYEKAKNVNFRPSGVTTTRQSITTMTNT
CVB4  FKPKHVKAYVPRPPRLCQYKKAKSVNFDVEAVTAERASLI    TT
CVB5  FKPKHVKTWVPRPPRLCQYQKAGNVNFEPTGVTESRTEITAMQTT
```

Fig. 4. Alignment of the amino acid sequence of the VP1 (1D) protein of the CVB. The consensus amino acid sequence, derived from all published full-length CVB sequences (Table 1), is shown underlined above the alignments of sequences for each of the prototype CVB serotypes. Amino acid positions at which diffferences occur among the sequenced CVB are designated by a dash.

acid identities among the CVB range from 80.1%–83.3%. As is clearly evident from the amino acid alignments of theVP2 proteins, a major region of diversity is located at amino acids 2135–2173, relative to CVB3 (Fig. 5). This region of significant amino acid diversity corresponds to the relative location of the E-F loop of the VP2 protein. The E-F loop is a major surface projection arising from one of the "walls" of a deep rift seen in the topography of all human picornaviruses whose capsid structure has been resolved at an atomic level (HOGLE et al. 1985; ROSSMAN and PALMENBERG 1988; ROSSMAN 1989; ARNOLD and ROSSMAN 1990; MUCKELBAUER et al. 1995). Variation in the size of this loop contributes to the differences in nucleotide length observed among the CVB genomes. Smaller regions of amino acid diversity, again relative to CVB3 are also observed in the B-C loop at amino acids 2066–2077, C-D loop at positions 2082–2089 and in the H-I loop at amino acid positions 2231–2239 of the VP2 protein alignment. The majority of the amino acid positions exhibiting variability are predicted to be surfce-exposed residues, based on the data obtained from CVB3.

Unlike the capsid proteins VP1 and VP2, no differences in the length of the VP3 (1C) protein are observed among the CVB. The VP3 protein intertypic nucleotide identity is similar to that observed for VP2, ranging from 70% (CVB1:4) to 73% (CVB1:5). The corresponding amino acid identities range from a low of 78.2% for CVB1 and 4 to a high of 86.1% for CVB1 and 5. Regions of nucleic acid diversity within the VP3 protein span amino acid positions immediate-

```
                 10        20        30        40        50        60        70        80
      SP--EECGYSDR-RSITLGNSTITTQECANVVVGYG-WP-YL-D-EAT-EDQPTQPDVATCRFYTL-SV-W---S-GWWW
CVB1  SPSAEECGYSDRVRSITLGNSTITTQECANVVVGYGVWPEYLKDNEATGEDQPTQPDVATCRFYTLESVQWMKNSAGWWW
CVB3  SPTVEECGYSDRVRSITLGNSTITTQECANVVVGYGVWPDYLKDSEATAEDQPTQPDVATCRFYTLDSVQWQKTSPGWWW
CVB4  SPTVEECGYSDRVRSITLGNSTITTQECANVVVGYGVWPDYLSDEEATAEDQPTQPDVATCRFYTLNSVKWEMQSAGWWW
CVB5  SPSAEECGYSDRVRSITLGNSTITTQECANVVVGYGTWPTYLKDEEATAEDQPTQPDVATCRFYTLESVMWQQSSPGWWW

                 90       100       110       120       130       140       150       160
      K-P-ALS--GLFGQNMQYHYLGR-GYT--VQCNASKFHQGCLLVVCVPEAEMGC----N-P----L--G--A--F-----
CVB1  KLPDALSQMGLFGQNMQYHYLGRTGYTIHVQCNASKFHQGCLLVVCVPEAEMGCSNLNNTPKFAELSGGDNARMFTDTEV
CVB3  KLPDALSNLGLFGQNMQYHYLGRTGYTVHVQCNASKFHQGCLLVVCVPEAEMGCATLDNTPSSAELLGGDSAKEFADKPV
CVB4  KFPDALSEMGLFGQNMQYHYLGRSGYTIHVQCNASKFHQGCLLVVCVPEAEMGCTNAENAPAYGDLCGGETAKSFEQNAA
CVB5  KFPDALSNMGLFGQNMQYHYLGRAGYTVHVQCNASKFHQGCLLVVCVPEAEMGCATLANKPDQKSLSNGETANTFDSQNT

                170       180       190       200       210       220       230       240
      -------VQ--V-NAGMGVGVGNLTI-PHQWINLRTNNSATIVMPY-NSVPMDNM-RHNN-TLM-IPF-PL-Y--G---Y
CVB1  GTSNDKKVQTAVWNAGMGVGVGNLTIFPHQWINLRTNNSATIVMPYINSVPMDNMYRHNNLTLMIIPFVPLNYSEGSSPY
CVB3  ASGSNKLVQRVVYNAGMGVGVGNLTIFPHQWINLRTNNSATIVMPYTNSVPMDNMFRHNNVTLMVIPFVPLDYCPGSTTY
CVB4  TGKTA  VQTAVCNAGMGVGVGNLTIYPHQWINLRTNNSATIVMPYINSVPMDNMFRHNNFTLMIIPFAPLDYVTGASSY
CVB5  TGQTA  VQANVINAGMGVGVGNLTIFPHQWINLRTNNSATIVMPYINSVPMDNMFRHNNFTLMIIPFAPLSYSTGATTY

                250       260
      -PIT-T-APM-AEYNGLRLA--Q
CVB1  VPITVTIAPMCAEYNGLRLASSQ
CVB3  VPITVTIAPMCAEYNGLRLAGHQ
CVB4  IPITVTVAPMSAEYNGLRLAGHQ
CVB5  VPITVTVAPMCAEYNGLRLAGKQ
```

Fig. 5. Alignment of the amino acid sequence of the VP2 (1B) protein of the CVB. The consensus amino acid sequence, derived from all published full-length CVB sequences (Table 1), is shown underlined above the alignments of sequences for each of the prototype CVB serotypes. Amino acid positions at which diffferences occur among the sequenced CVB are designated by a dash.

ly preceding and within the knob structure and the B-C loop (amino acids 3055–3082), the C-D loop (amino acids 3091–3096), the E-F loop (amino acids 3141–3147), the G-H loop (amino acids 3178–3192) and the H-I loop (amino acids 3202–3211) (Fig. 6). In CVB3 these regions contain multiple surface amino acid residues as described by Muckelbauer. The knob (3058–3069) is the major surface protrusion of the CVB3 VP3 protein. In the rhino- and polioviruses this structure has been shown to contain serotype-specific neutralization sites. (Sherry and Rueckert 1985; Minor et al. 1986; Sherry et al. 1986; Page et al. 1988; Reimann et al. 1991) This observation most likely explains why this area of the VP3 protein alignment exhibits such a significant degree of amino acid (and underlying nucleic acid) diversity among the CVB.

2.3 The Nonstructural Protein Coding Region (P2 and P3 Regions)

Unlike the capsid coding region, the nonstructural protein coding regions share a great degree of nucleic acid identity among the CVB. The intertypic amino acid identity for the nonstructural proteins is, with few exceptions (2A- CVB3:5, CVB4:5 and 3A- CVB1:4, CVB4:5), greater than 90% (and in most cases exceeds 95%). In all but two instances (2A- CVB1:4 and CVB4:5), the intertypic amino acid similarity is greater than 95% and in the overwhelming majority of cases approaches 100%. This high degree of conservation may be the result of a least

```
              10        20        30        40        50        60        70        80
      GLP-M-TPGS-QFLTSDDFQSP-AMPQ-DVTPEM-IPG-V-NLME-AEVDSVVP--N---------AY---V-S------
CVB1  GLPVMTTPGSTQFLTSDDFQSPSAMPQFDVTPEMQIPGRVNNLMEIAEVDSVVPVNNTDNNVNGLKAYQIPVQSNSDNRR
CVB3  GLPTMNTPGSCQFLTSDDFQSPSAMPQYDVTPEMRIPGEVKNLMEIAEVDSVVPVQNVGEKVNSMEAYQIPVRSNEGSGT
CVB4  GLPTMLTPGSTQFLTSDDFQSPSAMPQFDVTPEMNIPGQVRNLMEIAEVDSVVPINNLKANLMTMEAYRVQVRSTDEMGG
CVB5  GLPTMLTPGSNQFLTSDDFQSPSAMPQFDVTPEMAIPGQVNNLMEIAEVDSVVPVNNTEGKVSSIEAYQIPVQSNSTNGS

              90       100       110       120       130       140       150       160
      Q-FGFPL-PG---V--RTLLGEILNYYTHWSGS-KLTF-FCGSAMATGKFLLAYSP-GAG-P--R--AMLGTHV-WDVGL
CVB1  QVFGFPLQPGANNVLNRTLLGEILNYYTHWSGSIKLTFMFCGSAMATGKFLLAYSPPGAGVPKNRRDAMLGTHVIWDVGL
CVB3  QVFGFPLQPGYSSVFSRTLLGEILNYYTHWSGSIKLTFMFCGSAMATGKFLLAYSPPGAGAPTKRVDAMLGTHVVWDVGL
CVB4  QIFGFPLQPGASSVLQRTLLGEILNYYTHWSGSLKLTFVFCGSAMATGKFLLAYSPPGAGAPDSRKNAMLGTHVIWDVGL
CVB5  QVFGFPLIPGASSVLNRTLLGEILNYYTHWSGSIKLTFMFCGSAMATGKFLLAYSPPGAGAPTTRKEAMLGTHVIWDVGL

             170       180       190       200       210       220       230
      QSSCVLC-PWISQTHYR-V--D--TA-G---CWYQTN--VPA--Q--C-I-CF-SACNDFSVR-L-DT-FI-Q--FYQ
CVB1  QSSCVLCVPWISQTHYRYVVEDEYTAAGYVTCWYQTNIIVPADVQSTCDILCFVSACNDFSVRMLKDTPFIRQDNFYQ
CVB3  QSSCVLCIPWISQTHYRFVASDEYTAGGFITCWYQTNIVVPADAQSSCYIMCFVSACNDFSVRLLKDTPFISQQNFFQ
CVB4  QSSCVLCVPWISQTHYRYVVDDKYTASGFISCWYQTNVIVPAEAQKSCYIMCFVSACNDFSVRMLRDTQFIKQTNFYQ
CVB5  QSSCVLCIPWISQTHYRYVVVDEYTAGGYITCWYQTNIVVPADTQSDCKILCFVSACNDFSVRMLKDTPFIKQDSFYQ
```

Fig. 6. Alignment of the amino acid sequence of the VP3 (1C) protein of the CVB. The consensus amino acid sequence, derived from all published full-length CVB sequences (Table 1), is shown underlined above the alignments of sequences for each of the prototype CVB serotypes. Amino acid positions at which diffferences occur among the sequenced CVB are designated by a dash.

two factors. First, unlike the capsid proteins, the nonstructural proteins are not directly exposed to the host immune system and are, therefore, under much less pressure to mutate. Second, as nonstructural proteins they may be expected to have significant constraints on variability due to conformational requirements for interaction with host cell protein(s) or viral protein(s)/RNA.

Within the CVB P2 and P3 coding regions, genomic differences are due, in large part, to variations at the third nucleotide position of the codons comprising the P2 and P3 proteins. The protease 2A has the lowest mean intertypic nucleic and amino acid identities (77.7% and 90.7%, respectively) of all the P2 proteins. Differences are noted at 22 (14.7%) of the 150 amino acids that comprise this protein (Table 2). The intertypic nucleotide identity of the 2B protein ranges from 74.8% for CVB1 and 5 to 81.5% for CVB3 and 4, but maintains a significantly higher intertypic amino acid identity and similarity, ranging from 96% to 98%. Alignment of the amino acid sequences of the 2B protein demonstrates that the CVB differ at only five (5%) of the 99 amino acid positions of this protein (Table 3).

The 2C protein, the largest P2 protein (987 nucleotides, 329 amino acids), possesses the highest intertypic nucleic and amino acid identities of the P2 group of proteins, 81.7–82.9% and 97.3–98.5%, respectively. At the amino acid level, variation occurs at only nine (2.7%) of the 329 amino acids that comprise this protein (Table 4). A unique feature within the 2C coding region is a 41 base stretch, located from nucleotides 331 to 371, that is absolutely conserved among all the CVB sequenced to date (Fig. 7) (Iizuka et al. 1987; Jenkins et al. 1987; Lindberg et al. 1987; Klump et al. 1990; Tracy et al. 1992; Zhang et al. 1993; Chapman et al. 1994, Kang et al. 1994; Titchener et al. 1994). Among the polioviruses similar absolute nucleotide identity within this region is not observed (Toyoda et al. 1984).

Table 2. Alignment of the group B coxsackievirus 2A protein[a]

```
          10        20        30        40        50        60        70
G--G-QSGA---GNY-VVNRHLAT--DWQ-CVWE-YNRDLLVSTTT-HGCD-IARC-C-TGVYFCAS--KHYP-S

    80        90       100       110       120       130       140       150
FEGPGLVEVQESEYYP-RYQSHVLLA-GFSEPGDCGGILRCEHGV-G-VTMGGEGVVGFADVRDLLWLEDDAMEQ
```

Amino acid	CVB1	CVB3	CVB4	CVB5
2	A	A	P	V
3	F	F	Y	L
5	Q	Q	H	Q
10	V	V	V	I
11	Y	Y	Y	C
12	V	V	V	I
16	R	R	K	R
25	R	S	H	S
26	E	A	V	E
30	R	N	N	R
35	D	S	D	D
47	A	T	A	A
52	I	I	T	T
57	Q	Q	Q	R
59	T	T	T	S
68	R	K	K	R
69	N	N	S	N
74	V	I	V	V
92	K	R	K	K
102	A	A	T	A
121	V	I	I	I
123	I	I	L	L

[a]Amino acid differences among the different CVB prototype strains are designated for each amino acid position and listed below the consensus sequence.

Table 3. Alignment of the group B coxsackievirus 2B protein[a]

```
          10        20        30        40        50        60        70
GVKDYVEQLGNA-GSGFTNQ-CEQVNLLKESLVGQDSILEKSLKALVKIVSALVIVVRNHDDL-T-TATLALIGC

    80        90
TSSPWRWLK-KVSQYYGIPMAERQ
```

Amino acid	CVB1	CVB3	CVB4	CVB5
13	F	L	F	F
21	V	I	I	I
64	I	I	I	V
66	V	V	V	I
85	Q	Q	H	Q

[a]Amino acid differences among the different CVB protoype strains are designated for each amino acid position and listed below the consensus sequence.

Table 4. Alignment of the group B coxsackievirus 2C protein[a]

```
         10        20        30        40        50        60        70
N--WLKKFTEMTNACKGMEWIAIKIQKFI-WLKVKILPEV-EKHEFL-RLKQLPLLESQIATIEQSAPSQSDQEQ

     80        90       100       110       120       130       140       150
LFSNVQYFAHYCRKYAPLYAAEAKRVFSLEKKMSNYIQFKSKCRIEPVCLLLHGSPGAGKSVATNLIGRSLAEKL

         160       170       180       190       200       210       220
NSSVYSLPPDPDHFDGYKQQAVVIMDDLCQNPDGKDVSLFCQMVSSVDFVPPMAALEEKGILFTSPFVLASTNAG

     230       240       250       260       270       280       290       300
SINAPTVSDSRALARRFHFDMNIEVISMYSQNGKINMPMSVKTCD-ECCPVNFK-CCPLVCGKAIQFIDR-TQVR

         310       320
YSLDMLVTEMFREYNHRHSVG-TLEALFQ
```

Amino acid	CVB1	CVB3	CVB4	CVB5
2	S	N	N	N
3	G	S	G	N
30	E	E	E	D
41	K	R	K	K
48	N	N	S	N
271	E	D	E	E
280	K	K	R	R
296	R	R	K	R
322	A	T	A	A

[a]Amino acid differences among the different CVB prototype strains are designated for each amino acid position and listed below the consensus sequence.

The reason for the conservation of this area within the 2C encoding region of all CVB is unknown. Based on the amino acids encoded for within this area, nucleotide variance would clearly be permitted. If, similar to the polioviruses, the other non-CVB enteroviridae also fail to maintain an identical absolute sequence identity within the homologous nucleotide region, this site may serve as a target for the design of a "group-specific" primer for detection of the CVB.

Downstream of, and independent from, the previously discussed 41 base region are the sequences comprising the amino acids of the consensus nucleotide binding (NTP-B) motif of 2C (MIRZAYAN and WIMMER 1992, 1994). Similar to poliovirus the NTP-B motif is found to be conserved among all CVB for whom 2C amino acid sequence information exists. The conservation of this motif adds further support to the proposed ATPase activity of this protein.

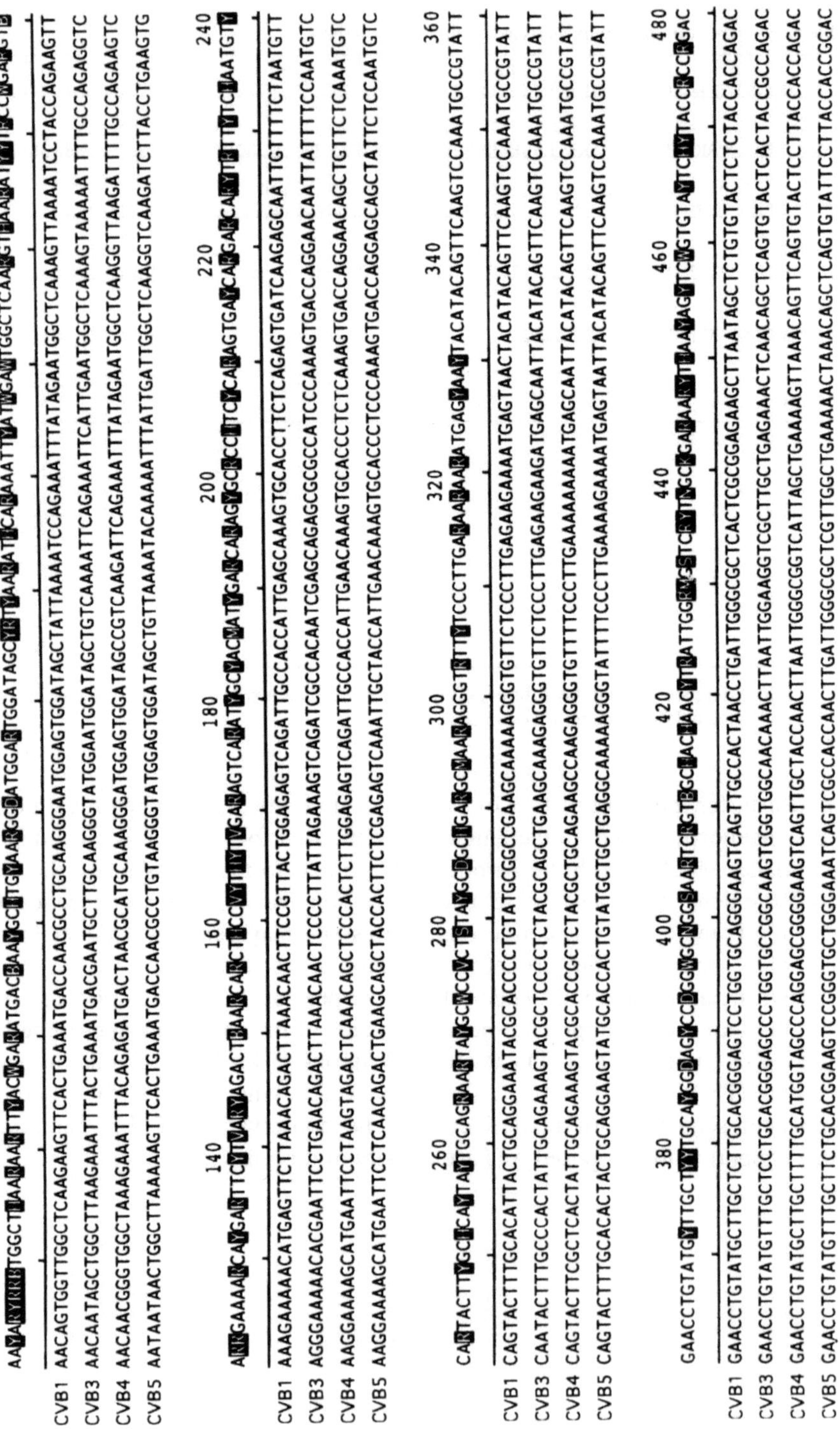

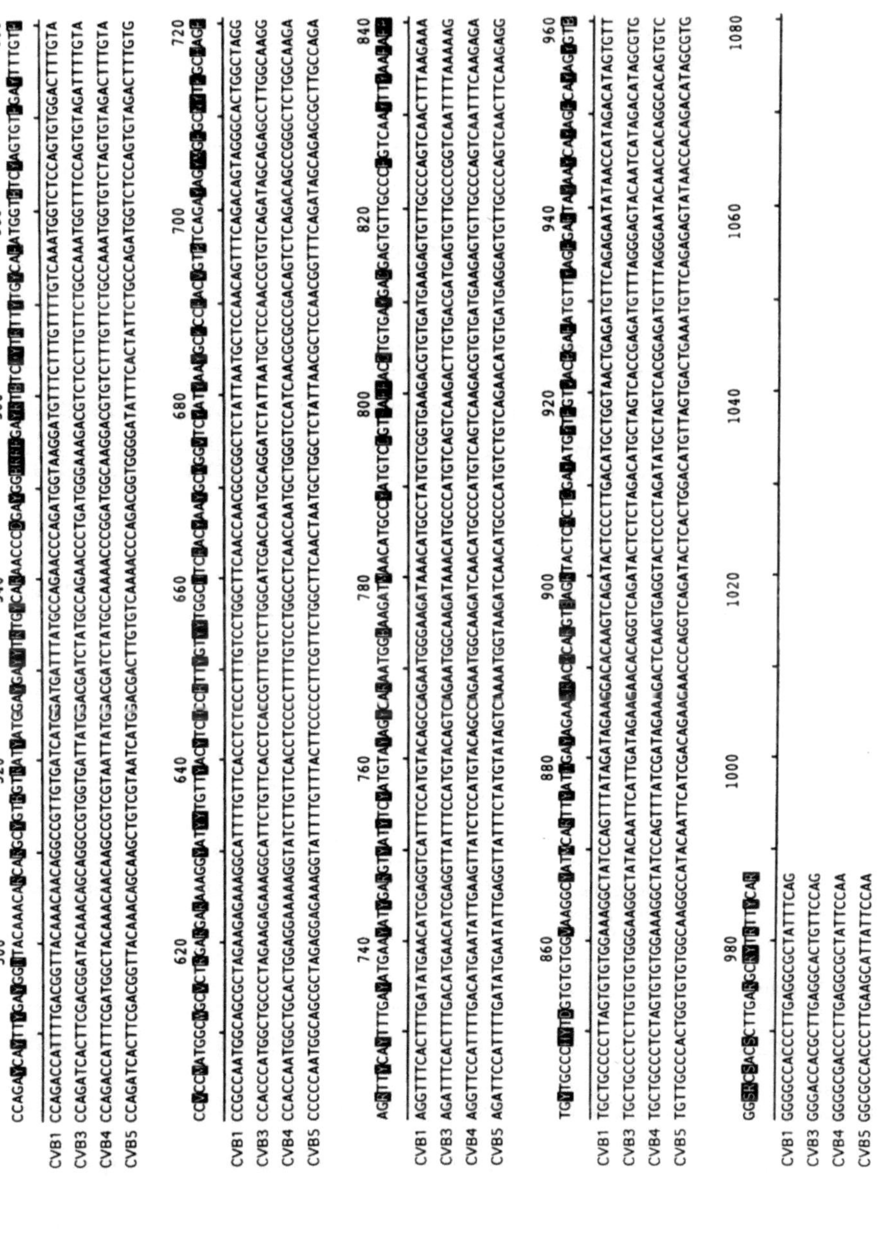

Fig. 7. Alignment of the nucleotide sequence of the 2C protein coding region of the CVB. The consensus sequence for the CVBs is shown underlined above the alignments of sequences from each of CVB serotypes. Nonconserved positions are designated by the single letter genetic code where R=G or A; K=G or T; Y=C or T; S=C or G; M=A or C; W=A or T; B=C, G or T; D=A, G or T; H=A, C or T; V=A, C or G; N=A, C, G or T.

Table 5. Alignment of the group B coxsackievirus 3B protein[a]

```
     10        20
GAYTG-PNQKP-VPTLRQAKVQ
```

Amino acid	CVB1	CVB3	CVB4	CVB5
6	M	V	M	M
12	K	R	K	K

[a]Amino acid differences among the different CVB prototype strains are designated for each amino acid position and listed below the consensus sequence.

Table 6. Alingment of the group B coxsackievirus 3C protein[a]

```
         10        20        30        40        50        60        70
GP-FEFAVAMMKRNSS-VKTEYGEFTMLGIYDRWAVLPRHAKPGPTILMNDQE-GV-DAKEL-D-DGTNLELTLL

     80        90       100       110       120       130       140       150
KLNRNEKFRDIRGFLAKEEVEVNEAVLAINTSKFPNMYIPVG-VT-YGFLNLGGTPTKRMLMYNFPTRAGQCGGV

         160       170       180
LMSTGKVLGIHVGGNGHQGFSA-LLKHYFNDEQ
```

Amino acid	CVB1	CVB3	CVB4	CVB5
3	V	A	A	A
17	T	R	T	T
54	I	V	V	V
57	L	L	L	V
63	V	V	I	V
65	K	K	R	K
118	Q	Q	R	Q
121	E	E	D	D
173	A	A	G	A

[a]Amino acid differences among the different CVB prototype strains are designated for each amino acid position and listed below the consensus sequence.

Table 7. Alignment of the group B coxsackievirus 3D protein[a]

```
         10        20        30        40        50        60        70
GEIEFIESSK-AGFPVINTPS-TKLEPSVFH-VFEGNKEPAVLR-GDPRL--NFEEAIF-KYIGNVNTHVDEYML

     80        90       100       110       120       130       140       150
EAVDHYAGQLATLDI-TEPM-LEDAVYGTEGLEALDLTTSAGYPYVALGIKKRDILSKKTRDLTKLKECMDKYGL

         160       170       180       190       200       210       220
NLPMVTYVKDELRS-EKVAKGKSRLIEASSLNDSVAMRQTFGNLYK-FHLNPG-VTGSAVGCDPD-FWSKIPVML

    230       240       250       260       270       280       290       300
DGHL-AFDYSGYDASLSPVWFACLK-LLEKLGY-HKETNYIDYLCNSHHLYRDKHYFVRGGMPSGCSGTSIFNSM

         310       320       330       340       350       360       370
INNIIIRTLMLKVYKGIDLDQFRMIAYGDDVIASYPWPIDASLLAEAG---GLIMTPADKG-CFNEVTW-NVTFL

    380       390       400       410       420       430       440       450
KRYFRADEQYPFLVHPVMPMKDIHESIRWTKDPKNTQDHVRSLCLLAWHNGEHEYEEFI-KIRSVPVG-CLTLPA

         460
FSTLRRKWLDSF
```

Amino acid	CVB1	CVB3	CVB4	CVB5
11	E	D	D	E
22	K	K	R	K
32	Q	Q	H	Q
45	N	S	N	N
51	R	K	K	K
52	A	A	V	A
60	S	S	F	S
91	S	S	N	S
95	R	K	K	K
165	A	I	A	A
197	T	T	A	T
204	I	V	I	I
216	L	L	V	L
230	V	I	I	I
251	L	M	L	L
259	S	T	T	T
349	R	K	K	K
350	D	G	D	D
351	C	Y	Y	Y
362	D	E	E	E
370	A	T	T	T
435	R	R	Q	K
444	A	R	R	R

[a]Amino acid differences among the different CVB prototype strains are designated for each amino acid position and listed below the consensus sequence.

Table 8. Alignment of the group B coxsackievirus 3A protein[a]

```
          10        20        30        40        50        60        70
GPP-YREIKIS--P-TPPPP-IADLLKSVD--A-REYCKEKGWLVPEI-S-LQIEKHVSRAFICLQALTTFVSVA

       80
GIIYIIYKLFAGFQ
```

Amino acid	CVB1	CVB3	CVB4	CVB5
3	I	V	V	I
12	I	V	V	V
13	A	A	T	A
15	E	E	E	D
21	A	A	V	A
31	S	S	R	S
32	E	E	Q	E
34	V	V	I	V
49	N	N	D	N
51	T	T	I	T

[a]Amino acid differences among the different CVB prototype strains are designated for each amino acid position and listed below the consensus sequence.

	20 40 60 80
	ANYYMSCCNYHRCCCAWWRHAACHTAGAAGYHHRNHHYWWWHGRYCAAYAGVYRGYYYRGYRHWCCARYYRRGYYYHGRY
CVB2 OHIO ▶	ATCTACCC.TTGCCCAATGCAACATAGAAGCTTAGTCCAAAAGGTCAATAGGCAGCCCAGTGCACCAACTGGGTCCAGAC
CVB2 2093 ▶	ATCCACCCTTTCCCCATTGTAACTTAGAAGCATGAATCAAACGGCCAATAGCTAGTTCGGTAAACCAACCGAGCTTTGGC
CVB2 HI ▶	ATTTACCC.TATCCCAAAGTAACTTAGAAGTACGCCACATATGATCAATAGGTGGCCCAGTACTCCAACTGGGTCTTGAT
CVB2 LO ▶	A.TCCGCC.CCCCCCAATGAAACCTAGAAGTCAGACTTAATCGGTCAACAGACAGCCTAGTATACCAACTAGGTTCCGAC
CVB2 WI ▶	ATTTACCC.TACCCCAAAATAACTTAGAAGCATGTCACATATGGTCAATAGGTGGCCCAGCAATCCAGTCGGGTCCCGAC
CVB2 BR ▶	ATTTACCC.TACCCCAAAGTAACTTAGAAGCATGTCACATATGGTCAATAGGTGGCCCAGCAATCCAGTCGGGTCCCGAC
CVB2 GR ▶	A.TCACCC.TCCCCCAATGTAACTTAGAAGTCAGACTTTAACGGTCAACAGACAGCCTAGTATACCAACTGGGTTCTGAC

	100 120 140 160
	CAAGYACYTCTGTTWNCCCCGGACYRAGTATCARTAGGCTGYYCRCRYGGCTGAARGAGAAARYGTTCGTTACCCGGCCA
CVB2 OHIO ▶	CAAGCACTTCTGTTA.CCCCGGACCAAGTATCAGTAGGCTGCTCACGCGGCTGAAGGAGAAAACGTTCGTTACCCGGCCA
CVB2 2093 ▶	CAAGCACCTCTGTTA.CCCCGGACTGAGTATCAATAGGCTGTTCACACGGCTGAAGGAGAAAACGTTCGTTACCCGGCCA
CVB2 HI ▶	CAAGCACTTCTGTTA.CCCCGGACCGAGTATCAATAGGCTGCTCGCGCGGCTGAAGGAGAAAGTGTTCGTTACCCGGCCA
CVB2 LO ▶	CAAGCACTTCTGTTTCCCCCGGACTGAGTATCAATAGGCTGCTCACGCGGCTGAAGGAGAAAACGTTCGTTACCCGGCCA
CVB2 WI ▶	CAAGCACTTCTGTTA.CCCCGGACCGAGTATCAATAGGCTGCTCGCGTGGCTGAAAGAGAAAGCGTTCGTTACCCGGCCA
CVB2 BR ▶	CAAGCACTTCTGTTA.CCCCGGACCGAGTATCAATAGGCTGCTCGCGTGGCTGAAAGAGAAAGCGTTCGTTACCCGGCCA
CVB2 GR ▶	CAAGTACTTCTGTTT.CCCCGGACTGAGTATCAATAGGCTGCCCACGCGGCTGAAGGAGAAAACGTTCGTTACCCGGCCA

	180 200 220 240
	AYTACTTCGARAAACCYAGTASCACCATGRARGTTGCGCRGYGTTTCGCTCMRCACRRCCCCAGTGYAGATCAGGYYGAT
CVB2 OHIO ▶	ACTACTTCGAGAAACCTAGTAGCACCATGAAGGTTGCGCAGTGTTTCGCTCCGCACAGCCCCAGTGTAGATCAGGTCGAT
CVB2 2093 ▶	ACTACTTCGAGAAACCTAGTACCACCATGAATGTTGCGCAGCGTTTCGCTCAGCACGACCCCAGTGCAGATCAGGCTGAT
CVB2 HI ▶	ACTACTTCGAGAAACCTAGTACCACCATGAAGGTTGCGCAGCGTTTCGCTCCGCACAACCCCAGTGTAGATCAGGCTGAT
CVB2 LO ▶	ATTACTTCGAAAAACCCAGTACCACCATGGATGTTGCGCAGTGTTTCGCTCCACACAACCCCAGTGTAGATCAGGCCGAT
CVB2 WI ▶	ATTACTTCGAGAAACCCAGTACCACCATGAAGGTTGCGCGGCGTTTCGCTCCACACAACCCCAGTGTAGATCAGGCCGAT
CVB2 BR ▶	ATTACTTCGAGAAACCCAGTACCACCATGAAGGTTGCGCGGCGTTTCGCTCCACACAACCCCAGTGTAGATCAGGCCGAT
CVB2 GR ▶	ATTACTTCGAAAAACCTAGTACCACCATGGACGTTGCGCAGTGTTTCGCTCCGCACAACCCCAGTGTAGATCAGGCTGAT

	260 280 300 320
	GAGYCWCCGCRWTCCCCACGGGYRACCGTGGCGGWGGCTGCGYTGGCGGCCTGCCCATGGGRYRACCCATGGACGCTTC
CVB2 OHIO ▶	GAGCCACCGCATTCCCCACGGGTGACCGTGGCGGTGGCTGCGTTGGCGGCCTGCCCATGGGGTAACCCATGGACGCTTC
CVB2 2093 ▶	GAGTCTCCGCATTCCCCACGGGCGACCGTGGCGGAGGCTGCGCTGGCGGCCTGCCCATGGGGCAACCCATGGACGCTTC
CVB2 HI ▶	GAGTCACCGCATTCCCCACGGGCGACCGTGGCGGTGGCTGCGCTGGCGGCCTGCCCATGGGGTAACCCATGGACGCTTC
CVB2 LO ▶	GAGTCACCGCAATCCCCACGGGCGACCGTGGCGGTGGCTGCGCTGGCGGCCTGCCCATGGGGTGACCCATGGACGCTTC
CVB2 WI ▶	GAGTCACCGCGTTCCCCACGGGCGACCGTGGCGGTGGCTGCGCTGGCGGCCTGCCCATGGACAACCCATGGACGCTTC
CVB2 BR ▶	GAGTCACCGCGTTCCCCACGGGCGACCGTGGCGGTGGCTGCGCTGGCGGCCTGCCCATGGACAACCCATGGACGCTTC
CVB2 GR ▶	GAGTCACCGCAATCCCCACGGGCAACCGTGGCGGTGGCTGCGCTGGCGGCCTGCCCATGGGGCAACCCATGGACGCTTC

	340 360 380 400
	AATRCYGACATGGTGYGAAGAGTCTATTGAGCTARTTGGTAKTCCTCCGGCCCCTGAATGCGGCTAAYCCYAACTGCGGA
CVB2 OHIO ▶	AATACTGACATGGTGTGAAGAGTCTATTGAGCTAGTTGGTAGTCCTCCGGCCCCTGAATGCGGCTAATCCTAACTGCGGA
CVB2 2093 ▶	AATACTGACATGGTGTGAAGAGTCTATTGAGCTAGTTGGTATTCCTCCGGCCCCTGAATGCGGCTAACCCTAACTGCGGA
CVB2 HI ▶	AATACTGACATGGTGCGAAGAGTCTATTGAGCTAGTTGGTAGTCCTCCGGCCCCTGAATGCGGCTAATCCTAACTGCGGA
CVB2 LO ▶	AATACTGACATGGTGCGAAGAGTCTATTGAGCTAATTGGTAGTCCTCCGGCCCCTGAATGCGGCTAATCCTAACTGCGGA
CVB2 WI ▶	AATACCGACATGGTGCGAAGAGTCTATTGAGCTAATTGGTAGTCCTCCGGCCCCTGAATGCGGCTAATCCCAACTGCGGA
CVB2 BR ▶	AATACCGACATGGTGCGAAGAGTCTATTGAGCTAATTGGTAGTCCTCCGGCCCCTGAATGCGGCTAATCCCAACTGCGGA
CVB2 GR ▶	AATGCTGACATGGTGCGAAGAGTCTATTGAGCTAATTGGTAGTCCTCCGGCCCCTGAATGCGGCTAATCCTAACTGCGGA

```
                            420                 440                 460                 480
            GCRKRYRCYYRCAHVCCAGTGRGYWGYMYGTCGTAAYGGGYAACTCTGCAGCGGRACCGACTACTTTGGGTGTCCGTGTT

CVB2 OHIO ▶ GCGTGCGCTTGCAACCCAGTGAGTAGCACGTCGTAATGGGTAACTCTGCAGCGGAACCGACTACTTTGGGTGTCCGTGTT
CVB2 2093 ▶ GCAGATACCCACATGCCAGTGGGTTGTCTGTCGTAACGGGCAACTCTGCAGCGGGACCGACTACTTTGGGTGTCCGTGTT
CVB2 HI ▶   GCAGATACCCACACACCAGTGGGCAGTCTGTCGTAACGGGCAACTCTGCAGCGGAACCGACTACTTTGGGTGTCCGTGTT
CVB2 LO ▶   GCAGATACCCACATGCCAGTGGGCAGTCTGTCGTAACGGGCAACTCTGCAGCGGAACCGACTACTTTGGGTGTCCGTGTT
CVB2 WI ▶   GCAGATACCCACATACCAGTGGGCAGTCTGTCGTAATGGGCAACTCTGCAGCGGAACCGACTACTTTGGGTGTCCGTGTT
CVB2 BR ▶   GCAGATACCCACATACCAGTGGGCAGTCTGTCGTAATGGGCAACTCTGCAGCGGAACCGACTACTTTGGGTGTCCGTGTT
CVB2 GR ▶   GCAGATACCCACATGCCAGTGGGCAGTCTGTCGTAACGGGCAACTCTGCAGCGGAACCGACTACTTTGGGTGTCCGTGTT
```

```
                            500                 520                 540                 560
            TCYTTTTATYYYHWNWYTGGCTGCTTATGGTGACAATTGARAGATTGTTRCCATATAGCTATTGGATTGGCCATCCGGTR

CVB2 OHIO ▶ TCCTTTTATTCCTTGTCTGGCTGCTTATGGTGACAATTGAGAGATTGTTGCCATATAGCTATTGGATTGGCCATCCGGTA
CVB2 2093 ▶ TCCTTTTATTTTTATACTGGCTGCTTATGGTGACAATTGAAAGATTGTTGCCATATAGCTATTGGATTGGCCATCCGGTG
CVB2 HI ▶   TCCTTTTATTCTAATTCTGGCTGCTTATGGTGACAATTGAGAGATTGTTACCATATAGCTATTGGATTGGCCATCCGGTG
CVB2 LO ▶   TCTTTTTATTCTCACATTGGCTGCTTATGGTGACAATTGAAAGATTGTTACCATATAGCTATTGGATTGGCCATCCGGTG
CVB2 WI ▶   TCCTTTTATTCTTAAACTGGCTGCTTATGGTGACAATTGAGAGATTGTTACCATATAGCTATTGGATTGGCCATCCGGTG
CVB2 BR ▶   TCCTTTTATTCTTAAACTGGCTGCTTATGGTGACAATTGAGAGATTGTTACCATATAGCTATTGGATTGGCCATCCGGTG
CVB2 GR ▶   TCTTTTTATCTCTACATTGGCTGCTTATGGTGACAATTGAAAGATTGTTACCATATAGCTATTGGATTGGCCATCCGGTG
```

```
                            580                 600                 620                 640
            WSWAAYAGRGCVATHRTHTWYYTDTTYRTTGGKTTYRTWCCVYTKRAYHWYRADRWYYWHAAHACCCTHMHYYWYATHWY

CVB2 OHIO ▶ TCTAACAGGGCGATCATTTACCTGTTTGTTGGTTTCATACCCTTGAATCTTGAGGTTTATAATACCCTCACTCTTATAAT
CVB2 2093 ▶ TCAAACAGAGCCATAATCTACTTGTTCGTTGGTTTTGTACCGCTGAATTATAAGGTCTTAAATACCCTTAATTACATTAC
CVB2 HI ▶   ACAAATAGAGCGATTGTCTACCTATTTGTTGGTTTTGTTCCATTGAATTTTAAAGTTCTAAAAACCCTTAATTTTATCTT
CVB2 LO ▶   AGTAACAGAGCAATTATATTTCTTTTTGTTGGGTTTATACCACTTGATATCAATAATTACAACACCCTACTCTATATTAT
CVB2 WI ▶   ACAAATAGAGCAATCGTTTACCTATTTATTGGTTTTGTTCCATTGAACTTCAAAGTTTTAAAAACCCTCAATTACATCTT
CVB2 BR ▶   ACAAATAGAGCAATCGTTTACCTATTTGTTGGTTTTGTTCCATTGAACTTCAAAGTTTTAAAAACCCTCAATTACATCTT
CVB2 GR ▶   AGTAACAGAGCAATTATATTCCTTTTTGTTGGGTTCATACCACTTGATTTCAATAATTATAACACCCTACTCTATATTAT
```

```
                            660                 680                 700                 720
            VVHAYTSAAYACRDSAARATG

CVB2 OHIO ▶ AAAACTCAATACATCAAGATG
CVB2 2093 ▶ AACACTCAACACATCAAAATG
CVB2 HI ▶   AGTACTCAATACAGCAAAATG
CVB2 LO ▶   CCAATTGAATACGAGAAGATG
CVB2 WI ▶   GATACTCAATACAGCAAAATG
CVB2 BR ▶   AATACTCAATACAGCAAAATG
CVB2 GR ▶   CCAATTGAATACGAGAAGATG
```

Fig. 8. Alignment of the nucleotide sequence of the 5' NTR of the CVB2. The consensus sequence for the CVBs is shown underlined above the alignments of sequences from each of CVB serotypes. Nonconserved positions are designated by the single letter genetic code where R=G or A; K=G or T; Y=C or T; S=C or G; M=A or C; W=A or T; B=C, G or T; D=A, G or T; H=A, C or T; V=A, C or G; N=A, C, G or T.

Four proteins are encoded for within the P3 region of the CVB genome. With the exception of the intertypic nucleotide identity between CVB1 and 3 for the 3A protein, all other CVB comparisons exhibit identities of between 71.2% and 77.8% In the case of the previously mentioned pair the nucleotide identity is higher at 84.3%. The intertypic nucleic acid identities of 3C (protease) and 3D (polymerase) are slightly higher, ranging from 79.8% to 86.9%. Attesting to the possible conformational constraints placed on the nonstructural proteins is the extremely conserved nature of 3B, 3C, and 3D amino acid sequences (Tables 5–7). In particular, the latter two proteins demonstrate only minor variations at <5% of the amino acid positions that comprise them. 3A exhibits somewhat greater variation but still remains significantly conserved among the prototypic CVB (Table 8). The intertypic amino acid identities of all four P3 proteins, with two exceptions (3A- CVB1:4 and CVB4:5), are greater than 90%.

3 Intratypic Genetic Divergence Among the Group B Coxsackieviruses

In the introduction to the first comparative analysis of the picornaviral capsid proteins, Palmenberg cites the major contribution of the discovery of reverse transcriptase to the expansion of our knowledge of the picornaviral genome (PALMENBERG 1989). In recent years the discovery of thermostable DNA polymerases and the development of the PCR has again led to a significant increase in the availability of new picornavirus sequences for analysis. The coupling of reverse transcription of the RNA enteroviral genome with PCR amplification of the RT product (RT-PCR) has dramatically facilitated the ability to obtain sequence information from these viruses.

3.1 Coxsackievirus B2

Although the complete genomic sequence of CVB2 has not yet been determined, RT-PCR technology has been employed to sequence 80% of the 5' NTR from seven clinical isolates (ROMERO et al. 1996). Using primers targeted to the VP4 coding region and to nucleotides 60–83 (relative to poliovirus) (OLIVE et al. 1990, ROMERO and ROTBART 1995) of the 5' NTR, an approximately 696 bp fragment, which includes the variable region, from six clinical isolates and the prototype strain of CVB2 (Ohio) has been amplified and directly cycle sequenced (ROMERO et al. 1996). These isolates and their year isolated are follows: Ohio-1947; 2093-1986; HI-1987; LO-1991; BR-1994; GR-1994 and WI-1994. Alignment of the seven CVB2 5' NTR sequences is shown in Fig. 8.

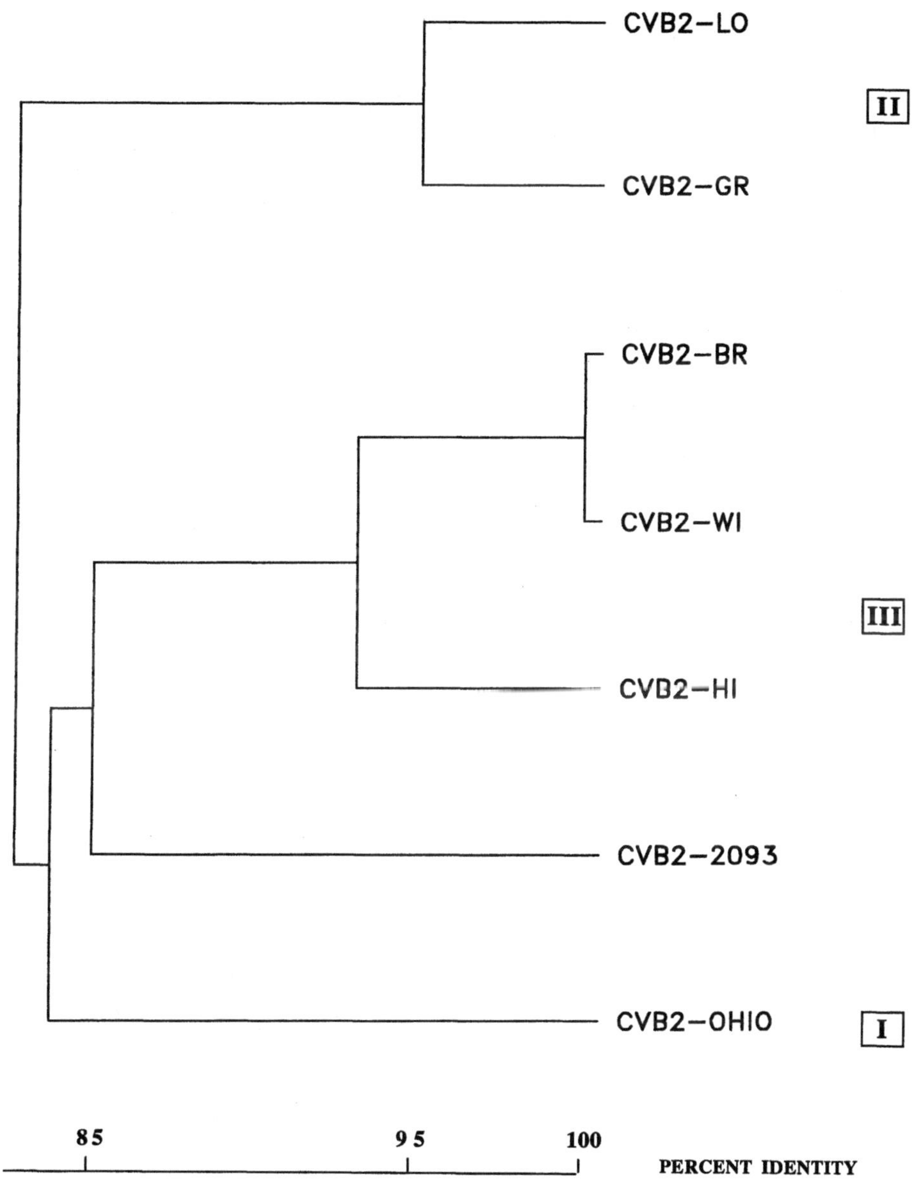

Fig. 9. Dendrogram of the relationships of the 5' NTR of CVB2 serotypes

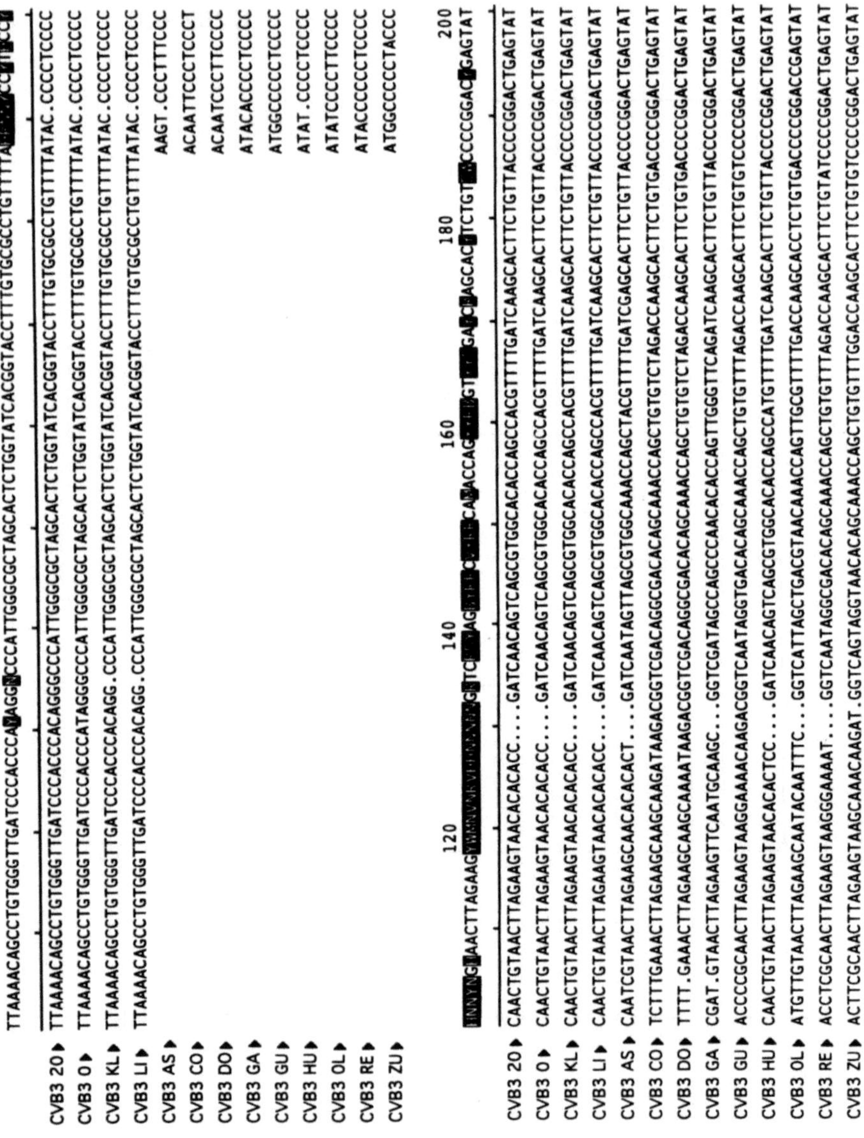

Fig. 10. Alignment of the nucleotide sequence of the 5' NTR of the CVB3. The consensus sequence for the CVBs is shown underlined above the alignments of sequences from each of CVB serotypes. Nonconserved positions are designated by the single letter genetic code where R=G or A; K=G or T; Y=C or T; S=C or G; M=A or C; W=A or T; B=C, G or T; D=A, G or T; H=A, C or T; V=A, C or G; N=A, C, G or T.

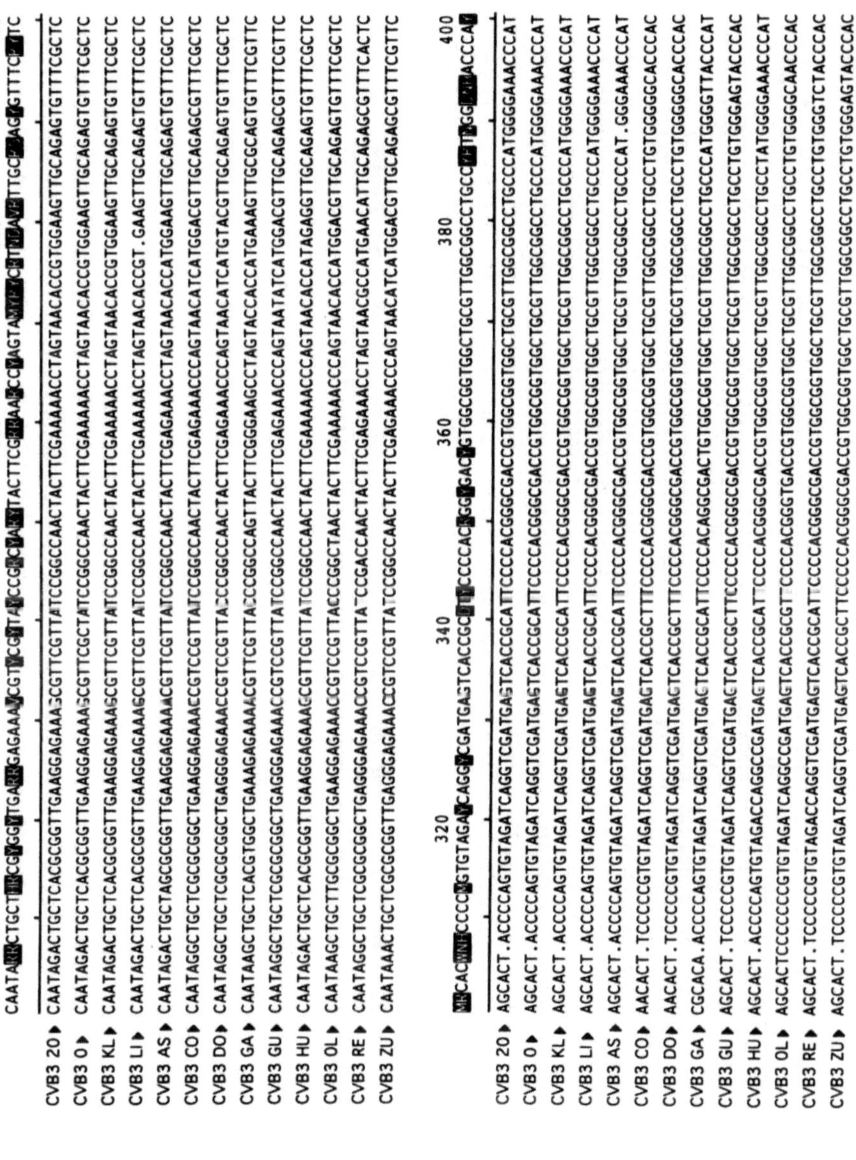

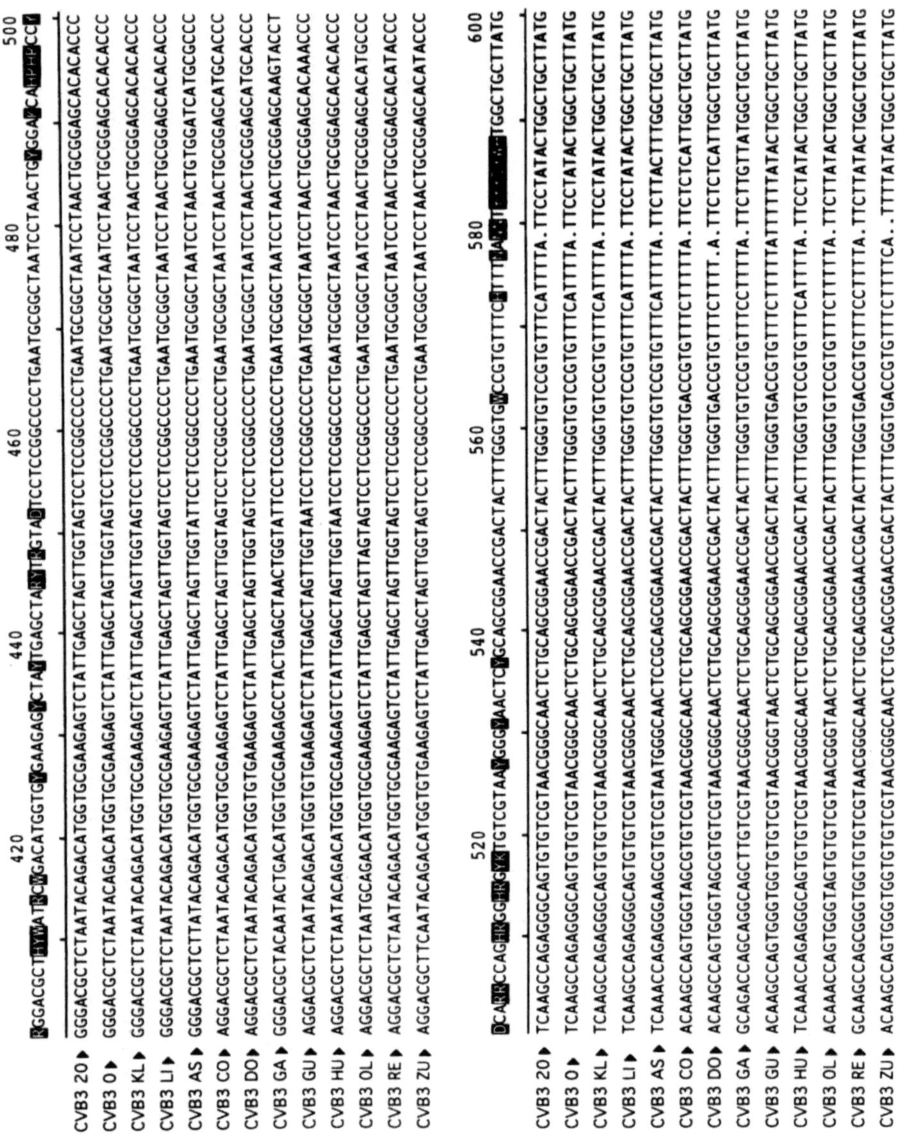
420
440
460
480
500
520
540
560
580
600
CVB3 20
CVB3 0
CVB3 KL
CVB3 LI
CVB3 AS
CVB3 CO
CVB3 DO
CVB3 GA
CVB3 GU
CVB3 HU
CVB3 OL
CVB3 RE
CVB3 ZU

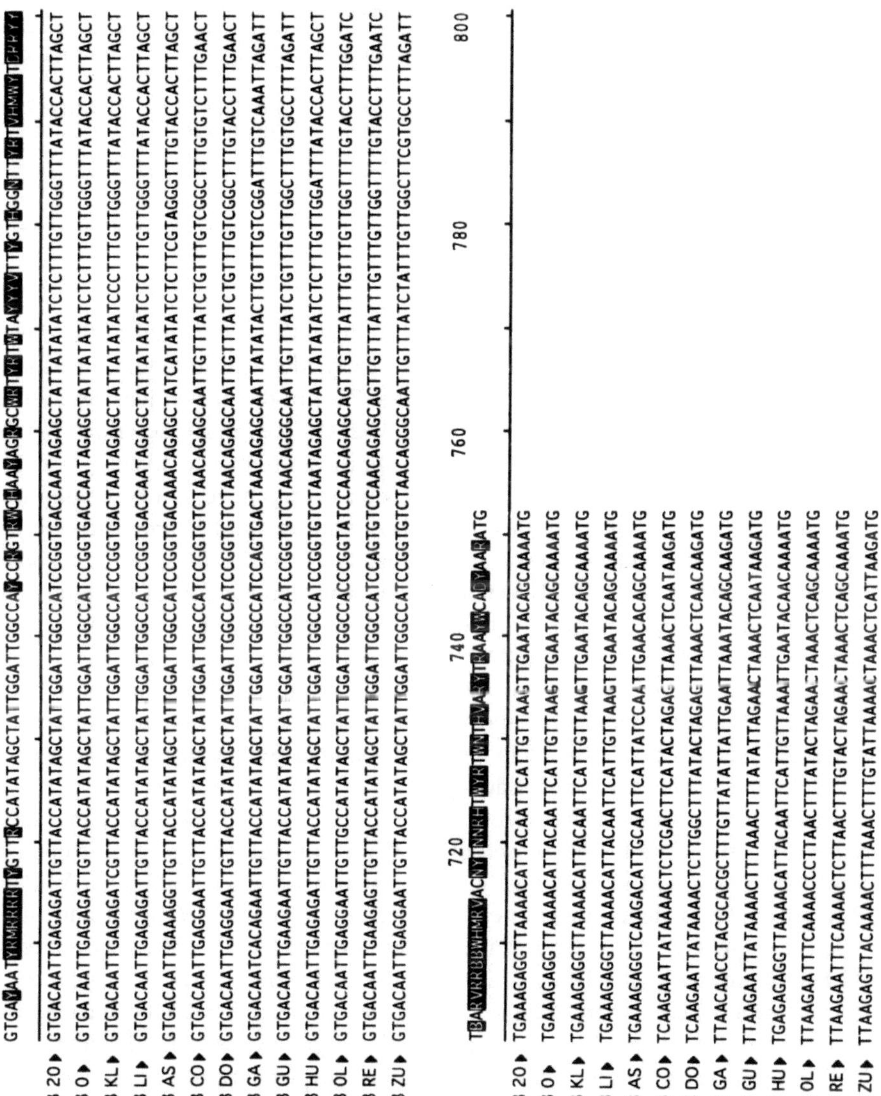

Table 9. Nucleotide identity[a] of the coxsackievirus B2 5' nontranslated region

Ohio:LO = 82.9	2093:LO = 84.7	LO:GR = 95.9	GR:WI = 85.0	WI:BR = 99.5
BR:HI = 93.7				
Ohio:GR = 83.5	2093:GR = 84.7	LO:WI = 85.2	GR:BR = 85.3	WI:HI = 93.2
Ohio:2093 = 84.0	2093:WI = 85.7	LO:BR = 85.5	GR:HI = 86.1	
Ohio:WI = 84.9	2093:BR = 86.1	LO:HI = 85.5		
Ohio:BR = 85.9	2093:HI = 88.1			
Ohio:HI = 86.7				

[a]Values given in percent. See text for discussion.

The amplified region of the CVB2 5' NTR from the 3' terminal of the downstream primer to the true initiation codon is nearly identical in length, measuring from 655 to 657 nucleotides. Within this region 72% of the nucleotides are absolutely conserved among all seven strains of CVB2. As previously described for the CVB intertypic comparisons of the 5' NTR, significant sequence diversity was primarily observed at two sites: (1) A downstream divergent region beginning at a variable length pyrimidine-rich tract immediately upstream of 5' amplification primer and extending downstream for approximately 60 nucleotides. (2) The known "variable region" of the 5' NTR which begins at nucleotide 560 of the alignment and extends to the true initiation codon (Fig. 8). The intratypic nucleotide identity of the CVB2 5' NTR ranges from 82.9% to 99.5% (Table 9). The prototypic strain, Ohio, is the most divergent of the seven CVB2 strains studied. This most likely relates to its being the temporally most distant of the isolates. Not surprisingly, the isolates BR and WI, both isolated in the same year and geographic area (Romero et al. 1996), have the highest nucleotide identity.

Based on the genomic divergence of the 5' NTR the CVB2 cluster into three groups (Fig. 9). Group I is composed of the Ohio strain alone. Group II is comprised of the LO and GR strains. Lastly, group III consists of the 2093, HI, WI, and BR strains. The nucleotide identities between the members of group I and II are sufficiently low (<85%) to suggest that, at the level of the 5' NTR, they represent different genotypes (Kew 1988) (Table 9). The differences between groups becomes less pronounced when groups I and II are compared with the members of group III. Confirmation of the genotypic relationships suggested by the analysis of 5' NTR these seven strains will need to await sequencing of additional CVB2 genomes and comparisons of sequences from the coding region (Ricco-Hesse et al. 1987)

3.2 Coxsackievirus B3

Four complete genomic sequences of CVB3 have been published: LI (Lindberg et al. 1987), KL (Klump et al. 1990), 20 (Tracy et al. 1992) and 0 (Chapman et al. 1994). In addition, the 5' NTR and capsid sequences of the M strain (M) of CVB3 (Gauntt et al. 1979; Lee et al. 1996) and capsid sequences of the RD variant of the Nancy strain of CVB3 (RD) (Reagan et al. 1984, Lindberg et al. 1992), a strain adapted to grow on RD cells, are also known. Additionally, using RT-PCR and cycle sequencing, the sequences of 80% of the 5' NTRs of nine clinical isolates of CVB3 have also been determined (ROMERO et al. 1994). These strains and their year of isolation are as follows: AS-1977, CO-1978, DO-1983, GA-1956, GU-1969, HU, OL-1965, RE-1962, and ZU-1974.

The length of the genomes of the completely sequenced CVBs isolates are nearly identical. The CVB3 20 and 0 strains each measure 7399 nucleotides, while the Nancy sequences reported by Lindberg and Klump are 7396 and 7399 nucleotides in length, respectively (LINDBERG et al. 1987; KLUMP et al. 1990; TRACY et al. 1992; CHAPMAN et al. 1994). Alignment of the complete and partial sequences of the CVB3 5' NTR is shown in Fig. 10. The length of the 5' NTRs of the CVB3 whose genomes have been completely determined (20, 0, KL, LI, and M) varies minimally; from 738 to 742 nucleotides. Evaluation of the partial 5' NTR sequences from the AS, CO, DO, GA, GU, HU, OL, RE and ZU strains indicates that they may be slightly longer, measuring up to 748 nucleotides. Variations in the length of the upstream variable region (discussed previously) are responsible for the differences in the size of the 5' NTRs among the CVB3. No differences in length were contributed by the variable region immediately upstream of the true initiation codon.

Table 10. Nucleotide identity[a] of the coxsackievirus B3 5' nontranslated region

AS:CO = 82.9	CO:DO = 97.0	DO:GA = 81.1	GA:GU = 79.7	GU:HU = 84.2	HU:KL = 97.0
KL:LI = 99.5	LI:OL = 83.3	OL:RE = 90.4	RE:0 = 84.3	0:20 = 99.7	
AS:DO = 83.2	CO:GA = 79.7	DO:GU = 91.3	GA:HU = 80.8	GU:KL = 83.4	HU:LI = 97.3
KL:OL = 83.4	LI:RE = 84.9	OL:0 = 83.6	RE:20 = 84.6	0:ZU = 82.8	
AS:GA = 80.7	CO:GU = 91.9	DO:HU = 84.4	GA:KL = 81.8	GU:LI = 83.5	HU:OL = 85.1
KL:RE = 84.2	LI:0 = 99.7	OL:20 = 83.9	RE:ZU = 91.3		
AS:GU = 81.9	CO:HU = 84.3	DO:KL = 84.1	GA:LI = 82.0	GU:OL = 88.7	HU:RE = 84.2
KL:0 = 99.2	LI:20 = 99.7	OL:ZU = 87.8			
AS:HU = 90.9	CO:KL = 83.9	FO:LI = 84.4	GA:OL = 80.3	GU:RE = 92.6	HU:0 = 96.8
KL:20 = 99.5	LI:ZU = 83.1				
AS:KL = 92.0	CO:LI = 84.0	DO:OL = 88.8	GA:RE = 80.6	GU:0 = 83.3	HU:20 = 97.1
KL:ZU = 82.9					
AS:LI = 92.3	CO:OL = 87.5	DO:RE = 89.7	GA:0 = 81.6	GU:20 = 83.6	HU:ZU = 83.2
AS:OL = 82.9	CO:RE = 89.6	DO:0 = 84.0	GA:20 = 81.9	GU:ZU = 96.1	
AS:RE = 82.6	CO:0 = 83.7	DO:20 = 84.3	GA:ZU = 78.9		
AS:0 = 92.0	CO:20 = 84.0	DO:ZU = 89.6			
AS:20 = 92.3	CO:ZU = 90.5				
AS:ZU = 81.4					

[a]Values given in percent. See text for discussion.

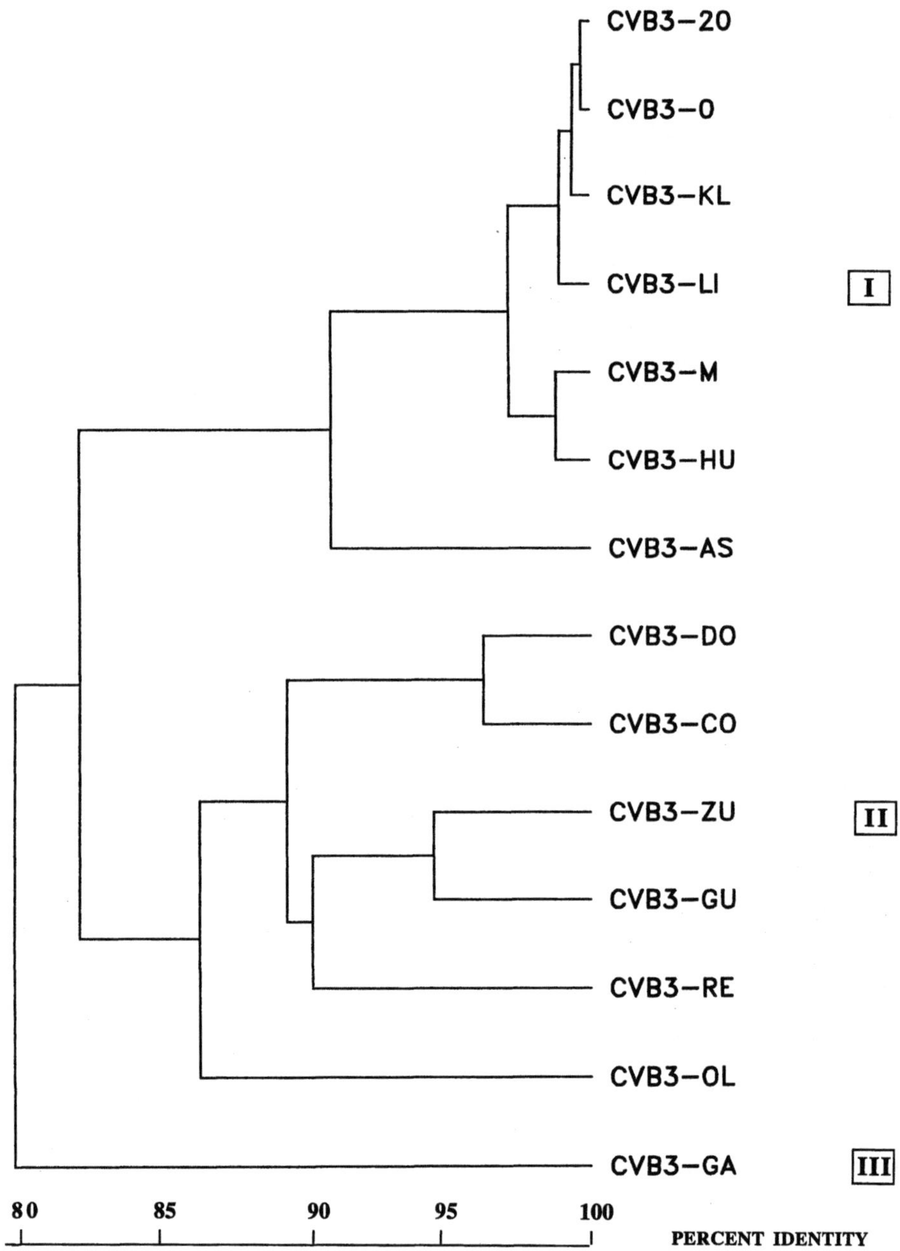

Fig. 11. Dendrogram of the relationships of the 5' NTR of CVB3 serotypes

The CVB3 intratypic nucleic acid identity of the 5' NTR ranges from 78.9% to 99.7% (Table 10). The GA strain is the most divergent of the 14 strains analyzed, exhibiting a nucleotide identity of 78.9%–82% with the other CVB3. The 20 and 0 strains share the greatest identity (99.7%). Sequence analysis indicates that the CVB3 5' NTR cluster into three groups (I–III) if sorted by a nucleotide identity of <85% (Fig. 11). Group I consists of strains whose sequences are the most similar to the prototype Nancy strain. This group can be further divided into two subgroups, Ia and Ib. Strains in the former subgroup (20, 0, KL, LI, M and HU) share an extremely high identity to one another (≤3.2% genomic divergence). The sole member of the Ib subgroup is the clinical isolate AS, which shares a 90.0%–93.2% identity to the preceding subgroup. The group II CVB3 5' NTR sequences can be subdivided into three subgroups: IIa (DO and CO), IIb (ZU, GU, and RE) and IIc (OL). The final group, III, is composed solely of the most divergent of the CVB3, GA.

From the preceding discussion it would appear that, for the 5' NTR, it is the CVB3 that exhibit the greatest plasticity among the CVB. However, this may be artifactual and merely reflective of the larger number of CVB3 strains that have been sequenced compared to the other CVB. As more strains of CVB are sequenced a clearer picture of the degree of variability tolerated within the 5' NTR of each serotype will become available.

The 3' NTR of the CVB3 varies minimally in length (98–100 nucleotides). The minor differences in the length of the CVB3 3' NTR are due to: (1) a relative deletion of a thymidine at position 7314 (relative to the CVB3 20 genome) in the CVB3 LI strain and (2) an addition of a single guanine at the terminal of the CVB3 KL strain. Of the 100 nucleotides comprising this region variation is observed at only four positions resulting in CVB3 intertypic nucleic acid identities of 95%–99%.

The intratypic nucleic acid and amino acid identities of the capsid region (P1) of the six CVB3 strains for whom data exist (20, 0, KL, LI, M, and RD) are extremely high (≥95%). Of the 2553 nucleotides encoding the four capsid proteins, variation is observed at only 118 positions (4.6%) (Table 11). The percentage of nucleotides demonstrating variation in each of the four capsid proteins is roughly equal: VP1, 4.3%; VP2, 4.4%; VP3, 5.0%; and VP4, 4.5%. In the P1 region the CVB3 M strain is the most divergent of the six strains at both the nucleotide and amino acid levels (Table 11). This strain alone is responsible for 83 (70.3%) of the 118 of the nucleotide variations and ten (41.7%) of 24 amino acid differences identified within the capsid coding region.

As a point of interest it should be noted that the recently submitted sequence of the CVB3 M strain (Lee et al. 1996) differs by five amino acid from the sequence reported in the CVB structural paper by Muckelbauer et al. (1995). These differences fall within the VP2 protein: 2084 D→Y, 2135 K→T, and 2245 V→I and VP3 protein: 3023 S→L and 3046 I→V. Of the six published sequences of CVB3, only the M strain exhibits an amino acid change at each of the positions listed.

Table 11. Nucleotide and amino acid differences in the P1 region of the B3 coxsackieviruses

Nucleotide[a]	20	0	KL	LI	M	RD	Amino acid[a]	20	0	KL	LI	M	RD
VP4 (1A)													
788	A	G	A	G	G	G	4016	R	G	R	G	G	G
826	A	A	A	A	G	A							
830	A	G	A	A	A	A	4030	I	V	I	I	I	I
844	G	G	G	G	A	G							
847	T	T	T	T	C	T							
889	C	C	C	C	T	C							
893	G	G	G	G	A	G	4051	G	G	G	G	S	G
904	A	A	A	A	G	A							
913	A	A	G	A	G	A							
VP2 (1B)													
987	T	C	C	C	T	C	2013	V	A	A	A	V	A
1001	T	C	T	T	C	T							
1048	G	G	G	G	A	G							
1060	A	A	A	A	G	A							
1075	A	A	A	A	G	A							
1084	T	T	T	T	C	T							
1090	A	A	A	A	T	A							
1156	G	G	G	G	A	G							
1198	C	C	C	C	T	C							
1199	G	G	G	G	T	G	2084	D	D	D	D	Y	D
1252	A	A	A	A	G	A							
1271–1273	GTA	GTA	GTA	GAT	ATA	GTA	2108	V	V	V	D	I	V
1294	T	T	T	T	C	T							
1361	G	G	G	G	A	G	2138	D	D	D	D	N	D
1401–1402	GC	GC	CG	CG	CG	GC	2151	S	S	T	T	T	S
1405	A	A	A	A	C	A							
1407–1408	AA	AA	AG	AA	CA	AA	2153	K	K	K	K	T	K
1429	C	C	C	C	T	C							
1501	C	C	C	C	T	C							
1513	C	C	C	C	T	C							
1519	A	A	A	A	G	A							
1528	C	C	C	C	T	C							
1540	T	T	T	T	C	T							
1576	T	T	T	T	C	T							
1660	C	C	C	C	T	C							
1678	T	T	T	T	C	T							
1682	G	G	G	G	A	G	2245	V	V	V	V	I	V
1717	G	G	G	G	A	G							
1718	T	T	T	T	C	T							
1726	A	A	A	A	G	A							
1729	A	A	A	A	C	A							
VP3 (1C)													
1756	T	T	T	T	C	T							
1805–1806	TC	TC	TC	TC	CT	TC	3023	S	S	S	S	L	S
1822	T	T	T	T	C	T							
1831	A	A	A	A	G	A							
1861	A	G	A	A	G	A							
1874	A	A	A	A	G	A	3046	I	I	I	I	V	I
1900	A	A	A	A	G	A							
1930	T	T	T	T	C	T							
1939	A	A	A	A	G	A							
1963	T	T	C	T	T	T							
1998	T	T	T	T	C	T							
2011	C	C	C	C	T	C							
2023	T	T	T	T	C	T							
2059	T	T	T	T	C	T							

2062	A	A	A	A	C	A							
2071	A	A	A	A	G	A							
2086	T	T	T	T	C	T							
2116	T	T	T	T	C	T							
2148	C	C	C	T	C	T	3137	P	P	P	L	P	L
2152	T	T	T	T	C	T							
2158	A	A	A	A	G	A							
2182	C	C	T	C	T	C							
2201	G	G	A	G	G	G	3155	V	V	I	V	V	V
2209	C	C	C	C	T	C							
2239	T	T	T	T	C	T							
2271	T	A	T	A	A	A	3178	F	Y	F	Y	Y	Y
2275	T	T	T	T	T	C							
2289	A	A	A	G	A	A	3184	Y	Y	Y	C	Y	Y
2335	G	G	G	G	C	G							
2344	G	G	G	G	A	G							
2379–2380	TG	TG	TG	TG	TA	AG	3214	V	V	V	V	V	A
2395	C	C	C	C	T	C							
2428	C	C	C	C	T	C							
2438	C	G	C	G	C	C	3234	Q	E	Q	E	Q	Q
VP1 (1D)													
2458	A	A	A	A	C	A							
2482	T	T	T	T	C	T							
2500	T	T	T	T	C	T							
2520	C	C	C	A	C	C	1023		T	T	T	N	TT
2530	A	A	A	A	G	A							
2560	A	A	G	A	A	A							
2569	G	G	G	G	G	A							
2572	A	A	A	A	G	A							
2585	G	G	G	G	A	G	1045	G	G	G	G	S	G
2593	C	C	T	C	C	C							
2632	T	C	C	C	T	C							
2638	A	A	A	A	G	A							
2683	G	G	G	G	A	G							
2690	A	G	A	G	G	A	1080	K	E	K	E	E	K
2701	T	T	T	T	C	T							
2726	T	T	T	T	A	T	1092	L	L	L	L	I	L
2743	A	A	A	A	G	A							
2788	C	C	C	C	T	C							
2848	A	A	A	A	G	A							
2851	C	C	T	C	T	C							
2881	C	T	T	T	T	C							
2896	A	A	A	A	G	A							
2902	A	A	A	A	G	A							
2908	T	T	T	T	C	T							
2914	T	T	T	T	C	T							
2920	A	A	A	A	T	A							
2983	C	C	C	C	T	A							
2990	A	A	A	A	G	A	1180	I	I	I	I	V	I
3016	C	C	C	C	T	C							
3034	C	C	C	C	T	C							
3037	A	A	A	A	G	A							
3061	A	A	A	A	G	A							
3070	C	C	C	C	T	T							
3079	G	G	G	G	A	G							
3118	C	C	C	C	T	C							
3133	T	T	T	T	A	T							
3169	G	G	G	G	T	G							
3243	G	A	A	A	A	A	1264	R	Q	Q	Q	Q	Q

[a]Nucleotide and amino acid positions are relative to the CVB3 20 strain (Tracy et al. 1992).

The majority of the nucleotide changes within the capsid coding region are of a silent nature. The differences observed at 118 nucleotide positions fall within 112 codons but result in changes of only 24 (2.8%) of the 851 capsid amino acids (Table 11). Six of the differences are located to residues comprising the β-structures of the capsid proteins; 1180 (βG1), 2013 (βA1), 2108 (βD), 3155 (Fβ), 3178 (βG2), and 3214 (βI). Nearly one third of the amino acid positions demonstrating variation (1080, 1264, 2084, 2138, 2151, 2153, 3184, and 3234) are known to be surface-exposed residues of CVB3 (MUCKELBAUER et al. 1995). Of note is that half of these surface residues are located in the VP2 protein and that all but one fall within its E-F loop, the largest of the surface loops.

Within the CVB3 nonstructural coding region variations are noted at 68 nucleotide positions in 54 codons and are distributed equally among the P2 and P3 regions (Tables 12, 13). These nucleotide differences result in changes of 35 amino acid residues, 18 in the P2 region and 17 in the P3 region. Unlike the nucleotide variations in the P1 region, where only 2.8% of the codons containing changes resulted in amino acid differences, in the P2 and P3 regions 66% and 63%, respectively, of the codons possessing nucleotide variations resulted in amino acid changes.

Table 12. Nucleotide and amino acid differences in the P2 region of the B3 coxsackieviruses

Nucleotide[a]	20	0	KL	LI	**Amino acid**[a]	20	0	KL	LI
2A									
3302	T	T	T	A	3	F	F	F	I
3305, 3307	GA	GA	GA	TG	4	G	G	G	W
3308–3309	CA	CA	CA	AC	5	Q	Q	Q	T
3311–3312	CA	CA	CA	AC	6	Q	Q	Q	T
3314–3316	TCA	TCA	TCA	ATC	7	S	S	S	T
3317	CA	CA	CA	GC	8	A	A	A	G
3321–3322	CA	CA	CA	GC	9	A	A	A	G
3323–3325	GTG	GCG	GTG	AGT	10	V	A	V	S
3326–3328	TAT	TAT	TAT	GTA	11	Y	Y	Y	V
3329–3331	GTG	GTC	GTG	TGT	12	V	V	V	C
3335	Â	A	A	G	14	N	N	N	D
3346	A	A	G	A					
3358	T	T	T	C					
3359	C	C	C	T	22	L	L	L	S
3434	A	G	G	G	47	T	A	A	A
2B									
3782	C	T	T	T	13	L	F	F	F
3804	A	G	A	A	20	Q	R	Q	Q
3913	G	G	G	A					
4928	A	T	A	A	95	M	L	M	M
4031	G	G	G	C	96	A	A	A	P
2C									
4078	G	G	A	G					
4165	G	G	A	G					
4177	A	A	G	A					
4306	A	G	A	A					
4582	G	G	G	T	180	Q	Q	Q	H
4585	C	T	T	T					
4948	T	T	T	G					

[a]Nucleotide and amino acid positions are relative to the CVB3 20 strain (Tracy et al. 1992).

Within the nonstructural protein coding region, the LI strain of CVB3 is the most divergent at both the nucleic and amino acid levels. Of the 68 nucleotide positions exhibiting differences among the CVB3 in the nonstructural coding region, 50 (73.5%) are due to LI alone. Some 80% (28 of 35) of the amino acid differences observed in the P2 and P3 regions are the result of variations in the LI sequence.

Among the structural proteins, the 2A protein exhibits the greatest divergence at both the nucleic and amino acid levels (94.4%–99.6% and 91.3%–99.3% identities, respectively). This divergence is disproportionately contributed to by the LI strain of CVB3 Nancy. Twenty (74%) of the 27 nucleotide differences and ten (76.9%) of the 13 amino acid differences in 2A were solely due to CVB3 LI (Table 12).

Table 13. Nucleotide and amino acid differences in the P3 region of the B3 coxsackieviruses

Nucleotide[a]	20	0	KL	LI	**Amino acid**[a]	20	0	KL	LI
3A									
5051	A	A	A	T	8	I	I	I	F
5119	C	T	C	C					
5249	G	G	G	A	74	V	V	V	M
3C									
5383	T	C	C	C					
5412	G	C	C	C	17	R	T	T	R
5560	C	C	C	T					
5588	A	A	A	G	76	K	K	K	E
5612	A	A	A	G	84	R	R	R	G
5621	A	A	A	G	87	R	R	R	G
5630	T	T	T	G	90	L	L	L	V
5735	T	C	C	C					
5894	T	T	T	A	178	Y	Y	Y	N
3D									
5943	A	A	A	T	11	D	D	D	V
6007	G	G	G	A					
6015	A	A	A	T	35	E	E	E	V
6055	A	A	A	T					
6142	A	A	A	G					
6212–6213	TG	TG	TG	AG	101	V	V	V	R
6247	A	A	A	T					
6341	T	T	T	C	144	C	C	C	R
6380	T	T	T	C	157	Y	Y	Y	H
6743	G	G	G	A	278	D	D	D	N
7026	T	T	C	T	372	V	V	A	V
7073	C	C	C	T					
7986–7087	TT	TT	TT	CC	392	V	V	V	A
7265	T	T	T	A	452	S	S	S	T
7274	C	C	C	G	455	R	R	R	G

[a]Nucleotide and amino acid positions are relative to the CVB3 20 strain (Tracy et al. 1992).

Alignment of the four fully sequenced CVB3 reveals that the variation in the 2A protein coding region of the LI strain is the result of the insertion of an adenine residue at position 10 of the 2A sequence (Fig. 12). The LI strain sequence downstream of the insertion to nucleotide 41 is identical to that of the 20 and KL strains at all positions and to the 0 strain, except at position 39, but out of frame by one nucleotide. The deletion of an adenine that should have occupied position 41 of the LI sequence corrects the frameshift and brings the sequence in line with those of the other CVB3. It is not clear whether the adenine insertion/deletion is the result of an error by the authors (LINDBERG et al. 1987) or due to true nucleotide variations within the LI strain sequence. Given the high intratypic nucleic acid identity (98.4%) among the CVB3 2B and 2C proteins the former may be a more likely possibility.

In the P3 region the CVB3 LI strain alone is responsible for 25 (86.2%) of 29 nucleotide and 15 (88.2%) of 17 amino acid differences (Table 13). The intratypic nucleic and amino acid identities for the four P3 proteins are 98.8%–100% and 96.7%–100%, respectively. The diminutive 3B (VPg) is the only protein absolutely conserved at both nucleic and amino acid levels among all four fully sequenced CVB3 strains.

The 3' NTR of the CVB3 varies in length from 97 to 99 nucleotides. The differences in size are due to: (1) the deletion, relative to CVB3 20, of a thymidine residue at position 7234 in CVB3 LI and (2) an additional terminal guanine in CVB3 KL. As in the previously discussed regions of the CVB3 genome, the LI strain is again the most divergent, containing four nucleotide differences. The intratypic nucleotide identity of the CVB3 3' NTR ranges from 93.8% to 99%.

While it appears that the CVB3 exhibit exceptionally high intratypic nucleotide and amino acid identities, it should be borne in mind that all four full-length published sequences are all derived from a common progenitor strain (Nancy). As is evident from the analysis of the partial sequences of the 5' NTRs from clinical isolates of CVB3, there exists considerable nucleotide variability among the wild-type strains. It is therefore likely that as sequences of non-Nancy strains of CVB3 become available the degree of nucleotide (and possibly amino acid) divergence among the CVB3 in the coding region will also increase.

```
           GGCGCNATTTGGACAACAATCAGGGGCAGYGTATGTGGGGNACTAC
CVB3 LI ▶  GGCGCAATTTGGACAACAATCAGGGGCAGTGTATGTGGGG.ACTAC
CVB3 20 ▶  GGCGC.ATTTGGACAACAATCAGGGGCAGTGTATGTGGGGAACTAC
CVB3 0 ▶   GGCGC.ATTTGGACAACAATCAGGGGCAGCGTATGTGGGGAACTAC
CVB3 KL ▶  GGCGC.ATTTGGACAACAATCAGGGGCAGTGTATGTGGGGAACTAC
```

Fig. 12. Alignment of the nucleotide sequence of the 2A protein coding region of the CVB3. The consensus sequence for the CVBs is shown underlined above the alignments of sequences from each of CVB serotypes. Nonconserved positions are designated by the single letter genetic code where R=G or A; K=G or T; Y=C or T; S=C or G; M=A or C; W=A or T; B=C, G or T; D=A, G or T; H=A, C or T; V=A, C or G; N=A, C, G or T.

Table 14. Nucleic and amino acid identities of the B4 coxsackieviruses

	JVB:E2		JVB:P-CB4		E2:P-CB4	
	Nucl	AA	Nucl	AA	Nucl	AA
5'NTR	87.9	NA	99.2	NA	87.9	NA
VP4(1A)	84.5	98.6	99	98.6	84.5	97.1
VP2(1B)	83.3	98.5	99.7	100	83.0	98.5
VP3(1C)	83.6	97.9	100	100	83.6	97.9
VP1(1D)	83.0	98.2	99.9	99.6	82.9	97.9
2A	74.9	90.0	99.6	100	75.1	90.0
2B	79.5	94.9	100	100	79.5	94.9
2C	81.5	98.8	98.4	100	81.4	98.8
3A	76.4	88.8	98.5	95.5	76.0	91
3B	78.8	95.5	100	100	78.8	95.5
3C	80.0	94.5	99.5	100	80.0	94.5
3D	79.7	94.6	99.9	99.8	79.7	94.6
3'NTR	87.4	NA	100	NA	87.4	NA

Nucl, nucleic acid identity; AA, amino acid identity; NA, not applicable. All values are expressed in percent.

3.3 Coxsackievirus B4

The complete genomic sequences of three strains of CVB4 (JVB Benschoten, E2 and P-CB4) have been reported (JENKINS et al. 1984; KANG et al. 1994; TITCHENER et al. 1994). JVB Benschoten (JVB) is considered the prototypic strain of CVB4. The E2 strain (KANG et al. 1994), a murine diabetogenic strain, was derived from the human Edwards strain of CVB4 (HARTIG et al. 1993). The third strain, P-CB4, is a β-cell tropic variant derived from JVB (TITCHENER et al. 1994). The total length of the genome of the JVB and P-CB4 strains is 7395 nucleotides. The E2 strain is a single nucleotide longer (7396), the result of an insertion of a thymidine at position 98, relative to JVB, of the E2 genome. No other insertions or deletions exist among the three strains. Among the three strains variations are noted at 1360 of the nucleotide positions that comprise the CVB4 genome. The highest nucleic acid identity (99.7%) is shared by the JVB and P-CB4 strains. The nucleic acid identity of the E2 strain with JVB and P-CB4 is 81.9% and 81.7%, respectively, thereby establishing it as a separate genotype within the CVB4 serotype (KEW 1988).

The intratypic nucleic acid identities of the nontranslated regions are shown in Table 14. A total of 106 nucleotide differences exist among the three CVB4 strains within the nontranslated regions. Of those differences, 93 are located within the 5' NTR, while 31 of the 5' NTR differences are clustered in the 100 nucleotides immediately upstream of the true initiation codon – a region known to be highly dissimilar among the enteroviruses (TOYODA et al. 1984). The remaining 13 NTR differences were located to the 3' NTR where only the E2 strain exhibits differences with JVB and P-CB4.

A comparison of the coding regions of the JVB and P-CB4 strains of CVB4 identified nucleotide differences within 1,147 codons (Table 15). The overwhel-

Table 15. Nucleotide and amino acid differences in the P1 region of the B4 coxsackieviruses

Nucleotide[a]	JVB	E2	P-CB4	Amino acid[a]	JBV	E2	P-CB4	Nucleotide[a]	JVB	E2	P-CB4	Amino acid[a]	JBV	E2	P-CB4
VP4(1A)															
750	G	G	A	4003	A	A	T	752	A	T	A				
755	G	A	G					758	G	T	G				
788	T	C	T					792	T	C	T				
802–803	GT	CG	GT	4020	S	T	S	806	A	C	A				
809	C	T	C					812	G	G	A				
815	T	C	T					818	T	C	T				
830	C	T	C					833	A	C	A				
845	G	A	G					851	T	C	T				
854	T	A	T					863	G	A	G				
866	C	G	C					869	T	C	T				
872	G	A	G					878	T	C	T				
881	T	C	T					887	A	G	A				
893	T	A	T					899	A	G	A				
902	C	T	C					914	A	C	A				
938	G	A	G					941	A	G	A				
944	G	A	G					947	C	T	C				
VP2(1B)															
953	C	A	C					956	G	A	G				
965	G	A	G					971	C	T	C				
974	A	G	A					980	C	T	C				
986	A	G	A					995	A	G	A				
1004	C	G	C					1007	G	A	G				
1010	C	T	C					1013	G	A	G				
1016	T	A	T					1022	G	A	G				
1031	G	A	G					1037	A	C	A				
1040	C	T	C					1043	C	T	C				
1055	T	C	T					1058	C	T	C				
1061C	A	C						1070	T	C	T				
1073	T	C	T					1079	C	T	C				
1085	A	G	A					1091	A	G	A				
1094	A	G	A					1097	G	A	G				
1100	A	G	A					1106	G	A	G				
1112	C	C	T					1118	T	A	T				

1121	T	C	T				
1130	G	A	G				
1145	G	A	G				
1151	T	G	T	2067	N	K	N
1206	T	C	T				
1220	G	T	G				
1250	C	T	C				
1262	A	G	A				
1271	A	G	A				
1277	T	C	T				
1292	A	G	A				
1298	A	G	A				
1307	A	G	A				
1313	T	C	T				
1319	T	A	T				
1328	T	C	T				
1349	A	G	A				
1355	C	T	C				
1361	A	G	A				
1370	A	G	A				
1376	G	A	G	2142	A	T	A
1388	G	A	G				
1397	A	G	A				
1409	G	A	G				
1415	C	T	C				
1429	C	T	C	2160	A	V	A
1442	A	T	A				
1454	G	A	G				
1466	T	C	T				
1481	G	A	G				
1490	T	C	T				
1499	T	C	T				
1514	A	G	A				
1526	A	G	A				
1538	T	C	T				
1550	A	T	A				
1562	C	T	C				
1571	C	T	C				

1124	G	T	G				
1133	T	C	T				
1146T	C	T					
1205	A	G	A				
1214	A	G	A				
1232	G	A	G				
1253A	T	A					
1268	C	T	C				
1274	T	C	T				
1280	G	A	G				
1295	C	T	C				
1304	C	T	C				
1310	T	C	T				
1314	C	T	C				
1322	G	A	G				
1340	T	C	T				
1352	T	C	T				
1358	T	C	T				
1367	C	T	C				
1374	G	A	G				
1386	T	C	T				
1391	T	C	T				
1400	G	A	G				
1410–1412	AGT	CAA	AGT	2154	S	Q	S
1421	G	A	G				
1437	A	G	A	2163	K	E	K
1451	G	A	G				
1463	C	T	C				
1472	T	C	T				
1484	G	A	G				
1494	T	C	T				
1508	T	A	T				
1523–1524	TT	CC	TT				
1532	A	T	A				
1547	C	T	C				
1559	A	T	A				
1568	T	C	T				
1583	C	T	C				

1598	T	C	T				
1611,1613	T,A	C,G	T,A				
1631	A	T	A				
1640	C	T	C				
1661	T	C	T				
1670	T	C	T				
1688	C	T	C				
1703	G	A	G				
1718	T	A	T				
1727	T	C	T				
VP3(1C)							
1745	T	C	T				
1760	C	T	C				
1778	G	A	G				
1805	T	C	T				
1814	G	A	G				
1841	T	C	T				
1853	G	A	G				
1871	T	A	T				
1889	G	A	G				
1895	A	G	A				
1905	T	C	T				
1921	T	A	T	3063	M	K	M
1931	G	A	G				
1955	C	T	C				
1961	C	T	C				
1970	A	T	A				
1976	G	A	G				
1988	C	T	C				
1997	G	A	G				
2003	G	A	G				
2016,2018	T,A	C,G	T,A				
2024	A	G	A				
2048	T	C	T				
2057	T	A	T				
2072	C	T	C				
2079	T	C	T				
2093	C	T	C				

1607	T	C	T				
1625	C	A	C				
1637	G	A	G				
1646	T	T	C				
1667	C	T	C				
1676	A	T	A				
1691	T	A	A				
1712	T	C	T				
1719,1721	C,A	T,G	C,A				
1730	T	C	T				
1751	T	C	T				
1766	G	C	G				
1787	T	T	C				
1811	A	G	A				
1829	A	T	A				
1850	A	G	A				
1860	C	T	C				
1877	A	G	A				
1892	A	G	A				
1901	T	C	T	3059	K	Q	K
1908	A	C	A				
1925	G	A	G				
1938	C	A	G				
1958	T	A	T				
1964	G	A	G				
1973	A	T	A				
1985	C	G	C				
1994	A	G	A				
2000	A	T	A				
2012	C	T	C				
2021	A	G	A				
2031	C	T	C				
2054	C	T	C				
2060	T	C	T				
2075	C	A	C				
2081	A	G	A				
2000	G	A	G				

2102	G	A	G				
2111	A	C	A				
2117	C	A	C				
2129	A	T	A				
2147	A	G	A				
2156	A	T	A				
2165	C	T	C				
2174	C	T	C				
2183	A	G	A				
2192	C	T	C				
2211	C	T	C				
2231	C	A	C				
2237	A	G	A				
2246	C	T	C				
2264	G	A	G				
2276	T	C	T				
2288	G	A	G				
2300	C	T	C				
2306	G	A	G				
2315	C	T	C				
2321	T	C	T				
2339	T	A	T				
2345	T	G	T				
32354	G	A	G				
2360	C	T	C				
2372	T	C	T				
2387	C	T	C				
2417	G	A	G				
2433–2435	ACA	GAC	ACA	3234	T	D	T
2447	G	A	G				
VP1(1D)							
2450	A	G	A				
2462	G	A	G				
2472	A	C	A				
2483	G	T	G				
2495	C	T	C				
2505,2507	C,C	A,A	C,C				
2516	G	A	G				

2105	A	C	A				
2114	T	C	T				
2126	A	G	A				
2135	C	T	C				
2153	G	A	G				
2162	C	T	C				
2168	G	A	G				
2177	T	C	T				
2189	C	T	C				
2204	C	T	C				
2225	T	C	T				
2234	T	C	T				
2240	G	A	G				
2249	C	T	C				
2273	T	G	T				
2282	G	A	G				
2297	T	G	T				
2303	T	A	T				
2309	T	C	T				
2318	A	G	A				
2330	A	C	A				
2342	A	G	A				
2351	A	G	A				
2357	C	T	C				
2369	C	T	C				
2381	A	G	A				
2396	T	A	T				
2426	T	C	T				
2437–2438	AC	CT	AC	3235	N	T	N
2459	A	G	A				
2465	C	T	C				
2480	G	G	A	1011	M	M	I
2486	A	G	A				
2498	G	C	G				
2510	C	T	C				
2522	T	C	T				

2528	A	G	A				
2537	T	C	T				
2543	A	G	A				
2558	A	G	A				
2567	C	T	C				
2573	G	A	G				
2579	A	T	A				
2609	T	C	T				
2621	C	A	C				
2630	A	G	A				
2547	A	G	A				
2665	C	T	C				
2684	C	T	C				
2706	C	T	C				
2720	T	C	T				
2735	C	T	C				
2744	G	A	G				
2753	A	G	A				
2757	A	C	A				
2780	T	C	T				
2789	G	T	G				
2801	G	A	G				
2807	C	A	C				
2816	T	C	T				
2822	C	T	C				
2840	C	T	C				
2854–2855	A,T	T,C	A,T	1136	D	V	D
2867	G	A	G				
2897	C	T	C				
2903	T	C	T				
2918	C	T	C				
2924	C	T	C				
2939	A	G	A				
2960	T	C	T				
2984	A	G	A				
2993	C	A	C				
3008	T	C	T				
3014	C	T	C				

2534	A	T	A				
2538,2540	C,G	T,A	C,G				
2552	G	A	G				
2561	T	C	T				
2570	G	A	G				
1576	T	C	T				
2588	G	T	G				
2612	C	T	C				
2624	A	G	A				
2649,2651	C,G	T,A	C,G				
2663	T	A	T				
2669	A	G	A				
2687	C	T	C				
2714	G	C	G				
2732	C	T	C				
2741	G	A	G				
2750	T	G	T				
2754,2756	C,A	T,G	C,A				
2762	A	G	A				
2783	T	C	T				
2792	C	T	C				
2803–2804	TT	AG	TT	1119	L	Q	L
2813	G	C	G				
2819	C	T	C				
2828	G	A	G				
2852	A	T	A				
2864	G	C	G				
2873	C	T	C				
2900	C	T	C				
2912	G	A	G				
2921	C	T	C				
2927	C	T	C				
2951	C	T	C				
2975	T	C	T				
2987	G	A	G				
2999	G	A	G				
3011	T	A	T				
3020	T	G	T				

3032	T	C	T				
3047	A	G	A				
3053	C	T	C				
3074	A	C	A				
3084,3086	T,A	C,G	T,A				
3104	A	T	A				
3110	G	A	G				
3131	C	T	C				
3140	A	G	A				
3167	C	T	C				
3179	C	T	C				
3185	A	G	A				
3191	T	C	T				
3199–3200	GC	CA	GC				
3205	C	G	C				
3215	T	C	T				
3242	C	T	C				
3260	T	C	T				
3264,3266	G,G	A,C	G,G	1272	A	T	A
3279	T	C	T				
3290	A	T	A				
3041	G	T	G				
3050	C	T	C				
3059	A	G	A				
3077	T	C	T				
3092	C	T	C				
3107	T	C	T				
3122	T	C	T				
3137	G	A	G				
3141	C	T	C				
3176	A	G	A				
3182	C	A	C				
3188	A	G	A				
3196–3197	CA	GC	CA				
3203	C	G	C				
3209	T	C	T				
3235–3236	GT	AC	GT	1263	S	N	S
3257	C	T	C				
3263	A	C	A				
3278	C	T	C				
3282	A	G	A	1279	I	V	I

[a]Nucleotide and amino acid positions are relative to the CVB4 JVBenschoten strain (Jenkins et al. 1987).

ming majority of which are of a silent nature. Only 88 amino acid position differences were observed among the CVB4 strains: 17 in the P1 region, 24 in the P2 region, 47 in the P3 region. Of these differences 82 (93.2%) are contributed solely by the E2 strain of CVB4. The total amino acid conservation within the entire coding region is 96%.

In the P1 region the VP1 protein contains the largest total number of amino acid changes, i.e., six, followed by five in the VP4 protein (Table 15). The intratypic nucleic and amino acid identities of the CVB4 capsid proteins is shown in Table 14. Of 15 amino acid differences found in the external capsid proteins (VP1–3), 11 may be located to surface-exposed residues based on the extrapolation of the data derived from the crystal structure of CVB3 (MUCKELBAUER et al. 1995). In the VP2 and VP3 proteins all amino acid differences are at possible surface-exposed residues. Within the former capsid protein, two of four amino acid differences (3059 and 3063) are located to the predicted position of the large E-F loop. In the VP3 protein two of the four differences are predicted to be located in the "knob" surface projection of this protein. In the case of the protein VP1, the two amino acids with possible surface exposure (1273 and 1279) are located at its carboxyl end. For the proteins of the P2 region the CVB4 intratypic amino acid identities are >90%. In all P2 proteins the JVB and P-CB4 strains share absolute identity (Tables 14, 16). In the P3 coding region the 3A protein exhibits the greatest variation among the three CVB4 strains (Tables 14, 17).

As has been postulated by other authors (MUCKELBAUER et al. 1995), the changes in the amino acids comprising the mature virion may effect cellular recognition and the ability to infect different tissues and/or hosts. 88% (15 of 17) of the observed amino acid differences within the capsid proteins are in the CVB E2 strain alone (Tables 14, 15). It is, therefore, tempting to speculate that the tissue (and host) specificity exhibited by these strains of CVB4, and in particular the E2 strain, may be due, in large part, to changes in the amino acid composition of the viral capsid.

As discussed earlier, the P-CB4 strain of CVB4 was derived from the JVB strain and exhibits β-cell tropism. The small number of differences observed between the JVB and P-CB4 strains, five nucleotides in the 5' NTR, one amino acid each in the VP4, VP1 and 3D, and four amino acids in 3D, provides a unique opportunity to further explore the CVB determinants of cell tropism. If infectious clones can be generated for these strains the possibility of mapping the location of the β-cell tropic determinant through the generation of chimeric constructs may be possible. The minimal number of changes present in the capsid and 5 'NTR of the P-CB4 strain would make mapping of a capsid and/or 5'NTR determinant relatively straightforward.

Table 16. Nucleotide and amino acid differences in the P2 region of the B4 coxsackieviruses

Nucleotide[a]	JVB	E2	P-CB4	**Amino acid**[a]	JBV	E2	P-CB4	**Nucleotide**[a]	JVB	E2	P-CB4	**Amino acid**[a]	JBV	E2	P-CB4
2A															
3293	C	A	C					3296	C	T	T				
3297	T	C	T	3	Y	H	Y	3302	A	T	A				
3305	T	G	T	5	H	Q	H	3308	A	G	A				
3311	A	T	A					3314	G	C	G				
3317	C	A	X					3320	G	A	G				
3332	T	C	T					3337–3338	AG	GA	AG	16	K	R	K
3341	A	G	A					3344	C	A	C				
3347	T	C	T					3350	G	A	G				
3353	C	T	C					3354	T	C	T				
3359	C	T	C					3362	G	A	G				
3365	C	T	C					3366,3368	G,G	T,T	G,G	26	V	F	V
3371	T	C	T					3377	A	G	A				
3382–3383	GC	AT	GC	31	C	Y	C	3384,3386	G,G	A,A	G,G	32	V	I	V
3398	T	C	T					3404	A	G	A				
3407	C	T	C					3410	T	C	T				
3413	A	G	A					3416	G	C	G				
3419	T	C	T					3422	T	A	T				
3425	C	T	C					3428	G	C	G				
3431	C	T	C					3437	G	A	G				
3440	C	T	C					3443	C	T	C				
3349	T	C	T					3461	A	G	A				
3464	C	T	C					3468,3470	A,A	T,T	A,A	60	T	S	T
3476	G	T	G					3482T	C	T					
3485	C	T	C	3487–3488	CC	TA	CC	66	A	V	A				
3491	C	A	C					3493–3494	AG	GA	AG	68	K	R	K
3496	G	A	G	69	S	N	S	3500	A	G	A				
3509	A	T	A					3512	T	G	T				
3513–3515	AGC	GTT	AGC	74	S	V	S	3518	T	C	T				
3524	A	C	C					3527	A	G	A				
3530	T	C	T					3531–3533	C	T	C				
3539	A	G	A					3542	C	T	C				

3548	A	G	A					3554	A	G	A				
3563	A	C	A					3585	T	C	T				
3593	T	A	T					3594–3596	ACA	CGG	ACA	102	T	R	T
3599	G	C	G					3608	A	G	A				
§&!$	A	G	A					3620	C	T	C				
3629	T	A	T					3630,3632	C,C	T,A	C,C				
3638	C	T	C					3641	A	G	A				
3644	C	T	C					3647	C	T	C				
3650	C	T	C					T	C	T					
3657,3659	C,T	A,A	C,T	123	L	I	L	3660–3662	GTC	CGA	GTC	124	V	R	V
3672,3674	G,T	C,G	G,T	128	G	R	G	3680	C	T	C				
3683	G	T	G					3689	T	C	T				
3692	C	T	C					3695	C	T	C				
3701	C	G	C					3704	T	C	T				
3710	G	T	G					3711	T	C	T				
3722	A	G	A					3728	T	C	T				
3731	A	G	A					3738	A	G	A				
2B															
3743	A	C	A					3746	G	C	G				
3752	T	C	T					3755	C	T	C				
3758	T	G	T					3761	G	A	G				
3762,3764	C,A	G,G	C,A	8	Q	E	Q	3770	T	C	T				
3774,3776	G,T	T,G	G,T	12	A	S	A	3779	T	C	T				
3785	A	T	A					3788	A	C	A				
3803	A	C	A					3806	C	T	C				
3815	T	C	T					3818	C	T	C				
3824	A	C	A					3830	A	G	A				
3839	A	G	A					3842	T	G	T				
3851	A	C	A					3854	C	T	C				
3855,3857	T,G	C,A	T,G					3872	A	G	A				
3876,3878	C,A	T,G	C,A					3881	T	C	T				
3893	T	A	T					3896	C	A	C				
3899	G	A	G					3902	G	C	G				
3911	A	G	A					3914	G	A	G				
3926	C	T	C					3929	G	T	G				
3930,3932	A,C	G,A	A,C	64	I	V	I	3941	A	T	A				

3947	A	T	A
3953	C	T	C
3959	T	A	T
3958	C	T	C
3974	T	A	T
3989	T	C	T
3998	G	A	G
4010	T	C	T
4032	C	A	C
2C			
4040	C	T	C
4052	A	C	A
4061	T	C	T
4076	C	T	C
4085	A	G	A
4103	C	T	C
4112	T	C	T
4133	C	G	C
4139	T	C	T
4145	T	A	T
4151	A	G	A
4176	C	T	C
4184	A	G	A
4190	A	G	A
4196	C	G	C
4200,4202	C,C	T,G	C,C
4214	G	A	G
4226	T	A	T
4244	T	A	T
4263	C	T	C
4283	C	T	C
4289	T	C	T
4298	C	T	C
4310	A	C	A
4316	C	G	C
4331	C	A	C

3950	C	G	C				
3956	T	A	T				
3965	C	T	C				
3969,3971	T,G	A,A	T,G	77	S	T	S
3983	A	C	A				
3994–3995	AC	GG	AC	85	H	R	H
4007	A	G	A				
4019	A	C	A				
4037	G	A	G				
4044–4046	GGG	AAC	GGG	3	G	N	G
4058	A	G	A				
4073	T	C	T				
4079	A	T	A				
4088	G	C	G				
4106	C	T	C				
4121	T	C	T				
4136	G	A	G				
4142	G	A	G				
4146,4148	T,G	C,C	T,C				
4160	G	A	G				
4180–4181	GT	AC	GT	48	S	N	S
4187	C	A	C				
4193	G	A	G				
4199	A	C	A				
4208	G	A	G				
4217	T	A	T				
4235	T	C	T				
4262	G	A	G				
4271	A	T	A				
4286	C	T	C				
4295	T	C	T				
4307	C	C	T				
4313	G	C	G				
4325	A	T	A				
4334	G	A	G				

4340	G	C	G	4343	T	C	T
4347	C	T	C	4349	T	A	T
4353	A	G	A	4355	A	G	A
4410,4412	T,G	C,A	T,G	4413	C	T	C
4415	T	G	T	4416,4418	T,G	C,A	T,G
4421	T	C	T	4424	T	G	T
4430	A	T	A	4433	A	T	A
4436	G	T	G	4446	A	T	A
4448	T	A	T	4451	T	G	T
4460	A	G	A	4469	G	T	G
4473	T	C	T	4475	A	T	A
4478	T	A	A	4484	G	A	G
4485	T	C	T	4490	C	T	C
4493	T	C	T	4496	A	T	A
4502	C	T	C	4505	C	G	C
4506,4508	T,A	C,G	T,A	4517	C	T	C
4520	A	T	A	4523	C	T	C
4526	T	C	T	4529	C	T	C
4544	A	G	A	4553	C	T	C
4556	A	G	A	4559	T	C	T
4568	T	C	T	4571	A	G	A
4577	A	G	A	4583	G	T	G
4589	C	G	C	4592	G	A	G
4595	C	T	C	4601	T	C	T
4602	T	C	T	4607	T	C	T
4610	C	T	C	4619	G	A	G
4622	T	A	T	4628	A	T	A
4631	C	T	C	4637	A	C	A
4649	T	G	T	4655	G	T	G
4667	T	C	T	4671,4673	T,G	C,C	T,G
4676	C	T	C	4682	C	A	C
4685	T	A	T	4694	G	T	G
4697	C	T	C	4700	A	C	A
4709	T	A	T	4718	C	T	C
4721	C	T	C	4724	G	C	G
4727	G	C	G	4730	A	T	A

4733	C	G	C					4743,4745	C,G	A,A	C,G
4751	G	T	G					4754	A	C	A
4763	C	T	C					4766	T	C	T
4769	T	C	T					4778	T	C	T
4787	T	C	T					4793	C	A	C
4805	G	A	G					4811	C	T	C
4814	G	A	G					4817	C	A	C
4826	C	G	C					4835	C	A	C
4841	G	A	G					4847	T	C	T
4856	T	C	T					4868	T	C	T
4876–4877	GA	AG	GA	280	R	K	R	4889	A	T	A
4892	G	T	G					4898	A	G	A
4910	G	A	G					4913	T	C	T
4916	C	T	C					4922	A	G	A
4924–4925	A,G	G,A	A,G	296	K	R	K	4928	T	A	T
4937	G	A	G					4943	C	A	C
4946	A	G	A					4949	T	C	T
4955	A	T	A					4958	C	G	C
4961	G	T	G					4964	G	A	G
4973	G	A	G					4988	G	A	A
4994	T	C	T					4997	C	G	C
5000	G	A	G					5003	G	A	G
5006	C	A	C					5015	G	G	C
5018	A	C	A					5024	A	G	A

[a]Nucleotide and amino acid positions are relative to the CVB4 JV Benschoten strain (Jenkins et al. 1987).

Table 17. Nucleotide and amino acid differences in the P3 region of the B4 coxsackieviruses

Nucleotide[a]	JVB	E2	P-CB4	Amino acid[a]	JBV	E2	P-CB4	Nucleotide[a]	JVB	E2	P-CB4	Amino acid[a]	JBV	E2	P-CB4
3A															
5030	G	C	G	3	V	I	V	5033	A	G	A				
5034,5036	G,G	A,A	G,G					5042	A	G	A				
5048	T	C	T					5050–5051	AA	GG	AA	9	K	R	K
5057	T	C	T					5060	C	T	C				
5061	A	G	A	13	T	A	T	5066	T	G	T				
5075	A	T	A					5078	A	T	A				
5086	T	C	C	21	V	A	A	5090	C	T	C				
5093	A	C	A					5097,5099	T,G	C,A	T,C				
5104–5105	AG	GA	AG	27	K	R	K	5108	A	G	A				
5114	C	T	C					5117	G	T	G	31	R	S	R
5118	C	G	A	32	Q	E	Q	5123	T	A	T				
5124	A	G	G	34	I	V	V	5129	A	G	A				
5138	T	C	T					5141	G	A	G				
5147	G	G	T	41	K	K	N	5154	C	T	C				
5159	T	G	T					5168	C	A	C				
5169–5170	GA	AG	GA	49	D	S	D	5174	T	C	T				
5176–5177	TT	CC	CT	51	I	T	T	5180	C	A	C				
5186	T	C	T					5192	G	A	G				
5195	T	C	T					5198	C	G	C				
5201	T	C	T					5207	A	T	A				
5220	T	C	T					5229	C	G	C				
5222	A	G	A					5228	G	C	G				
5231	A	T	A					5240	G	C	G				
5243	T	C	T					5246	G	A	G				
5249	C	A	X					5252	A	G	A				
5255	T	C	T					5258	T	A	T				
5261	T	C	T					5267	T	C	T				
5273	A	G	A					5274	C	T	C				
5285	G	C	G					5291	G	A	G				
3B															
5294	T	A	T					5297	A	T	A				
5303	G	T	G					5306	G	A	G				

5312	T	C	T				
5321	A	G	A				
5333	T	A	T				
5337,5339	C,A	T,T	C,A	16	L	F	L
5345	G	A	G				
3C							
5360	T	A	T				
5375	C	T	C				
5400,5402	T,C	G,T	T,C	15	S	A	S
5408	G	T	G				
5435	C	A	C				
5441	A	T	A				
5462	T	A	T				
5471	A	C	A				
5477	C	T	C				
5486	C	T	C				
5492	G	T	G				
5498	T	C	T				
5507	T	C	T				
5516	G	A	G				
5522	T	C	T				
5526,5528	C,G	G,A	C,G	57	L	V	L
5541	T	C	C				
5551	G	A	G	65	R	K	R
5563–5564	AT	CC	AT				
5579	A	C	A				
5592,5594	C,G	A,A	C,A				
5609	G	A	G				
5624	T	C	T				
5631–5633	AAG	CGA	AAG	92	K	R	K
5639	A	G	A				
5648	T	C	T				
5663	A	C	A				
5669	C	T	C				
5684	T	C	T				
5699	C	T	C				
5705	A	T	A				

5315	C	T	C				
5324	T	C	T				
5336	A	T	A				
5342	G	A	G				
5366	T	C	T				
5384	C	A	C				
5405	T	T	C				
5417	A	C	A				
5439	T	C	T				
5444	C	T	C				
5465	C	G	C				
5474	C	G	C				
5480	T	C	T				
5489	G	A	G				
5495	T	C	T				
5499,5501	C,T	T,G	C,T				
5510	C	T	C				
5519	C	G	C				
5525	G	A	G				
5534	C	T	C				
5544	A	G	A	63	I	V	I
5561	A	C	A				
5576	A	G	A				
5580	C	T	C				
5603	A	G	A				
5616	A	C	A				
5625,5627	C,A	T,G	C,A				
5636	G	A	G				
5641–5642	TG	CC	TG	95	V	A	V
5660	C	G	C				
5566	A	C	A				
5675	T	C	T				
5696	C	T	C				
5702	C	A	C				
5709–5711	AGG	CAA	AGG	118	R	Q	R

5714	C	T	C					5720	C	G	C	121	D	E	D
5732	A	C	A					5736	C	T	C				
5744	T	C	T					5753	A	T	A				
5757,5759	A,A	C,C	A,A					5765	C	A	C				
5780	T	C	T					5783	A	T	A				
5784,5786	A,G	C,T	A,G					5789	T	A	T				
5801	C	T	C					5804	T	C	T				
5897	T	C	T					5810	C	A	C				
5816	C	A	C					5819	T	A	T				
5828	G	T	C					5831	A	T	A				
5834	G	A	G					5837	C	A	C				
5849	G	A	G					5855	T	A	T				
5861	G	A	G					5864	T	A	T				
5870	A	G	A					5874–5875	GG	TC	GG	173	G	S	G
5879	C	A	C					5880,5882	C,T	T,G	C,T				
5883–5884	AA	CG	AA	176	K	R	K	5894	T	C	T				
3D															
5899	G	T	G					5905	C	A	C				
5927	A	G	A					5930	C	T	C				
5932	G	C	G					5936	A	G	A				
5939	T	A	T	11	D	E	D	5942	A	T	A				
5945	T	C	T					5948	C	T	C				
5954	C	T	C					5960	T	C	T				
5963	A	C	A					5966	A	C	A				
5968–69	GT	AC	GT	21	S	N	S	5971–5972	GA	AG	GA	22	R	K	R
5975	T	A	T					5978	G	A	G				
5981	A	T	A					5984	A	G	A				
5987	A	C	A					6003,6005	G,C	A,A	G,C	33	V	I	V
6011	A	G	A					6014	A	G	A				
6017	C	T	C					6020	G	A	G				
6023	A	G	A					6026	A	G	A				
6029	A	G	A					6032	C	G	C				
6035	C	A	C					6041	C	T	C				
6044	C	T	C					6048,6050	C,G	A,C	C,G	48	P	T	P
6051,6053	C,C	A,A	C,C					6054,6056	C,T	T,A	C,T				
6059	A	G	A					6080	A	C	A				
6084–6085	TT	TC	CT	60	F	S	L	6089	A	G	A				

6095	A	T	A				
6104	C	G	C				
6122	G	A	G				
6129,6131	C,A	A,G	C,A	75	L	M	L
6143	T	C	T				
6164	C	A	C				
6170	T	C	T				
6178–6179	AC	GT	AC	91	N	S	N
6185	G	A	G				
6193	A	G	A	96	K	R	K
6200	A	G	A				
6218	G	T	G				
6236	T	A	T				
6248	A	G	A				
6260	C	T	C				
6269	T	C	T				
6273,6275	T,A	C,C	T,G				
628	G	A	G				
6317	C	T	C				
6323	C	A	C				
6354	T	C	T				
6365	G	C	G				
6374	A	T	A				
6380	G	A	G				
6392	T	G	T				
6419	G	C	G				
6428	A	G	A				
6437	A	G	A				
6446	C	T	C				
6452	C	T	C				
6461	T	G	T				
6473	A	G	A				
6479	T	C	T				
6495	G	A	G				
6512	G	A	G				
6518	T	A	T				
6524	G	T	G				

6098	A	T	A				
6119	C	T	C				
6125	C	T	C				
6140	G	T	G				
6149	T	C	T				
6167	T	C	T				
6173	C	T	C				
6182	T	A	T				
6188	A	C	A				
6197	G	T	G				
6212	C	T	C				
6224	G	C	G				
6245	A	C	A				
6251	T	C	T				
6263	A	C	A				
6270,6272	G,A	A,C	G,A	122	A	T	A
6281	T	A	T				
6290	G	A	G				
6318	C	T	C				
6350	C	T	C				
6360,6362	T,G	C,G	T,G				
6371	G	C	G				
6377	C	T	C				
6383	G	A	G				
6407	G	A	G				
6425	T	C	T				
6431	C	G	C				
6443	C	G	C				
6447,6449	T,G	C,A	T,G				
6458	T	G	T				
6464	G	A	G				
6476	A	T	A				
6482	T	G	T				
6495	G	A	G	197	A	T	A
6515	G	T	G				
6521	A	G	A				
6536	C	T	C				

6539	G	T	G				
6548	A	C	A				
6563	T	C	T				
6569	A	C	A				
6576	G	C	G				
6587	A	C	A				
6593	T	C	T				
6599	C	T	C				
6605	C	T	C				
6614	T	A	T				
6626	T	C	T				
6650	T	C	T				
6657,6659	T,G	C,A	T,G				
6665	T	A	T				
6674	C	G	C				
6681,6683	A,A	T,C	A,A	259	T	S	T
6692	G	A	G				
6707	C	T	C				
6719	C	T	C				
6728	C	T	C				
6756,6758	C,T	A,A	C,T				
6770	C	A	C				
6779	C	T	C				
6788	C	T	C				
6800	C	T	C				
6809	C	T	C				
6815	C	T	C				
6821	T	C	T				
6830	C	A	C				
6839	G	C	G	311	L	F	L
6848	C	T	C				
6854	T	A	T				
6860	C	T	C				
6869	A	G	A				
6884	A	T	A				
6899	G	C	G				
6911	T	C	T				

6542	C	T	C				
6552,6554	G,T	T,G	G,T	216	V	L	V
6566	A	G	A				
6572	T	G	T				
6581	A	T	A				
6590	C	T	C				
6595	T	G	T	230	I	R	I
6602	C	T	C				
6609–6611	TCC	AGT	TCC				
6620	C	T	C				
6635	C	A	C				
6653	A	G	A				
6660	C	T	C				
6671	A	G	A				
6680	C	T	C				
6689	A	G	A				
6704	A	T	A				
6713	A	G	A				
6725	C	T	C				
6755	A	G	A				
6761	C	G	C				
6773	A	T	A				
6785	T	A	T				
6797	C	T	C				
6803	A	G	A				
6812	T	C	T				
6818	C	A	C				
6827	G	A	G				
6833	A	G	A				
6842	G	A	C				
6851	A	G	A				
6847	T	A	T				
6861, 6863	T,G	C,A	T,G				
6881	T	A	T				
6887	T	C	T				
6905	A	G	A				
6914	T	A	T				

6915–6917	TGG	CTA	TGG	337	W	L	W	6920	C	T	C				
6923	A	T	A					6932	T	A	T				
6938	C	A	C					6941	T	A	T				
6950	T	A	T					6955	A	G	A	350	D	G	D
6959	C	T	C					6962	G	C	G				
6963,6965	C,A	T,G	C,A					6968	C	T	C				
6980	G	T	G					6989	A	G	A				
6995	T	C	T					7004	A	G	A				
7005,7007	G,C	C,A	G,C	367	V	L	V	7028	T	C	T				
7034	G	A	G					7040	T	C	T				
7055	A	G	A					7058	A	G	A				
7068,7070	T,G	C,T	T,G					7088	C	A	C				
7091	G	C	G	395	M	I	M	7097	C	T	C				
7103	C	T	C					7106	G	A	G				
7109	T	A	T					7121	C	T	C				
7124	A	G	A					7130	A	T	A				
7133	G	A	G					7139	T	C	T				
7143,7145	G,T	T,C	G,T	413	D	Y	D	7148	T	C	T				
7154	C	A	C					7158	C	T	C				
7166	A	G	A					7169	G	A	G				
7172	T	C	T					7178	C	T	C				
7190	C	A	C	428	H	Q	H	7193	A	G	A				
7199	G	A	G					7205	C	G	C	433	F	L	F
7206–7208	ATC	TAT	ATC	434	I	Y	I	7210	A	C	A	435	Q	P	Q
7214	G	A	G					7220	A	G	A				
7226	C	T	C					7232	T	G	T				
7241	C	T	C					7247	T	C	T				
7253	C	T	C					7259	T	C	T				
7262	G	T	G					7265	C	T	C				
7271	T	C	T					7277	A	G	A				
7281,7283	T,G	C,A	T,G					7286	T	C	T				
7287,7289	T,C	G,T	T,C	461	S	A	S								

[a]Nucleotide and amino acid positions are relative to the CVB5 JV Benschoten strain (Jenkins et al. 1987).

4 Summary

As documented in the preceding discussion, the noncoding regions, and in particular the 5' NTR, of the CVB are tolerant of a substantial degree of nucleotide diversity while still being capable of fulfilling the life cycle requirements for these viruses. While diversity among the CVB is observed in the sequences encoding for the capsid proteins, it tends to involve predominantly those regions coding for amino acids located at the surface of the virus and not those responsible for the structural integrity of the mature virion, i.e., β-barrels and α-helices. It is these capsid surface differences that define the six serotypes of the CVB and subdivide them antigenically into strains. Additionally, these proteins most likely play the major role in determining host and cellular tropism. The most conserved of the CVB proteins and, therefore those with the least diversity in their coding sequences, appear to be the nonstructural proteins. Perhaps, as speculated earlier, it is a conformational requirement imposed by the necessity to interact with host or viral substrates that maintains the high degree of amino acid identity of this group of viral proteins.

References

Arnold E, Rossman MG (1990) Analysis of the structure of a common cold virus, human rhinovirus 14, refined at a resolution of 3.0 Å. J Mol Biol 211:763–801

Chapman NM, Tu Z, Tracy S, Gauntt CJ (1994) An infectious cDNA copy of the genome of a non-cardiovirulent coxsackievirus B3 strain: its complete sequence analysis and comparison to the genomes of cardiovirulent coxsackieviruses. Arch Virol 135:115–130

Chumakov KV, Agol VI (1976) Poly(C) sequence islocated near the 5' end of the encephalomyocarditis virus RNA. Biochem Biophys Res Commun 71:551–557

Drebot MA, Nguan CY, Campbell JJ, Lee SHS, Forward KR (1994) Molecular epidemiology of enterovirus outbreaks in Canada during 1991–1992: identification of echovirus 30 and coxsackievirus B1 strains by amplicon sequencing. J Med Virol 44:340–347

Gauntt CJ, Pallansch MA (1996) Coxsackievirus B3 clinical isolates and murine myocarditis. Virus Research 38:125–136

Gauntt CJ, Trousdale MD, LaBadie DRL, Paque RE, Nealon T (1979) Properties of coxsackievirus B3 variants which are amyocarditic of myocarditic for mice. J Med Virol 3:207–220

Harber J, Wimmer E (1993) Aspects of the molecular biology of picornaviruses. In: Carrasco L, Sonenberg N, Wimmer E (ed.), Regulation of Gene Expression in Animal Viruses. Plenum Press, New York

Harris TJR, Brown F (1976) The location of the poly(C) tract in the RNA of foot-and-mouth disease virus. J Gen Virol 33:493–501

Hartig PC, Madge GE, Webb SR (1993) Diversity within a human isolate of coxsackie B4: relationship to viral-induced diabetes. J Med Virol 11:23–30

Hogle JM, Chow M, Filman DJ (1985) Three-dimensional structure of poliovirus at 2.9 Å resolution. Science 229:1358–1365N, Kuge S, Nomoto A (1987) Complete nucleotide sequence of the genome of coxsackievirus B1. Virology 156:64–73

Jenkins O, Booth JD, Minor PD, Almond JW (1987) The complete nucleotide sequence of coxsackievirus B4 and its comparison to other members of the picornaviridae. J Gen Virol 63:1835–1848

Kang Y, Chatterjee NK, Nodwell MJ, Yoon JW (1994) Complete nucleotide sequence of a strain of coxsackie B4 virus of human origin that induces diabetes in mice and its comparison with nondiabetogenic coxsackie B4 JBV strain. J Med Virol 44:353–361

Kew OM (1988) Applications of molecular epidemiology to the surveillance of poliomyelitis. Proceedings of the National Academy of Medicine, Institute of Medicine conference "Poliomyelitis Vaccines: Re-evaluating Policy Options." Washington DC: National Academy of Sciences

Klump WM, Bergmann I, Muller BC, Ameis D, Kandolf R (1990) Complete nucleotide sequence of infectious coxsackievirus B3 cDNA: two initial 5' uridine residues are regained during plus-strand RNA synthesis. J Virol 64:1573–1583

Kopecka H, Brown B, Pallansch M (1995) Genotypic variation in coxsackievirus B5 isolates from three different outbreaks in the United States. Virus Research 38:125–136

Lee C, Maull E, Chapman N, Tracy S, Wood J, Gauntt C (1996) Generation of an infectious cDNA of a highly cardiovirulent coxsackievirus B3 (CVB3m) and comparison to other infectious CVB3 cDNAs (Submitted)

Lindberg AM, Stalhandske POK, Pettersson U (1987) Genome of coxsackievirus B3. Virology 156:50–63

Lindberg MA, Crowell RL, Zell R, Kandolf R, Pettersson U (1992) Mapping of the RD phenotype of the Nancy strain of coxsackievirus B3. Virus Res. 24:187–196

Melnick JL (1996) Enteroviruses: polioviruses, coxsackieviruses, echoviruses and newer enteroviruses. In: Fields BN, Knipe DM, Howley PM (eds), Virology (3rd Edn) Lippincott-Raven Press, Philadelphia

Minor PD, Ferguson M, Evans DM, Almond JW, Iconogle JP (1986) Antigenic structure of polioviruses of serotypes 1, 2, and 3. J Gen Virol 67:1283–1291

Mirzayan C, Wimmer E (1992) Genetic analysis of an NTP-binding motif in poliovirus polypeptide 2C. Virology 189:547–555

Mirzayan C, Wimmer E (1994) Biochemical studies on poliovirus polypeptide 2C: Evidence for ATPase Activity. Virology 199:176–187

Muckelbauer JK, Kremer M, Minor I, Diana G, Dutko FJ, Groarke J, Pevear DC, Rossmann MG (1995) The structure of coxsackievirus B3 at 3.5 Å resolution. Structure 3:653–667

Olive DM, Al-Mufti S, Al-Mulla W, Khan MA, Pasca A, Stanway G, Al-Nakib W (1990) Detection and differentiation of picornaviruses in clinical samples following genomic amplification. J Gen Virol 71:2141–2147

Page GS, Mosser AG, Hogle JM, Filman DJ, Rueckert RR, Chow M (1988) Three-dimensional structure of poliovirus serotype 1 neutralizing determinants. J Virol 63:1781–1794

Palmenberg AC (1989) Sequence alignments of picornaviral capsid proteins. In: Semler BL, Ehrenfeld E (eds) Molecular aspects of picornavirus infection and detection. American Society for Microbiology, Washington DC, pp 211–241

Perez-Bercoff R, Gander M (1977) The genomic RNA of mengovirus. I. location of the poly(C) tract. Virology 80:426–429

Reagan KJ, Goldberg G, Crowell RL (1984) Altered receptor specificity of coxsackievirus B3 after growth in rhabdomyosarcoma cells. J Virol 49:635–640

Reimann BY, Zell R, Kandolf R (1991) Mapping of a neutralizing antigenic site of coxsackievirus B4 by construction of an antigen chimera. J Virol 64:3475–3480

Rico-Hesse R, Pallansch M, Nottay B, Kew O (1987) Geographic distribution of wild poliovirus type 1 genotypes. Virology 160:311–322

Romero JR, Tracy S, Chapman N, Gauntt C (1994) Genetic variation in coxsackievirus B3 genomes. 13th Annual American Society for Virology Scientific Program and Abstracts, pp 47–11

Romero JR, Rotbart HA (1995a) Sequence analysis of the downstream 5' nontranslated region of seven echoviruses with different neurovirulence phenotypes. J Virol 69:1370–1375

Romero JR, Leser JS, Tracy S, Chapman N, Winters D (1995b) Amino acid sequence variation in the major surface loop (VP2 E-F) of the cell receptor site from 8 clinical isolates of coxsackievirus B3. Pediatric Research 37:187A

Romero JR, Tracy SM, Pong, AL, Hinrich SH, Leser JS, Dunn JJ (1996) Genetic diversity and molecular epidemiology among coxsackievirus B2 (CVB2) during a community outbreak. Pediatric Research 39:184A

Rossman MG (1989) The canyon hypothesis. J Biol Chem 264:14587–14590

Rossman MG, Palmenberg AC (1988) Conservation of the putative receptor attachment site in picornaviruses. Virology 164:373–382

Rowlands DJ, Harris TJR, Brown F (1978) Moore precise location of the polycytidylic acid tract in foot and mouth disease virus RNA. J Virol 26:335–343

Rueckert RR, Wimmer E (1984) Systematic nomenclature of picornaviral proteins. J Virol 50:937–959

Rueckert RR (1996) Picornaviridae: The viruses and their replication. In: Fields BN, Knipe DM, Howley

PM (eds), Virology (3rd edn) Lippincott-Raven, Philadelphia, pp 609–654

Sherry B, Mosser AG, Colonno RJ, Reuckert RR (1986) Use of monoclonal antibodies to identify four neutralization immunogens on a common cold picornaviruses of serotypes 1, 2, and 3. J Virol 57:246–257

Sherry B, Rueckert RR (1985) Evidence for at least two dominant neutralization antigens on human rhinovirus 14. J Virol 53:137–143

Titchener PA, Jenkins O, Szopa TM, Taylor KW, Almond JW (1994) Complete nucleotide sequence of a beta-cell tropic variant of coxsackievirus B4. J Med Virol 42:369–373

Toyoda H, Kohara M, Kataoka Y, Suganuma T, Omata T, Imura N, Nomoto A (1984) Complete nucleotide sequences of all three poliovirus serotype genomes: implication for genetic relationship, gene function and antigenic determinants. J Mol Biol 174:561–585

Tracy S, Chapman NM, Tu Z (1992) Coxsackievirus B3 from an infectious cDNA copy of the genome is cardiovirulent in mice. Arch Virol 122:399–409

Zhang G, Wilsden G, Knowles NJ, McCauley JW (1993) Complete nucleotide sequence of a coxsackie B5 virus and its relationship to swine vesicular disease virus. J Gen Virol 74:845–853

Coxsackievirus B5 and the Relationship to Swine Vesicular Disease Virus

N. J. KNOWLES[1] AND J. W. MCCAULEY[2]

1 Introduction

Coxsackieviruses are divided into two groups (A and B) based largely on their pathogenicity and replication in newborn mice. Both groups have been associated with a wide variety of neurological, cardiovascular, gastroenteric, ophthalmic, and exanthematic diseases. The main natural hosts of coxsackieviruses are humans but several studies have established that coxsackievirus B5 (CVB5) and swine vesicular disease virus (SVDV) are antigenically closely related.

Coxsackievirus B5 was first isolated in 1952 from the stools of a patient with mild paralytic disease (Faulkner strain). Swine vesicular disease (SVD) is a highly contagious disease of pigs which was first recognized in Italy during 1966 (NARDELLI et al. 1968) and subsequent outbreaks have been reported in Hong Kong, Japan, and Europe between 1970 and 1996. In 1968, the causative agent, SVDV, was characterized as an enterovirus in the family Picornaviridae (NARDELLI et al. 1968). No antigenic relationships were found between SVDV and the eight porcine

[1]Institute for Animal Health, Pirbright Laboratory, Ash Road, Pirbright, Woking, Surrey, GU24 0NF, UK

[2]Institute for Animal Health, Compton Laboratory, Compton, Newbury, Berkshire, RG20 7NN, UK

enterovirus (PEV) serotypes which were recognized at that time and it was tentatively designated serotype 9 (DUNNE 1975). However, GRAVES (1973) had already shown that SVDV and the human enterovirus CVB5 were antigenically related. This was confirmed by others (BROWN et al. 1973), and it has been proposed that SVDV should be considered a porcine strain of CVB5 (KNOWLES et al. 1979).

Swine vesicular disease is not fatal and usually results in only minor delays to fattening schedules; however, it is clinically similar to other more serious vesicular diseases of pigs: foot and mouth disease (FMD), vesicular stomatitis and vesicular exanthema of swine. Some pigs develop a serological response to SVDV but show no clinical signs (NARDELLI et al. 1968; BURROWS et al. 1973). This subclinical infection can be produced by exposure of pigs to small amounts of virus and there is evidence to suggest that SVD can spread in a subclinical form (MANN et al. 1975).

The only authenticated cases of human infections with SVDV have been found in laboratory workers (BROWN et al. 1976a; MCKERCHER and GRAVES 1981; LOXAM and HEDGER 1983; SELLERS 1984). In a survey of veterinary officers engaged in the control of disease in the field, no evidence of clinical or subclinical disease was found (WATSON 1981). The incubation period in the cases of human infections varied from 1 to 2 weeks or possibly 5 weeks. The pathogenesis is not known. People who come into contact with infected pigs may inhale virus aerosols emitted by these animals (SELLERS and HERNIMAN 1974) and similarly, laboratory workers may breath in virus aerosols generated from virus preparations of high titer. Virus may also enter through cuts and abrasions. In some individuals virus was isolated from nasal washings, but no serum antibodies were detected. Others developed a serological response to SVDV but did not become ill. In a third group illness occurred and serum antibody was present. It was presumed in the second and third groups that virus entered the body and multiplied in the organs and tissues (BROWN et al. 1976a; MCKERCHER and GRAVES 1981; SELLERS 1984).

The clinical symptoms in humans ranged from a mild influenza-like illness to a severe illness with fever, generalized muscle pain, weakness, lethargy, stomach pain and diarrhea. There was one case of aseptic meningitis. Vesicular lesions were not seen, unlike in FMD and hand, foot, and mouth disease in humans. Loss of weight occurred in some patients, but all recovered without sequelae (SELLERS 1984). Workers who became ill with SVD virus infection developed high levels of SVD virus neutralizing antibodies several weeks after their illnesses (BROWN et al. 1976a).

The serological relationship between the porcine pathogen SVDV and the human pathogen CVB5 raised the question of the possibility of a zoonosis. GARLAND and MANN (1974) attempted to infect pigs with CVB5; the results showed that the antibody titers attained were proportional to the dose of virus administered and the neutralizing antibody titer was very similar with both CVB5 and SVDVs. Prior exposure to CVB5 sensitized both inoculated and in-contact pigs so that subsequent exposure to SVDV resulted in a typical anamnestic response. Serum antibody was found to both viruses after cross-challenge but the response to SVDV was higher than that to CVB5. Furthermore, CVB5 did not produce clinical disease in pigs; however, CNS lesions were observed microscopically in infected animals which were similar to lesions seen in SVDV infected pigs (GARLAND and MANN 1974; MONLUX et al. 1975; KODAMA 1976; LAI et al. 1980).

The host range of CVB5 and SVDV appear to be limited to humans, pigs, and infant mice (NARDELLI et al. 1968; BROWN et al. 1973). It has been suggested that CVB5 gave rise to the agent of SVDV (Brown et al. 1973; Graves 1973) or SVDV arose from a recombination of human and animal enteroviruses. Experimental infection of pigs with CVB5 has been reported only for five virus isolates, the prototype Faulkner strain (GARLAND and MANN 1974; MONLUX et al. 1975; KODAMA 1976) and four isolates from New York in 1972 (LAI et al. 1980). However, these studies are not sufficient to prove that CVB5 cannot produce clinical lesions in pigs.

2 Serological Reactivity Between CVB5 and Swine Vesicular Disease Virus

Early work showed that SVDV can be neutralized by antisera directed against CVB5 and vice versa, although the two viruses are clearly not identical since in some reciprocal cross-neutralization tests the viruses are distinguishable and they exhibit differences in immunodiffusion tests. The virus neutralization tests which originally demonstrated that SVDV was antigenically related to CVB5 utilized enterovirus horse serum polls from the National Institute for Allergy and Infectious Disease (prepared by the Lim Benyesh-Melnick scheme; SCHMIDT et al. 1971) in which two pools neutralized SVDV both in tissue culture and in new-born mice, and serum from a pig 56 days convalescent following inoculation with SVDV (Hong Kong/36/71) neutralized CVB5 (Faulkner) replication in tissue culture (GRAVES 1973). Immunodiffusion tests carried out at the same time (BROWN et al. 1973) showed that CVB5 (Faulkner) could be differentiated from all the SVDVs examined at that time: a line of precipitin common to CVB5 and SVDV was seen but a specific spur was obtained for each virus and its homologous antiserum. In the former virus neutralization tests binding of antibody to antigen has a functional consequence to the biology of the virus since reactivity between neutralizing epitopes and neutralizing antibody only is detected, whilst in immunoprecipitation tests there is no necessity for such a functional interaction between antigen and antibody. In both cases analysis of cross-reactivity is focused on the capsid: firstly, because that is the location of the epitopes associated with neutralization by antibody, and secondly, purified virus preparations were used in the immunoprecipitation experiments. In neither case would antibody against nonstructural virus polypeptides be detected.

3 Antigenic Sites of Swine Vesicular Disease Virus Recognized by Mouse Monoclonal Antibodies

Recently mouse monoclonal antibodies (Mabs) have been used to localize antigenic sites within the capsid of SVDV. KANNO et al. (1995) used eight neutralizing antibodies directed against the H3'76 strain of SVDV and selected mutant viruses capable of overcoming the neutralizing effects of individual Mabs. Nucleotide sequence analysis showed amino acid substitutions in these resistant viruses in VP1, VP2 and VP3.

Located within VP1, residues 87 and 88 were associated with resistance to one of the Mabs. This region is equivalent to site 1 of poliovirus (PV) and NIm-I of rhinoviruses (SHERRY et al. 1986; PAGE et al. 1988). Also within VP1, substitutions at residues 272 and 275 confer resistance to neutralization and are located at a site equivalent to PV site 3. Poliovirus site 3 of SVDV is also associated with changes in VP2 (residues 70 and 233) and in VP3 (residues 60, 73, and 76). Viruses with amino acid substitutions in the site equivalent to antigenic site 2 of PV have been found in SVDV at amino acid residues 154 and 163 of VP2.

These results indicate that the immune system of the mouse recognizes SVDV in a manner similar to its recognition of PV and defines similar antigenic sites. The sites are all located on loops between β-strands or to NH_2- or COOH- terminal

Fig. 1. Location of amino acid residues involved in SVDV (strain H3'76) antigenic sites as mapped onto the 3-D structure of coxsackievirus B3 (MUCKELBAUER et al. 1995). The three major capsid polypeptide are shown in cyan (VP1), green (VP2) and red (VP3); residues involved in N-MAb resistance are shown in blue (VP1), yellow (VP2) and orange (VP3). Distinct antigenic sites are labeled 1–3 (KANNO et al. 1995). The figure was created using the program RasMol version 2.6 (SAYLE and MILNER-WHITE 1995)

VP2

```
                                βA2
                   ┌βA1┐     ┌──┐                                ┌αZ┐
CVB5      SPSAEECGYSDRVRSITLGNSTITTQECANVVVGYGTWPTYLKDEEATAEDQPTQPDVAT  60
SVDV-UK   ------------------------------------V-----------------------  60
SVDV-Jap  ------------------------------------V-----------------------  60
CVB1      ------------------------------------V--E----N---G-----------  60
CVB3      --TV--------A-----------------------V--D----S---------------  60
CVB4      --TV--------------------------------V--D--S-----------------  60
CVA9      --TV--------------------------------R-----R-D---------------  60
PV1       --NI-A------LQL------------A--S--A--R--E--R-S--NPV----E----A  60

                          βC
           ┌─ βB─┐        ┌─┐          ┌──αA──┐┌─────βD─────┐        ┌─
CVB5      CRFYTLESVMWQQSSPGWWWKFPDALSNMGLFGQNMQYHYLGRAGYTVHVQCNASKFHQG  120
SVDV-UK   -----------------------------------------------I------------  120
SVDV-Jap  -----------------------------------------------I------------  120
CVB1      ---------Q-MKN-A-----L-----Q---------------T---I------------  120
CVB3      ------D--Q--KT-------L------L--------------T----------------  120
CVB4      ------N--K-EMQ-A-----------E---------------S---I------------  120
CVA9      ------D-IK-EKG-V-------E---D-------------------I------------  120
PV1       ------DT-S-TKE-R-----L----RD--------Y-----S-----------------  120

          ──βE──┐┌──────────────────puff──────────────────────────
CVB5      CLLVVCVPEAEMGCATLANKP::DQKSLSNGETANTFDSQ::NTTGQTAVQANVINAGM:  175
SVDV-UK   ---------------------::-P----K--I--M-E--::-S--E------------:  175
SVDV-Jap  ---------------------::-P----K--I--M-E--::-S--E------------:  175
CVB1      --------------SN-N-T-::KFAE--G-DN-RM-TDTEVG-SNDKK--TA-W----:  177
CVB3      -----------------D-T-::SSAE-LG-D--KE-ADKPVASGSNKL--RV-Y----:  177
CVB4      --------------TNAE-A-::AYGD-CG----KS-EQN::AA--K----TA-C----:  175
CVA9      -------------G-VVGQAF::SATAMA--DK-YE-T-A::TQSD--K--TAIH----:  175
PV1       A-G-FA---MCLAGDSNTTTMHTSYQNANP--KGG--TGTFTPDNN--SPARRFCPVDYL  180

          ─────┐┌─αB─┐  ┌─βF┐      ┌─βG1┐      ┌βG2┐    ┌────βH────┐
CVB5      :::GVGVGNLTIFPHQWINLRTNNSATIVMPYINSVPMDNMFRHNNFTLMIIPFAPLSYS  232
SVDV-UK   :::-----------------------------------------------V---------  232
SVDV-Jap  :::-----------------------------------------------V---------  232
CVB1      :::--------------------------------------Y----L-------V--N--  234
CVB3      :::-----------------------------T-------------V---V---V--D-C  234
CVB4      :::---------Y--------------------------------------------D-V  232
CVA9      :::---------Y-------------------------------Y-----V---VK-D-A  232
PV1       LGN-TLL--AFV----I-------C--L-L--V--LSI-S-VK---WGIA-L-L---NFA  240

                ┌──────βI──────┐
CVB5      TGATTYVPITVTVAPMCAEYNGLRLAGKQ     261
SVDV-UK   -----------------------------     261
SVDV-Jap  -----------------------------     261
CVB1      E-SSP-------I-------------SS-     263
CVB3      P-S---------I--------------H-     263
CVB4      ---SS-I---------S----------H-     261
CVA9      DT-S----------------------QA-     261
PV1       SESSPEI---L-I----C-F----NITLPRLQ  272
```

Fig. 2a–c. Alignment of the three outer capsid polypeptides of CVB5/1954/UK/85 (Zhang et al. 1993a), SVDV/UKG/27/72 (SVDV-UK; Seechurn et al. 1990), SVDV/H3'76 (SVDV-Jap; Inoue et al. 1989), CVB1/Japan (Iizuka et al. 1987), CVB3/Nancy/Conn/47 (Klump et al. 1990), CVB4/JVB/NY/51 (Jenkins et al. 1987), CVA9/Griggs (Chang et al. 1989) and PV1/Mahoney/Ohio/41 (Kitamura et al. 1981; Racaniello and Baltimore 1981). Colons represent pad characters introduced to produce optimal alignments and dashes indicate that the amino acid is the same as that of CVB5. a VP2; b VP3; c VP1. The approximate positions of selected structural details deduced from the 3-D structure of CVB3 (Muckelbauer et al. 1995) are indicated above the alignments

```
VP3

                                                    ┌─αZ─┐             ┌──
CVB5          GLPTMLTPGSNQFLTSDDFQSPSAMPQFDVTPEMAIPGQVNNLMEIAEVDSVVPVNNTEG  60
SVDV-UK       ----LS----------------------------D-------------------------  60
SVDV-Jap      ----LS----------------------------D-------------------------  60
CVB1          ---V-T----T-----------------------Q---R-------------------DN  60
CVB3          -----N----C----------------Y------R---E-K--------------Q-VGE  60
CVB4          ----------T-----------------------N-----R-------------I--LKA  60
CVA9          -----N----T-----------C-L-------S-N---E-K----------------VQD  60
PV1           ---V-N------Y--A-N----C-L-E-----PID---E-K-M--L--I-TMI-FDLSAT  60

              ─Knob──┐┌βB┐        ┌─βC─┐            ┌─αA─┐  ┌─────βD─────┐
CVB5          KVSSIEAYQIPVQSNSTNGSQVFGFPLIPGASSVLNRTLLGEILNYYTHWSGSIKLTFMF  120
SVDV-UK       --M------------P-----------T---N---------------A------------  120
SVDV-Jap      --M------------P-----------T---N---------------A------------  120
CVB1          N-NGLK----------D-RR-------Q---NN---------------------------  120
CVB3          --N-M-------R--EGS-T-------Q--Y---FS------------------------  120
CVB4          NLMTM---RVQ-R-TDEM-G-I-----Q-------Q-----------------L----V-  120
CVA9          TTDQM-MFR---TI-APLQQ----LR-Q--LD--FKH----------A-----M----V-  120
PV1           -KNTM-M-RVRLSDKPHTDDPILCLS-S-ASDPR-SH-M-----------A--L-F--L-  120

                      ┌─βE1─┐         ┌αB┐   ┌─βF─┐     ┌βG1─┐       ┌βG2┐
CVB5          CGSAMATGKFLLAYSPPGAGAPTTRKEAMLGTHVIWDVGLQSSCVLCIPWISQTHYRYVV  180
SVDV-UK       ------------------------------------------------------------  180
SVDV-Jap      ------------------------------------------------------------  180
CVB1          --------------------V-KN-RD--------------------V------------  180
CVB3          -----------------------K-VD------------------------------F-A  180
CVB4          ----------------------DS--N--------------------V------------  180
CVA9          -----------I-------NP-K---D------I---I---------V---------L-Q  180
PV1           --FM-----L-VS-A----DP-KK-------------I------TMVV----N-T--QTI  180

                      ┌──βH──┐          ┌───βI───┐
CVB5          VDEYTAGGYITCWYQTNIVVPADTQSDCKILCFVSACNDFSVRMLKDTPFIKQDSFYQ  238
SVDV-UK       M----------------------A------------------------------N-F-  238
SVDV-Jap      M----------------------A---------A--------------------N-F-  238
CVB1          E-----A--V--------I----V--T-D----------------------R--N---  238
CVB3          S-------F--------------A--S-Y-M------------L-------S-QN-F-  238
CVB4          D-K---S-F-S------VI---EA-KS-Y-M--------------R--Q----TN---  238
CVA9          Q----SA--V------GMI--PG-PNSSS-M--A-----------R-----S--NKL-  238
PV1           D-SF-E----SVF---R----LS-PREMD--G-----------L-R--TH-E-KALA-  238
```

```
VP1

CVB5          ::GPPGEAVERAIAR::VADTISS:::::::::GPVNSESIPALTAAETGHTSQVVPADT  47
SVDV-UK       ::-----VMG-----::-----G-:::::::::------------------------S--  47
SVDV-Jap      ::-----V-------::-----G-:::::::::------------------------S--  47
CVB1          ::--VE-S----MV-::----V--:::::::::K-T---------------------S--  47
CVB3          ::--VED-ITA--G-::----VGT:::::::::--T---A-----------------G--  47
CVB4          ::--TE-S----MG-::-----AR:::::::::--S---Q------V--------D-S--  47
CVA9          ::-DVE--I---VVH::----MRT:::::::::--S--A-V-----V--------T-S--  47
PV1           GL-QML-SMIDNTV-ET-GAAT-RDALPNTEAS--TH-KE------V---A-NPL--S--  60

                              ┌αZ─┐   ┌────βB────┐      ┌─βC──┐       ┌───
CVB5          MQTRHVKNYHSRSESTVENFLCRSACVYYTTYK:NHGTDG:DN::FAYWVINTRQVAQL: 102
SVDV-UK       ---------------------------F-----:--DS--:--::--------------: 102
SVDV-Jap      ---------------------------F-----:--DS--:--::--------------: 102
CVB1          ---------------SI------------A--:::NNNSE:KG::Y-E-----------L 101
CVB3          ----------------I-----------F-E-:::KNSGA:KR::Y-E--LTP--A---: 100
CVB4          ------H--------SI----------I-IK-S:SAESNNLKR::Y-E-----------: 103
CVA9          ---------------------G------MEE--:TTDKHVNKK::-VA-P---K-MV-M: 103
PV1           V-----VQHR-----SI-S-FA-G---TIM-V:D-PASTTNKDKL--V-K-TYKDTV--: 118

              ─αA─┐ ┌───────βD───────┐             ┌─βE1─┐            ┌αB┐
CVB5          RRKLEMFTYARFDLELTFVITSTQEQSTIQGQDSPVLTHQIMYVPPGGPVPTKINSYSWQ 162
SVDV-UK       --------------------------P-VR---A-------------------V------ 162
SVDV-Jap      --------------------------P-VR---T-------------------V------ 162
CVB1          -----:---L------------A--P--ATSV-A--Q-Q--------------VTD-A-- 160
CVB3          -----F---V--------------QP--T-N--AQI---------------D-VD--V-- 160
CVB4          ---M-----I-C-M--------H--M--ATNS-V--Q---------------SV-D-V-- 163
CVA9          ---------L---M-V------R-DPG-TLA--M----R----------I-A-VDD-A-- 163
PV1           -----F---S---M-----V-A::NFTETNNGHALNQVY--------A---E-WDD-T-- 176

                    ┌βF┐     ┌βG1┐       ┌βG2┐
CVB5          TSTNPSVFWTEGSAPPRISIPFISIGNAYSMFYDGWAKFD:KQ:::::G::TYGINTLNN 214
SVDV-UK       -----------------M-----G-------------R--:--:::::-::----S---- 214
SVDV-Jap      -----------------M-----G-------------R--:--:::::-::---TS---- 214
CVB1          ------------N----M------------C-----TQ-S:RN:::::-::V-------- 212
CVB3          ------------N----M----L-------N-----SE-S:RN:::::-::V-------- 212
CVB4          ------I-----N----M----M------T------SN-S:RD:::::-::I--Y-S--- 215
CVA9          ------I-----N--A-M------------N-----SN--QR::::::-::S--Y----- 215
PV1           --S---I-Y-Y-T--A---V-YVG-S----H----FS-VPL-DQSAAL-DSL--AAS--D 236

                ┌─βH─┐        ┌───────βI────────┐
CVB5          MGTLYMRHVNDGSPGPIVSTVRIYFKPKHVKTWVPRPPRLCQYQKAGNVNFEPTGVTESR 274
SVDV-UK       ------------G---------------------------------------------G- 274
SVDV-Jap      ------------G---------------------------------------------G- 274
CVB1          ----------EAGQ---K-------------A-----------E-QK----N-----TT- 272
CVB3          -----A----A--T---K--I----------A-I---------E--K----Q-S---TT- 272
CVB4          ---I-A-----S---GLT--I----------AY----------K--KS---DVEA--AE- 275
CVA9          L-HI-V---SGS--H--T--I-V------TRA-----------K--FS-D-T--PI-DT- 275
PV1           F-I-AV-V---HN-TKVT-KI-V-L----IRV-C-----AVA-:YGPG-DYKDGTL-PLS 295

CVB5          TE:::ITAMQTT                 283
SVDV-UK       -D:::--T-K--                 283
SVDV-Jap      -D:::--T-K--                 283
CVB1          SN:::--T:::-                 278
CVB3          QS:::--T-TN-                 281
CVB4          ASL::--T                     281
CVA9          KDINTV-TVAQSRRRGDMSTLNTH     299
PV1           -K:D:L-TY                    302
```

extensions. SVDV site 1 (site 1 of PV, site NIm-IA of rhinovirus) is the BC loop of VP1 and is located near the fivefold axis of the virion. Site 2 can be subdivided into site 2a and site 2b which are located on the EF loop of VP2 and are located approximately in the middle of the protomer. Site 3 has also been subdivided: site 3a is identical to site 3a of PV and comprises the COOH- terminal of VP1 and the NH_2-terminal of VP3; site 3b is similarly made up of components of two polypeptide chains, the BC loop of VP2, the HI loop of VP2 and the BC loop of VP3. Figure 1 shows the location of these amino acid residues mapped onto the three-dimensional structure of coxsackievirus B3 (CVB3; MUCKELBAUER et al. 1995) and all the amino acid substitutions can be seen to be located on the surface of the virus capsid. It is likely that the structure of the CVB3 virion is very similar in overall topology to that of CVB5 and SVDV. To align the capsid polypeptides of CVB3 and CVB5/SVDV only two insertions are required; an insertion of two amino acid residues in the BC loop of VP1 of CVB5/SVDV and an insertion of two amino acids in the VP2 puff of CVB3 (Fig. 2). These differences seem unlikely to affect the capsid structure outside the direct vicinity of the insertion/deletion sites, but this cannot be proved until the structure of either CVB5 or SVDV is determined.

4 Comparison of the CVB5 and Swine Vesicular Disease Virus Capsid Structures

There are a total of 43 amino acid substitutions in the three outer capsid polypeptides when CVB5/1954/UK/85 (ZHANG et al. 1993a) and SVDV/UKG/27/72 (SEECHURN et al. 1990) are compared, with no insertions or deletions; 22 occur in VP1, 9 in VP2 and 12 in VP3 (Fig. 2). The two amino acid insertion in CVB5/SVDV relative to CVB3 is invariant in the former viruses. The amino acid differences between CVB5 and SVDV can be visualized by mapping them onto the three-dimensional structure of the capsid of CVB3 (MUCKELBAUER et al. 1995; Fig. 3). Less than half of the amino acid substitutions are exposed on the surface of the virion (Fig. 3b), although several lie in the same areas as antigenic sites (compare Fig. 3b with Fig. 1). Changes in the antigenic sites may be the result of immune selection.

A cellular receptor used by CVB1, CVB3, CVB5 and a number of echoviruses (EV) has been shown recently to be the cell surface marker CD55 (decay-accelerating factor; DAF) by a number of groups (BERGELSON et al. 1994; WARD et al. 1994; BERGELSON et al. 1995; SHAFREN et al. 1995). It is not known whether SVDV also uses CD55 as a receptor. Differences in the capsid proteins between CVB5 and SVDV may have resulted from adaptation of SVDV to receptor molecules expressed on the surface of pig cells.

Fig. 3a, b. Differences between CVB5/1954/UK/85 (ZHANG et al. 1993a) and SVDV/UKG/27/72 (SEECHURN et al. 1990) mapped onto the 3-D structure of CVB3 (MUCKELBAUER et al. 1995). The three major capsid polypeptide are shown in cyan (VP1), green (VP2) and red (VP3); the differences are shown in blue (VP1), yellow (VP2) and orange (VP3); insertions/deletions are shown in white; a ribbon diagram with differences shown in spacefill; b spacefill diagram

5 SDS-Polyacrylamide Gel Electrophoresis of CVB5/SVDV Structural Polypeptides

In the past SDS-polyacrylamide gel electrophoresis (PAGE) has been used to distinguish different isolates of CVB5 and SVDV (HARRIS and BROWN 1975; HARRIS et al. 1979). In the case of SVDV the migration of polypeptides VP1, VP3, and VP4 was similar in most of the isolates, but variation was observed in the migration of VP2. The polypeptide patterns obtained for four CVB5s analyzed by HARRIS and BROWN (1975) were clearly different from those of several SVDV isolates. The polypeptide migration patterns of two CVB5 isolates from Birmingham, UK in 1973 (8068, 9954) and one from Hong Kong (HKB5) were identical but differed from that obtained for the Faulkner/USA/52 strain. The FRA/1/73 SVD virus, isolated in the Bordeaux area in January 1973, had a polypeptide pattern more like that of the recent CVB5 isolates; however, another SVD virus, 2862, isolated from Annecy in the Haute Savoie area of eastern France at about the same time, gave a polypeptide pattern indistinguishable from the SVD viruses isolated in England, Austria and Italy during the winter of 1972–1973.

Recent studies suggest that SVD viruses belonging to different polypeptide pattern groups also belong to different genetic sublineages of SVDV; however, these sublineages are not more closely related to CVB5 (Zhang G, Knowles NJ, Wilsden G, McCauley JW, unpublished data).

6 RNA Hybridization and RNase T_1 Oligonucleotide Mapping

RNA-RNA hybridization assays have been used to compare isolates of CVB5 and SVDV (BROWN and WILD 1974; BROWN et al. 1976b; HARRIS et al. 1977). The RNAs of four SVDV isolates showed a high degree of similarity, although small differences were apparent. Greater differences were observed between SVDV and isolates of CVB5; however, the Faulkner strain of CVB5 was reported to be more closely related to SVDV than to two CVB5 isolates from 1973.

Rnase T_1 mapping showed that SVDV/UKG/27/72 RNA shares more oligonucleotides with SVDV/HKN/36/71 RNA than with the earlier SVDV/ITL/1/66 RNA. The conservation of several long oligonucleotides by the SVDV RNAs suggested that there were continuous regions of common sequence (HARRIS et al. 1977). The RNA fingerprints also showed that CVB5/Faulkner RNA and CVB5/8068/73 RNA were as different from each other as they were from the SVDV/UKG/27/72 RNA (HARRIS et al. 1977), supporting the hybridization analyses.

Thus these early data suggested that considerable variation existed within the CVB5 species, but SVDV was much more homogeneous.

7 Genetic Relationships Between Swine Vesicular Disease Virus and Other Coxsackie B Viruses

A study of the nucleotide sequence of the VP1/2A gene junction of nine CVB5 viruses isolated from three distinct outbreaks in the United States during 1967, 1972, and 1982–1983 showed that virus isolates from the same outbreak were more closely related to each other (>92.8% nucleotide identity) than they were to viruses isolated during temporally unrelated outbreaks (84%–91.6%), despite their wide geographic distribution (KOPECKA et al. 1995). The CVB5 isolate 1954/UK/85 (ZHANG et al. 1993a), from the United Kingdom, was related to the 1982–1983 US epidemic strains while an isolate from Honduras in 1985 was distinct from all the other strains. The SVDV strain examined was more distantly related to the CVB5s than they were to each other.

Another recent study examined a panel of 54 SVDV isolates which included representative viruses from 1966 until the present day and from Europe and Asia. Two regions of the genome were analyzed (the regions encoding VP1 and 3BC) and both formed a single genetic phylum. All SVDVs differed in nucleotide sequence diversity by no more than 16% in VP1 and 12% in 3BC. It is therefore likely that there was a single common ancestor to all SVD viruses. A small number of CVB5s were also analyzed but in both regions of the genome much greater genetic diversity was observed (Zhang et al. 1993b; Zhang G, Haydon DT, Knowles NJ, McCauley JW, manuscript in preparation). The genetically diverse population of CVB5, as discussed above, has been proposed as a likely source of the ancestor of SVDV.

The first complete genome sequence of CVB5 that was determined gave support to the notion that CVB5 gave rise to SVDV since the regions of the genome encoding the capsid and the virus nonstructural proteins were colinear and the degree of genetic variation was similar throughout the genome (ZHANG et al. 1993a). An analysis of the genetic relationships between SVDV and CVB5 with other coxsackie B viruses, for which complete genome sequences were known, revealed no greater nucleotide sequence homology outside of the capsid region between SVDV and CVB5 than between SVDV and CVB1, CVB3 or CVB4. Similarly, CVB5 was as equally related to the other CVB as it was to SVDV. A striking result occurred when SVDV was compared to echoviruses. A partial nucleotide sequence of EV9 (Barty), which encoded 3B and 3C, showed much a much higher degree of homology to SVDV than to other CVB viruses and the homology was higher than that observed between SVDV and any of the CVB viruses. This lead us to speculate that EV9 viruses might have a close genetic relationship with SVDV and that SVDV might have arisen, at least in part, from an EV9-like parent.

The complete genome sequences of two EV9 viruses (Hill and Barty) have now been reported (Zimmermann et al. 1995; Zimmermann H, Eggers H, Nelsen-Salz B, unpublished data; EMBL/GenBank accession numbers X84981 and X92886). Analysis of the relationship between CVB5, SVDV and these two echo-

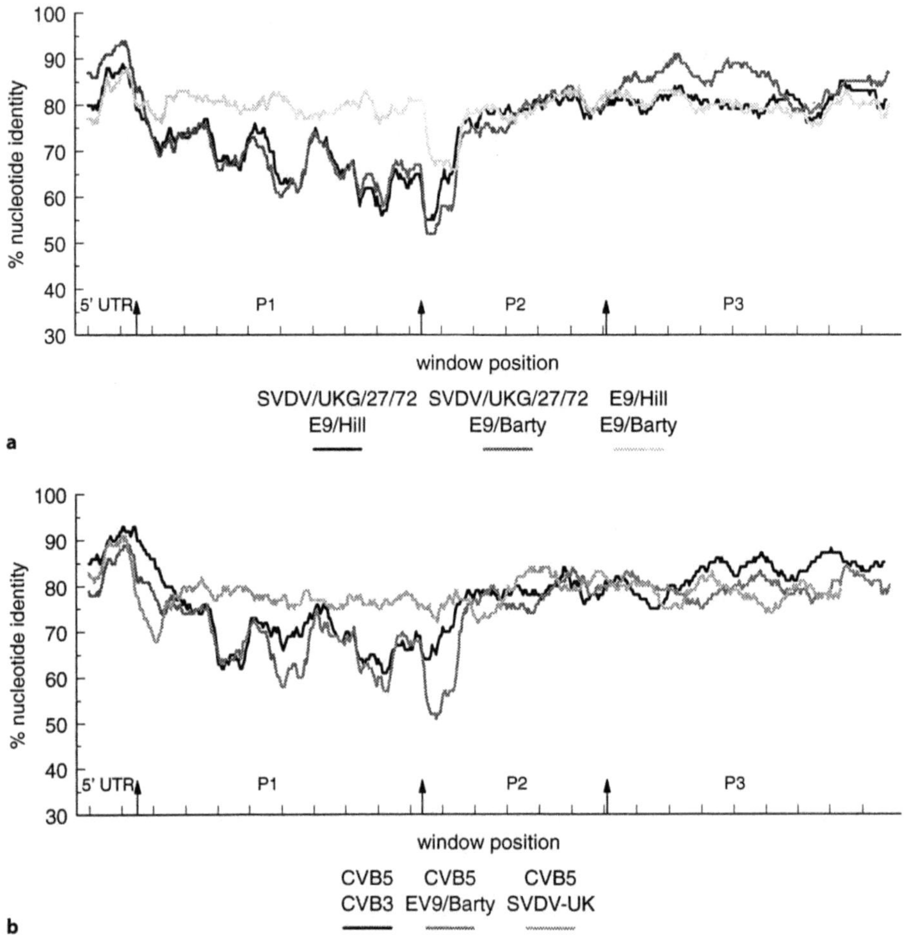

Fig. 4a, b. The genetic identity between CVB5/1954/UK/85 (Zhang et al. 1993a), SVDV/UKG/27/72 (Seechurn et al. 1990), CVB3/Nancy (Klump et al. 1990), EV9/Hill (Zimmermann et al. 1995) and EV9/Barty (EMBL/GenBank accession number X92886). The percentage nucleotide sequence identity measured over 300 bases between pairs of viruses is plotted on the Y-axis as the window of comparison was moved in 13 nucleotide steps. The divisions on the ordinate represent the passage by the comparison window of 300 nucleotides starting with positions 1–300 and terminating at approximately 7103–7402. The location of the coding-regions for the precursor polypeptides P1, P2 and P3 are also indicated. a The genome of SVDV/UKG/27/72 compared in turn against EV9/Hill and EV9/Barty, also EV9/Hill compared against EV9/Barty. b The genome of CVB5/1954/UK/85 compared in turn against optimally aligned SVDV/UKG/27/72, CVB3/Nancy; and EV9/Barty

viruses are shown in Fig. 4a, b. It is evident from Fig. 4a that SVDV is more closely related to EV9/Barty in the 3BC region of the genome than EV9/Barty is to EV9/Hill (or any other CVB/EV that have been sequenced) but the homology is restricted to this region. CVB5 does not share this relationship (Fig. 4b). The two EV9 virus strains show a similar genetic relatedness over the whole genome (except for an insertion at the COOH-terminal of VP1 in EV9/Barty) in a similar qualitative and quantitative fashion as do CVB5 and SVDV (Fig. 4b) and as does PV3/Leon and PV3/Finland (as described in ZHANG et al. 1993a). It is noteworthy, that all the CVB/EV show a similar genetic relationship outside of the capsid-coding genes (in the 5' noncoding region and in the P2 and P3 nonstructural protein coding region) which is exemplified in Fig. 4b by the comparison of CVB5 and CVB3. Recently, partial sequence analysis has shown that most echoviruses, with the notable exceptions of types 22 and 23, are genetically homogeneous with the CVB (HUTTENEN et al. 1996). The closer relationship of SVDV and EV9/Barty in the 3BC genome region may be an artifact of sequence conservation seen in the noncapsid coding regions when compared to the capsid-coding region and not be the result of genetic recombination.

8 Conclusions

CVB5 and SVDV are clearly members of the same virus species; however, all SVD viruses form a single genetic lineage within this species. The two viruses are nevertheless distinct entities since in nature they infect different host speciesm humans and pigs. We propose therefore that SVDV be considered a subspecies of CVB5 which infects pigs and belongs to a well defined genctic lineage of viruses.

References

Bergelson JM, Chan M, Solomon KR, St. John NF, Lin H, Finberg RW (1994) Decay-accelerating factor (CD55), a glycosylphosphatidylinositol-anchored complement regulatory protein, is a receptor for several echoviruses. Proc Natl Acad Sci USA 91:6245–6248

Bergelson JM, Mohanty JG, Crowell RL, St. John NF, Lublin DM, Finberg RW (1995) Coxsackievirus B3 adapted to growth in RD cells binds to decay-accelerating factor (CD55). J Virol 69:1903–1906

Brown F, Talbot P, Burrows R (1973) Antigenic differences between isolates of swine vesicular disease virus and their relationship to coxsackie B5 virus. Nature 245:315–316

Brown F, Wild F (1974) Variation in the coxsackievirus type B5 and its possible role in the etiology of swine vesicular disease. Intervirology 3:125–128

Brown F, Goodridge D, Burrows R (1976a) Infection of man by swine vesicular disease virus. J Comp Path 86:409–414

Brown F, Wild TF, Rowe LW, Underwood BO, Harris TJR (1976b) Comparison of swine vesicular disease virus and coxsackie B5 virus by serological and RNA hybridization methods. J Gen Virol 31:231–237

Burrows R, Greig A, Goodridge D (1973) Swine vesicular disease. Research in Veterinary Science 15:141–144

Chang KH, Auvinen P, Hyypiä T, Stanway G (1989) The nucleotide sequence of coxsackievirus A9; implications for receptor binding and enterovirus classification. J Gen Virol 70:3269–3280

Dunne HW (1975) Porcine enteroviruses. 4 edn. In: Dunne HW, Leman AD (eds), Diseases of Swine. Iowa State University Press, Ames, Iowa pp 353–368

Garland AJM, Mann JA (1974) Attempts to infect pigs with coxsackie virus type B5. J Hyg Cambs

73:85–96

Graves JH (1973) Serological relationship of swine vesicular disease virus and coxsackie B5 virus. Nature 245:314–315

Harris TJR, Brown F (1975) Correlation of polypeptide composition with antigenic variation in the swine vesicular disease and coxsackie B5 viruses. Nature 258:758–760

Harris TJR, Doel TR, Brown F (1977) Molecular aspects of the antigenic variation of swine vesicular disease and coxsackie B5 viruses. J Gen Virol 35:299–315

Harris TJR, Underwood BO, Knowles NJ, Crowther JR, Brown F (1979) Molecular approach to the epidemiology of swine vesicular disease: correlation of variation in the virus structural polypeptides with serological properties. Infect Immun 24:593–599

Huttenen P, Santti J, Pulli T, Hyypiä T (1996) The major echovirus group is genetically coherent and related to coxsackie B viruses. J Gen Virol 77:715–725

Iizuka N, Kuge S, Nomoto A (1987) Complete nucleotide sequence of the genome of coxsackievirus B1. Virology 156:64–73

Inoue T, Suzuki T, Sekiguchi K (1989) The complete nucleotide sequence of swine vesicular disease virus. J Gen Virol 70:919–934.

Jenkins O, Booth JD, Minor PD, Almond JW (1987) The complete nucleotide sequence of coxsackievirus B4 and its comparison to other members of the Picornaviridae. J Gen Virol 68:1835–1848

Kanno T, Inoue T, Wang YF, Sarai A, Yamaguchi S (1995) Identification of the location of antigenic sites of swine vesicular disease virus with neutralization-resistant mutants. J Gen Virol 76:3099–3106

Kitamura N, Semler BL, Rothberg PG, Larsen GR, Adler CJ, Dorner AJ, Emini EA, Hanecak R, Lee JJ, Van der Werf S, Anderson CW, Wimmer E (1981) Primary structure, gene organization and polypeptide expression of poliovirus RNA. Nature 291:547–553

Klump WM, Bergmann I, Mueller BC, Ameis D, Kandolf R (1990) Complete nucleotide sequence of infectious coxsackievirus B3 cDNA: two initial 5' uridine residues are regained during plus-strand RNA synthesis. J Virol 64:1573–1583

Knowles NJ, Buckley LS, Pereira HG (1979) Classification of porcine enteroviruses by antigenic analysis and cytopathic effects in tissue culture: description of 3 new serotypes. Arch Virol 62:201–208

Kodama M (1976) Outline of studies on swine vesicular disease in Japan. Bull Off Int Epiz 86:423–432

Kopecka H, Brown B, Pallansch M (1995) Genotypic variation in coxsackievirus B5 isolates from three different outbreaks in the United States. Virus Res 38:125–136

Lai SS, McKercher PD, Moore DM, Gillespie JH (1980) Response of pigs to recent isolates of coxsackievirus B5. Comp Immun Microbiol Infect Dis 2:459–468

Loxam JG, Hedger RS (1983) Swine vesicular disease: clinical signs, diagnosis, epidemiology and control. Rev sci tech Off Int Epiz 2:11–24

Mann JA, Burrows R, Goodridge D (1975) Mild and sub-clinical infections with swine vesicular disease virus. Bull Off Int Epiz 83:117–122

McKercher PB, Graves JH (1981) Swine Vesicular Disease. In: Beran GW (ed), Handbook of Zoonoses. Section B: Viral Zoonoses, Vol. 2. CRC Press, Boca Raton pp 161–167

Monlux WS, McKercher PD, Graves JH (1975) Brain and spinal cord lesions in pigs inoculated with swine vesicular disease (UKG strain) virus and coxsackievirus B5. Am J Vet Res 36:1745–1749

Muckelbauer JK, Kremer M, Minor I, Diana G, Dutko FJ, Groarke J, Pevear DC, Rossmann MG (1995) The structure of coxsackievirus B3 at 3.5 Angstrom resolution. Structure 3:653–667

Nardelli L, Lodetti E, Gualandi GL, Burrows R, Goodridge D, Brown F, Cartwright B (1968) A foot-and-mouth disease syndrome in pigs caused by an enterovirus. Nature 219:1275–1276

Page GS, Mosser AG, Hogle JM, Filman SJ, Rueckert RR, Chow M (1988) Three-dimensional structure of poliovirus serotype 1 neutralizing determinants. J Virol 62:1781–1794

Racaniellio VR, Baltimore D (1981) Molecular cloning of poliovirus cDNA and determination of the complete nucleotide sequence of the viral genome. Proc Natl Acad Sci USA 78:4887–4891

Sayle RA, Milner-White EJ (1995) RasMol: biomolecular graphics for all. Trends in Biochemical Sciences 20:374-376

Schmidt NJ, Melnick JL, Wenner HA, Ho HH, Burkhardt MA (1971) Evaluation of enterovirus immune horse serum pools for identification of virus field strains. Bull Wrld Hlth Org 45:317–330

Seechurn P, Knowles NJ, McCauley JW (1990) The complete nucleotide sequence of a pathogenic swine vesicular disease virus. Virus Res 16:255–274

Sellers RF. (1984). Vesicular Diseases. In: Brown F, Wilson G (eds), Topley and Wilson's Principles of Bacteriology, Virology and Immunity, Vol 4, Virology. Edward Arnold, London pp 213–232

Sellers RF, Herniman KAJ (1974) The airborne excretion by pigs of swine vesicular disease virus. J Hyg Cambs 72:61–65

Shafren DR, Bates RC, Agrez MV, Herd RL, Burns GF, Barry RD (1995) Coxsackievirus B1, B3, and B5 use decay-accelerating factor as a receptor for cell attachment. J Virol 69:3873–3877

Sherry B, Mosser AG, Colonno RJ, Rueckert RR (1986) Use of monoclonal antibodies to identify four neutralization immunogens on a common cold picornavirus, human rhinovirus 14. J Virol 57:246–257

Ward T, Pipkin PA, Clarkson NA, Stone DM, Minor PD, Almond JW (1994) Decay-accelerating factor CD55 is identified as the receptor for echovirus 7 using CELICS, a rapid immuno-focal cloning method. EMBO J 13:5070–5074

Watson WA (1981) Swine vesicular disease in Great Britain. Can Vet J 22:195–200

Zhang G, Wilsden G, Knowles NJ, McCauley JW (1993a) Complete nucleotide sequence of a coxsackie B5 virus and its relationship to swine vesicular disease virus. J Gen Virol 74:845–853

Zhang G, Wilsden G, Knowles NJ, McCauley JW (1993b) Nucleotide sequence diversity in a range of swine vesicular disease virus isolates and coxsackievirus B5 isolates. Abstract of a poster presented at the IXth International Congress of Virology, Glasgow, August 8–13, P28-24 p 221

Zimmermann H, Eggers HJ, Kraus W, Nelsen-Salz B (1995) Complete nucleotide sequence and biological properties of an infectious clone of prototype echovirus 9. Virus Res 39:311–320

Higher Order Structures of Coxsackievirus B 5' Nontranslated Region RNA

K. M. CURREY[1] AND B. A. SHAPIRO[2]

1 Introduction

A complete understanding of the function of RNA molecules requires a knowledge of their higher order structure (secondary and tertiary) as well as the characteristics of the sequence (primary structure). The two- and three-dimensional structure of RNA is important for many functions, including regulation of transcription and translation, catalysis, and transport of proteins across membranes. Molecules with the same function have the potential to fold into similar structures although they might differ in primary structure, a fact that helps to illustrate the importance of secondary and tertiary structure in relation to function. Examples of such constancy

[1]SIDS Institute, Department of Pediatrics, University of Maryland Medical System, 22 South Greene Street, Baltimore, Maryland, 21201; USA and Laboratory of Mathematical Biology, Division of Basic Sciences, National Cancer Institute, Frederick Cancer Research and Development Center, Frederick, Maryland, 21702, USA

[2]Image Processing Section, Laboratory of Mathematical Biology, Division of Basic Sciences, National Cancer Institute, Frederick Cancer Research and Development Center, Frederick, Maryland, 21702, USA

in secondary structure exist in tRNAs, 5sRNAs, 16sRNAs and viroid RNAs. Secondary and tertiary structure of $tRNA^{Phe}$ (KIM et al. 1974) and of a hammerhead ribozyme (PLEY et al. 1994) have been shown by their crystal structure. Currently little is known of tertiary interactions, but studies on tRNA indicate these are weaker than secondary structure interactions (RIESNER and ROMER 1973; CROTHERS and COLE 1978; JAEGER et al. 1990). It is however very difficult to crystallize and/or get nuclear magnetic resonance spectrum data for large RNA molecules. Therefore, a logical place to start in determining the 3-D structure of RNA is determination of the secondary structure.

In the enteroviruses, it has been shown that RNA secondary structure in the 5' nontranslated region (5' NTR) is important in determining neurovirulence in the polioviruses and a common higher order structure is implicated for transcription and in the internal ribosome entry site (IRES), also known as the ribosomal landing pad (RLP). More recently it was shown that a single nucleotide change was responsible for attenuating a cardiovirulent coxsackievirus B3 phenotype; RNA structure was implicated in this functional change (TU et al. 1995). Although coxsackievirus and poliovirus are both enteroviruses and share many viral mechanisms, the virulence "switch" in poliovirus is primarily related to translational efficiency whereas the mechanism in coxsackievirus may be related to transcriptional efficiency. This difference is also reflected in the effect on higher order structure of the RNA.

Currently accepted models of the RNA secondary structure for the 5' NTR in enteroviruses assume the same structure for all enteroviruses including coxsackievirus and poliovirus. In this model, the local region containing the nucleotide responsible for attenuation of virulence in CVB3 is devoid of structure. However, recent studies place secondary structure elements in this region and demonstrate alteration of these structures with mutation of this nucleotide (TU et al. 1995). Thus a secondary structure map for CVB3 that contains elements of the currently accepted structure yet has unique features is probable. The features unique to CVB3 occur in the region identified as determining cardiovirulence and affecting RNA transcriptional efficiency.

2 RNA Structure Prediction Concepts and Methods

The primary structure of an RNA molecule is relatively easy to determine. Because experimental methods for determining RNA secondary and tertiary structure (when the primary sequence folds back on itself and forms base pairs) have not kept pace with the rapid discovery of RNA molecules and their function, computer predictions of secondary and tertiary structures have increasingly been utilized and a methodology has been developed. Accurate prediction of RNA structure requires an understanding of basic interactions such as hydrogen bonding, stacking, and hydration in differing structural contexts.

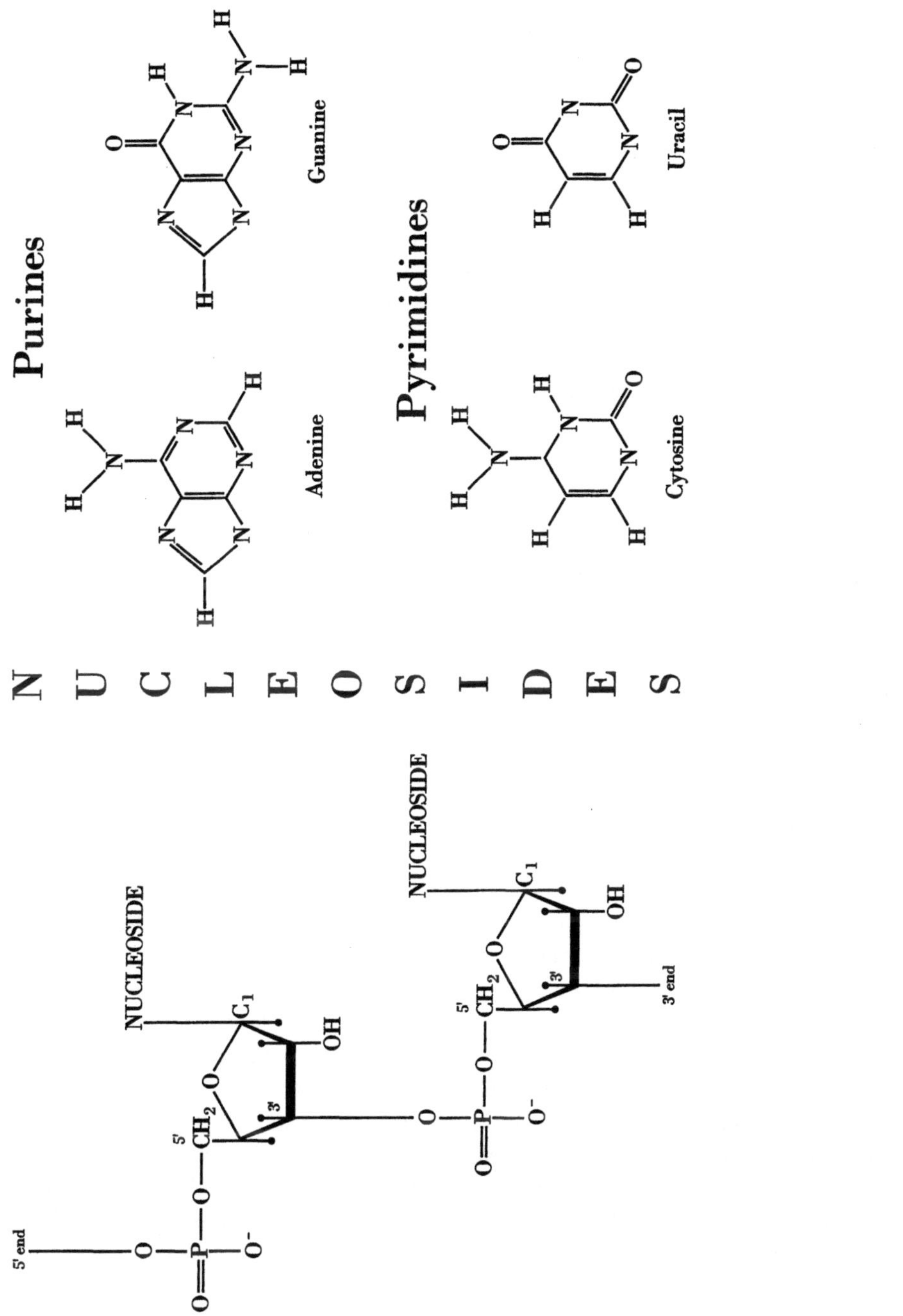

Fig. 1. The chemical structure for RNA with its phosphodiester linkages. Note that the nucleosides are connected to the ribose sugars at the anomeric carbon – C1. Depicted also are the chemical structures for the individual nucleosides

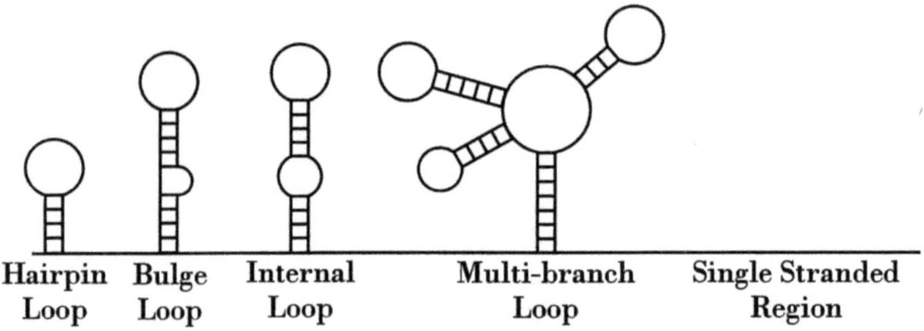

Fig. 2. Examples of the secondary structural elements. The cross-hatching indicates hydrogen bonds within a helix. Circular regions are loops which consist of unpaired nucleotides. The single-stranded region shown also consists of unpaired nucleotides

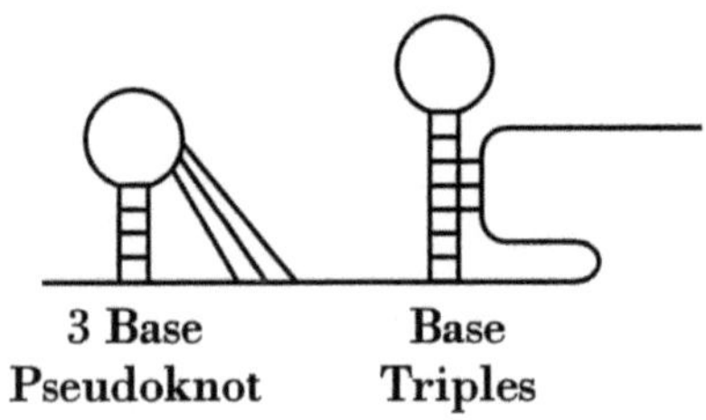

Fig. 3. Two examples of tertiary interactions. On the left is a typical pseudoknot and on the right are base triples

2.1 Fundamental Interactions

RNA molecules are polynucleotides containing ribose sugars connected by 3'-5' phosphodiester linkages. The bases are connected to the ribose sugars in the position of the anomeric carbon (C1) (Fig. 1). The glycosidic torsion angle specifies the rotation of the base relative to the sugar. Other torsion angles specify the folding of the sugar-phosphate backbone. An RNA chain can fold back upon itself to form hydrogen bonds between bases. Most commonly, Watson-Crick base-pairs between A-U and G-C are formed, which involve two and three hydrogen bonds, respectively. The consecutive formation of base pairs results in the formation of a double helix of the A form and these paired regions define the secondary structure. Secondary structural elements formed by RNA include bulges, hairpins, internal loops, and junctions or multi-branched loops (Fig. 2). Interactions between helices or loop regions are termed tertiary interactions and include pseudoknots and base triples (Fig. 3).

2.2 Methods of Structure Analysis

Classically, phylogenetic analysis has been felt to be the most accurate method for secondary structure determination. Natural phylogenetic analysis assumes that RNAs of similar function share common structural elements. These elements are strongly inferred by compensatory mutation of bases in a double helix, for example a G-C to an A-U pair. Phylogeny cannot predict standard base pairs well, since the rules for these pairs are not precise.

A phylogenetic analysis compares sequences and looks for similar secondary structures. Alignments are helpful for studying sequence homology and usually contain gaps to compensate for base deletions and insertions. Secondary structures determined from the alignment must show at least one compensatory base pair mutation per helix in order to be confirmed by this method. Tertiary interactions such as pseudoknots and triple base interactions have been proposed once the secondary structural elements were established. Tertiary base pairings are inferred from compensatory changes, just as is secondary structure.

Secondary structure prediction programs often use thermodynamic parameters derived from oligonucleotides which form simple base-paired structures. RNA foldings are usually computed to minimize free energy. At a given temperature, the molecule is assumed to be in equilibrium and a minimum energy solution can be regarded as a most probable folding. Energy is assigned to hydrogen bonds between bases and to the stacking of one base over another. Various loops are given destabilizing energies. A principle of additivity is assumed, so that the overall free energy of a folding is the sum of the energies of the stacked base pairs and the loops. This assumption is essential since better experimental data are not yet available to provide more suitable rules.

2.3 Energy Rules

Free-energy data come from melting studies performed on small oligonucleotides (ZUKER 1989a and references therein). This information was summarized by SALSER (1977). Standard base pairs such as G-C and A-U are allowed. In addition, the nonstandard G-U base pair is allowed, although the nonstandard G-U base pair has sometimes not been allowed to occur at either end of a helix. Loop destabilizing energies depend on the size (number of single-stranded bases) of the loop and on the nature of the base pairs which close them. The folding temperature implicit in "Salser's rules" is 25°C.

Various alternatives and modifications to the above rules have appeared. Tinoco's rules no longer depend on closing base pairs for determining hairpin and interior loop destablilizing energies (CECH et al. 1983). Ninio adjusted the energy rules so that the phylogenetically determined structure would also be a minimum energy structure as frequently as possible (NINIO 1979). The nonstandard pairs were expanded to include G-G, when considering stacking involving nonstandard base pairs, while the Salser rules do not.

More recently, new experimental data on RNA duplex stability have resulted in energy rules which supersede Salser's rules. A new rule adjusts the energies of hairpin and interior loops depending on the nature of the terminal mismatched pair(s) (FREIER et al. 1986). These newer rules are referred to as Turner's rules have 37°C as the folding temperature and the only nonstandard base pair allowed is G-U (TURNER et al. 1988). Later, thermodynamic data for tetraloops were also included, further improving the ability to predict known RNA structures (JAEGER et al. 1989).

Thermodynamic calculations allow one to alter energy rules for folding at different temperatures, using either Salser's rules (STEGER et al. 1984) or Turner's rules (JAEGER et al. 1993) for folding at arbitrary temperatures.

2.4 Types of Secondary Structure Prediction Programs

There are several different approaches to predicting RNA secondary structure. Because no single program as of yet has consistently yielded definitive results, the programs are often used in combination to gain insight into the secondary structure of an RNA molecule.

2.4.1 Hairpin Algorithms

Some programs predict only hairpins or helices. Energy considerations may or may not be used. These programs do not predict multibranched loops and limit the distance along the sequence between base pairs. These programs are good for finding local folding motifs but the output may include hairpins that are incompatible with one another either because they involve hydrogen bonding of the same bases or because pseudoknots are created (SEQL, KANEHISA and GOAD 1982).

2.4.2 Multiple Hairpin Prediction

Another kind of secondary structure program generates a list of all possible helices and then pieces them together in all possible ways to form secondary structures. The computer memory required by these programs is proportional to the number of helices used, which grows as the square of the sequence size. For large sequences, computation time more than memory is the limiting factor (CRUSOE, GOUY 1987).

2.4.3 Recursive Programs

Recursive programs build optimal foldings one base at a time. They comprise two distinct parts. The first part, called the fill, computes and stores minimum folding energies for all fragments based on minimum folding energies of small fragments, starting with pentanucleotides. The final part, called the traceback, computes a structure by searching though the matrix of folding energies. The fill algorithm requires the bulk of computing time, while the time for a single traceback is negligible. If multiloops are treated in a simple manner, such algorithms can execute in a time proportional to the cube of sequence length. Computer storage requirements grow as the square of the sequence length; memory, not computational time, usually limits this kind of algorithm. Pseudoknots cannot be handled, and the additivity assumption for free-energy contributions is necessary. This type of recursive program is also referred to as a dynamic programming algorithm. By their nature, recursive algorithms predict the optimal folding of all subfragments. This means, for example, that optimal foldings of a growing RNA sequence can be simulated without any additional cost, once the entire sequence is folded. Recursive algorithms are designed to yield a single solution; but with some effort, the traceback algorithm can be extended to yield multiple solutions (FOLD, ZUKER and STIEGLER 1981; MFOLD, ZUKER 1989a, b).

2.4.4 RNA Folding Simulation

Another kind of algorithm simulates the folding of a RNA molecule in time. The program first compiles a list of stems that can form which may contain small bulge or interior loops. The free energies of these stems are then computed, and the stems given weights or Boltzman probabilities. One stem is then chosen at random using the probabilities. Subsequent stems are then selected, removing stems incompatible with the ones already chosen. The process terminates when no further stems can be added. Pseudoknots are not permitted, although pseudoknots could be allowed. The algorithm requires a modest amount of storage and is also very fast. Large RNA molecules can be folded quickly, but there is no guarantee of achieving a minimum energy folding. The strategy is to refold the molecule over and over again and to compute statistics on which stems tend to occur repeatedly (MONTE-CARLO, MARTINEZ 1984, 1988).

2.4.5 Genetic Algorithms

A recently developed algorithm for secondary structure prediction is a genetic algorithm. When the genetic algorithm is used, an initial stem "pool" (thousands of possible base pairing regions) is created. The selection operator then chooses a set of best fit structures from the current population. From this selected set a new generation is created. Mutations induce random changes in the RNA structure at each generation. Crossover allows structural information within the population of structures to be exchanged, creating new structures based on the mutations and exchanged structural characteristics. Each generation iterates through this process of selection, mutation, and crossover. Eventually the population converges to structures that are highly fit. The fitness criterion used is free energy. Pseudoknots can be allowed (GA, SHAPIRO and NAVETTA 1994; GULTYAEV et al. 1995; SHAPIRO and WU 1996b).

2.5 Poliovirus and Coxsackievirus Secondary Structure Prediction

Computer prediction of higher order structure of RNA from the primary sequence remains a challenging problem. Confidence in the results exists only when there are biologic data to confirm the results or when phylogenetic comparison of related sequences lends credence to the results. The most commonly used algorithm has been MFOLD (JAEGER et al. 1989; ZUKER 1989; ZUKER et al. 1991), a dynamic programming algorithm which calculates a single best free energy and employs a suboptimal option to compute n suboptimal solutions in addition to the most optimal solution. Genetic algorithms have more recently been utilized for RNA structure prediction (SHAPIRO and NAVETTA 1994; GULTYAEV et al. 1995; SHAPIRO and WU 1996a). Genetic algorithms utilize selection, mutation, and crossover operators. The genetic algorithm has performed better than MFOLD when predicting secondary structure from single or individual sequences even when MFOLD employs its suboptimal option to incorporate related sequences in the analysis (SHAPIRO and WU 1996a; CURREY and SHAPIRO 1996).

Although there is increasing confidence in the genetic algorithm to perform well at predicting RNA secondary structure from individual sequences, MFOLD is also utilized since it is still the most commonly used, and thus accepted, RNA secondary structure prediction program. MFOLD has classically required phylogenetic comparison to arrive at a "correct" structure. Previous secondary structure analysis of poliovirus has had to rely on supporting biochemical data or comparative sequence analysis of related sequences to verify the model.

MFOLD is a recursive or dynamic programming algorithm (DPA) and works in two stages (Sect. 2.4.3). The first part, called the fill algorithm, computes and stores minimum folding energies for all fragments of the sequence. The second algorithm, called the traceback, computes a minimum energy structure by searching systematically through the matrix of stored energies. It examines possible base pairs in the context of what neighboring base pairs might be. Representative suboptimal foldings are generated by selecting all predicted base pairs one at a time and computing n best foldings that contain them within an energy range. A distance criterion that ensures that no two structures are "too close" is used to avoid multiple generation of similar structures. When using MFOLD for secondary structure prediction of poliovirus and coxsackievirus, the optimal and the top 9999 suboptimal solutions are collected based on minimum free energy, not to exceed 10% above optimal energy since the "real" solution is felt to be within 10% of the optimal energy. A distance of 0 is allowed between folds so as not to eliminate potential structures. MFOLD is implemented on a CONVEX 3240, VAX 6220, MasPar MP-2, and CRAY YMP computer in the Frederick Biomedical Supercomputing Center, Frederick Cancer Research and Development Center, National Cancer Institute.

The genetic algorithm is implemented on a massively parallel MasPar-2 supercomputer which is a SIMD (single instruction multiple data) machine with 16384 processors and which iterates over a three-step evolution -like procedure: selection, mutation, and crossover across all processors in parallel at each generation (an iteration constitutes one generation), using minimal free energy as a criterion to improve secondary structures. Initially, during a preprocessing stage, the GA generates a stem pool consisting of all possible fully zipped (or partially zipped) stems from the given sequence. At each generation, in each processor, the GA selects two RNA structures from itself and its eight neighbors (assuming the 8-way interconnected mesh architecture of the MasPar) using a ranked rule (biased towards the better free energies) and takes them as parents P_1 and P_2. Then the GA mutates the RNA structures by randomly picking stems, according to an annealing mutation operator from the stem pool, to form two child structures (C_1 and $C_{2)}$ excluding conflicting structures. Next the GA does a crossover operation between (P_1,P_2) and (C_1,C_2) by distributing stems from P_1 and P_2 to C_1 and C_2 to complete the two new structures. Between these two structures, the one with the better energy is selected. Thus, 16384 new structures are created in parallel at each generation. This process is repeated for each generation and 400–600 generations are allowed per run for poliovirus and coxsackievirus as a function of the convergence properties of the run with multiple solutions obtained to yield a representative structure.

Data from both MFOLD and the genetic algorithm are analyzed with the dot matrix histogram which represents the stems and structures graphically. This allows us to collect lists of stems according to percent occurrence. It is part of Structurelab, a Computer Workbench software system (SHAPIRO and KASPRZAK 1995) that integrates numerous folding and analysis programs. These programs include MFOLD, GA, DRAW (used to visualize the secondary structures (SHAPIRO et al. 1982), 2 D STEM HISTOGRAM (the dot matrix histogram), and others and can be implemented on various host (VAX 6220, MASPAR MP-2, CONVEX 3240, CRAY-YMP, and SPARC, and SILICON GRAPHICS (SGI) work stations.

2.6 Tertiary Interactions

Both the primary and secondary structures are known to contain signals for gene regulation and expression. Tertiary structure elements such as pseudoknots have also been implicated as signals for gene regulation and expression (PLEIJ 1990). A pseudoknot is defined as a structure formed when a single-stranded loop region pairs with a complementary sequence outside this loop. Consequently, a pseudoknot is always characterized by at least two stems. If all four type of single-stranded loop regions are taken into account, it can be seen that, in principle, 14 different types of pseudoknots are possible, although not all are sterically possible. Examples include base pairing between a hairpin loop and a single-stranded region, a hairpin loop and bulge loop, a multibranch loop and multibranch loop, etc. Almost all known pseudoknots involve stems of three base pairs or more and many of the pseudoknots described to date are H-type or hairpin pseudoknots. These are formed when hairpin loops interact with single-stranded regions not within other loops. Pseudoknots now appear to be a common structural motif in a number of RNAs, including viral RNAs (PLEIJ 1990; PLEIJ et al. 1985; JACOBSON et al. 1993).

Triple base interactions have also been described and occur when nucleotides in single-stranded positions interact with base pairs. Determining triple base interactions at this time is not feasible with computer prediction methods although methodology using comparative sequence analysis has been reported to help identify potential basetriples (GAUTHERET et al. 1995).

Prediction of RNA higher order structure is a computationally intense problem. The time requirement for predicting RNA secondary structure using MFOLD increases approximately as a function of the length cubed on a single processor computer. The availability of the MasPar computer with parallel architecture changes the time relation to length from a cubed to a squared function. For example, folding a long sequence such as HIV requires five times as much cpu time on the CRAY-YMP than on the MasPar (SHAPIRO et al. 1995). Computer configurations and time required for tertiary structure prediction are even more demanding. Deterministic algorithms such as MFOLD are not able to predict tertiary structure; rather they are solved in nondeterministic ways such as with the genetic algorithm implemented on the MasPar.

Even with supercomputers (massively parallel machines and Cray-like machines), the only tertiary structure elements that can be predicted to date are pseudoknots. Three representative tertiary structure programs available each have their own characteristics. A tertiary interaction program available in Structurelab has the ability to identify all possible tertiary interactions between unpaired regions in a secondary structure. A second program predicts all possible pseudoknot base pairings (MARTINEZ 1990). It does not eliminate overlapping structures and does not calculate free energies of the structures. Thirdly, the genetic algorithm, in addition to predicting RNA secondary structure, can predict H-type pseudoknots (SHAPIRO and WU 1996b). The genetic algorithm for pseudoknots calculates the free energy of the individual pseudoknots in addition to the free energy for the entire structure containing the pseudoknot(s). Competing or overlapping pseudoknots are not allowed. Visualization of the pseudoknots in 3-D can be accomplished with a computer program provided by H. Martinez (personal communication).

Although methodology for RNA structure prediction has continued to improve, the presence of supporting biological or biochemical data remains useful. The biological data can confirm predicted structural elements and guide some structural analyses by incorporation into the programs. Conversely, predicted structures can suggest and guide further biological experiments. Used together, they become very powerful tools for understanding RNA structure-function relationships.

3 The Polio Story

The body of knowledge about the genetics and mechanisms of enteroviral pathogenesis is primarily derived from poliovirus studies and has been extrapolated to

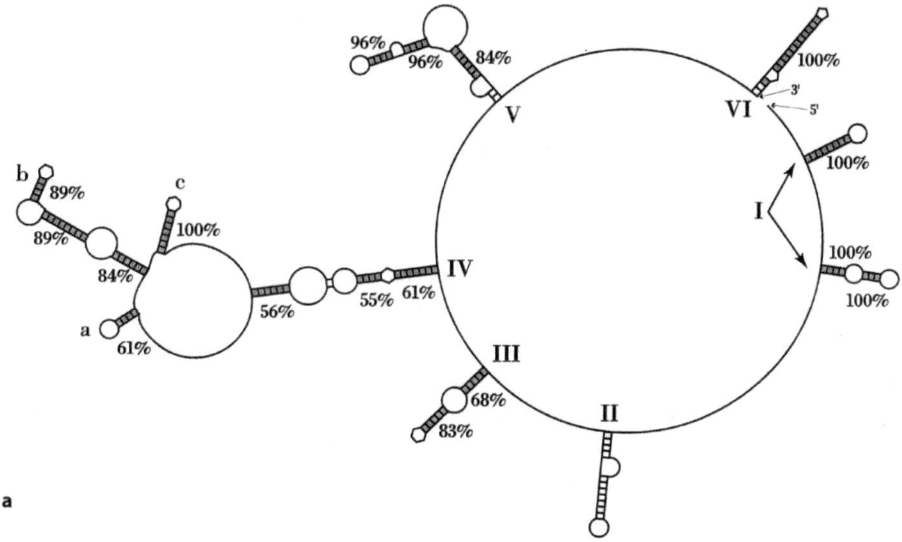

a

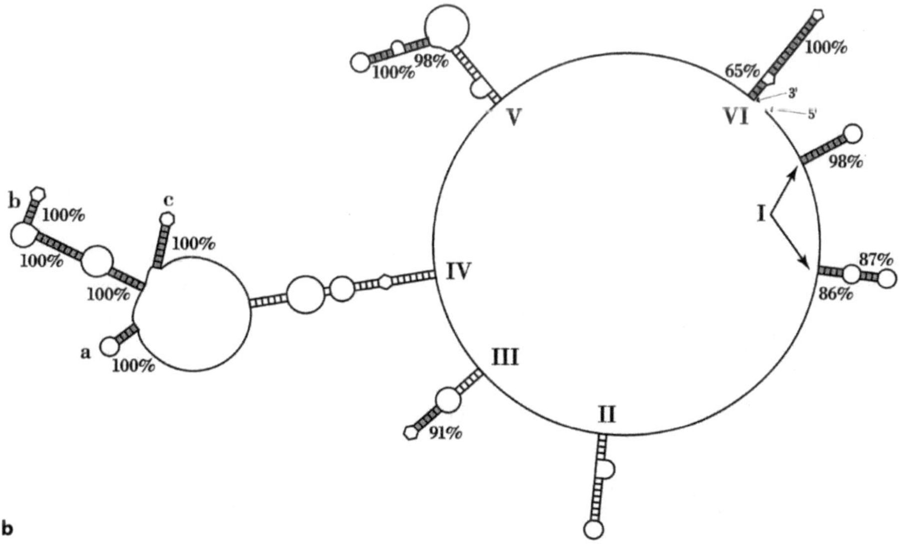

b

Fig. 4A, B. The outline of the currently accepted model of poliovirus 3 (Skinner et.al.) which covers the first 620 nucleotides of the 5' NTR. The stems predicted by the GA (A) and the DPA (B) at the 55% threshold are the shaded stems on the outline structure. Structure 10–34 is indicated by I, 51–78 by II, 126–164 by III, 188–222 by IV, 236–443 by V, 471–538 by VI, and 585–620 by VII. The GA predicts some longer range interactions at the beginning of the sequence and the DPA produces multiple stems not found in the model.

other viruses within the enterovirus family. Likewise, the biological and biochemical data supporting the secondary structure model for the enteroviruses come from poliovirus studies. The poliovirus genome consists of an approximately 7440 nucleotide positive sense mRNA which codes for a single polypeptide that is later cleaved into its functional units. It has an unusually long (approximately 750 nucleotide in length) NTR which has long been known to be essential for the virus' survival and long believed to have secondary structure of its RNA as an important feature. The 5' NTR of the poliovirus genome is now known to contain sequences and structural elements necessary for viral replication and for cap-independent translation and includes the region primary responsible for the artificial attenuation of neurovirulence.

RNA secondary structure, in addition to the primary structure, has been implicated in the multiple functions of the 5' NTR which is predicted to form six structured domains (I-VI) (reviewed in BLYN et al. 1995) (Fig. 4). Both replication and internal ribosome binding processes involve interactions with proteins that depend on RNA structure. Andino and colleagues (1990) proposed a cloverleaf secondary structure within the first 100 nucleotides of the 5' NTR and related it to replication. This structure appears to require a specific fit with viral replication proteins (ROHLL et al. 1994; XIANG et al. 1995). PELLETIER and SONNENBERG (1988) demonstrated that the region necessary for internal ribosome binding and translation initiation in poliovirus was between nucleotides 140 and 631 and termed this the ribosomal landing pad (RLP). PERCY et al. (1992) identified specific structural elements within this region required for internal ribosome entry. As a result, this region is also termed the IRES, or internal ribosome entry site. A number of studies have demonstrated that small deletions, insertions, or substitutions within the IRES are detrimental to viral translation (KUGE and NOMOTO 1987; TRONO et al. 1988a, b; DILDINE and SEMLER 1989; HALLER and SEMLER 1992) and that compensatory mutations which restore disrupted domain structures tend to also restore IRES activity (KUGE and NOMOTO 1987; HALLER and SEMLER 1992). Studies using mutants in which various domain structures were either partially or completely deleted demonstrated that the requirements for translation include only domains II, IV, and V and sequences in domain VI (DILDINE and SEMLER 1989; PERCY et al. 1992; HALLER et al. 1993; MEEROVITCH et al. 1993; JACKSON et al. 1994). Further studies have shown that within domain IV, substructure IVa is important but not essential, substructure IVb is essential for the virus' survival and necessary for translation, and substructure IVc is not necessary for translation (BLYN et al. 1995).

Previous work with poliovirus defined single nucleotide changes in each poliovirus strain as attenuating sites (EVANS et al. 1985; LAMONICA et al. 1986, 1987; OMATA et al. 1986). These sites occur within the region of nucleotides 472–481 (domain V; Fig. 4) in the respective poliovirus genomes. Transitions at these sites correlate closely with reversion from the vaccine (attenuated) phenotype to neurovirulent (disease-causing) phenotype. MACADAM et al. (1992) demonstrated the importance of the region 464–542 in temperature sensitive mutants and the absolute necessity of stems from base positions 471–538, 477–534, 485–513, and 491–506 in the determination of neurovirulence in poliovirus 3. The single nucleotide change in the 5' NTR described above has been shown to markedly decrease translational efficiency in neuronal cells while replication is marginally affected (LAMONICA and RACANIELLO 1989).

A short NTR (approximately 65 nucleotides in length) extends between the termination codon and the 3' terminal poly (A) tract. ROHLL et al. (1995) examined the role of this region in replication and demonstrated a stem loop structure. JACOBSON et al. (1993) reported evidence for a pseudoknot in this region and described its role in viral RNA amplification. PIERANGELI et al. (1995) demonstrated that point mutations which either destabilized the stem and loop structure or altered the sequence of the loop abolished both the infectivity and viral RNA synthesis. These were not restored by compensatory mutations that reconstructed the native secondary structure of the domain, suggesting that the secondary/tertiary folding of the 3'-NTR is not the only determinant for template recognition and initiation of RNA synthesis, but rather that a specific primary structure is also required.

Several investigators have previously reported models for the RNA secondary structure of the 5'NTR in enteroviruses using computer prediction methods. RIVERA et al. (1988) used comparative sequence analysis to arrive at a consensus structure for the enteroviruses (including poliovirus) and rhinoviruses. PIILPENKO et al. (1989), in addition to comparative sequence analysis, used biochemical and genetic data to support their models for poliovirus. SKINNER et al. (1989) used the dynamic programming algorithm FOLD, which yields a single most optimal structure, and compared the predicted structures of poliovirus and related enteroviruses to arrive at a consensus structure for poliovirus. They supported their model with biologic data from poliovirus type 3. LE and ZUKER (1990) used purely computer based methods comparing optimal and suboptimal structures of the enteroviruses and rhinoviruses to arrive at a consensus model for the enteroviruses. The program used was MFOLD. There has been good agreement among Pilipenko, Skinner, and Le and Zuker. Biologic data has continued to support the consensus model first proposed in its entirety by SKINNER et al. (1989). Figure 4 shows the secondary structure model for poliovirus reported by Skinner on which the functional domains are superimposed.

More recently, the 5' NTR of poliovirus has been folded with the genetic algorithm which found the structural elements present in the currently accepted model of the 5' NTR of poliovirus at a high rate (CURREY and SHAPIRO 1996). The elements for which there are biological or biochemical data support were predicted from a single sequence without requiring biological data or phylogenetic sequence comparison. A secondary structure model, very similar to the one predicted by SKINNER et al. (1989), for poliovirus 3, was predicted using the genetic algorithm. This result was obtained without utilizing related genomic sequences as required by the DPA for a "consensus" model. All of the known structures which have been shown to be biologically significant are present in the structure predicted by the GA (Fig. 4). Using a 55% threshold, the GA finds 18 of the 21 stems present in the model including all 12 of the stems predicted by the DPA. Also at the 55% threshold, the GA only predicts 11 stems not found in the model while the DPA predicts 24 stems not found in the model. The selected threshold was 55%, which allows as many stems as possible to be selected but eliminates competing or overlapping structures. A case could be made for the existence of three of those "extra" stems for the DPA and for five of the "extra" stems predicted by the GA. Stems 2–88 and 35–45 were reported by ANDINO et. al. (1990) and ROHLL et. al. (1994) and comprise the "cloverleaf" structure which has biological support. Stem 90–142 could exist instead of stem 126–164 and still be compatible with the protein binding data. The remaining GA predicted stems either occur within a multibranch loop or form the base of known stems and thus may exist. The algorithm also performed well at predicting secondary structures for poliovirus 1 and 2 and significantly better than MFOLD.

The GA thus may be very informative in predicting structures for genomic elements and does not require the use of phylogenetically related sequences for comparison or consensus structure analysis. GULTYAEV et al. (1995) also report that genetic algorithm produced structures are more consistent with phylogenetic proven structures than minimum energy solution and attribute this to the folding pathway simulated by genetic algorithm methods. The DPA is a deterministic algorithm working to arrive at the most optimal solution which is always the same, whereas the GA is a non-deterministic algorithm that mutates and recombines stems which may be more biologically realistic. Thus the GA may be capturing important properties of folding which work in concert with free energy but does not use free energy as the sole criterion. Perhaps the innate characteristics of the genetic algorithm, stochastically searching for, and selecting, the fittest stems lends itself to more biologic relevance than the dynamic programming algorithms. Because of its accuracy and independence from requiring supporting data, it may offer insights into secondary structures of other genetic elements. For example, the GA may be able to aid in elucidating unique structural elements within the enteroviruses. TU et. al. (1995) reported that the cardiovirulent phenotype in CVB3 is related to a single base change in the 5' NTR and potentially to the secondary structure involving this nucleotide. This was not apparent from the consensus RNA secondary structure model for the enteroviruses; rather it was suggested by the GA methodology described here.

4 The Coxsackievirus B3 Story

The coxsackievirus B RNA is a typical enterovirus genome. It has a relatively long 5' NTR preceding the single, long, open reading frame (ORF) followed by a 3' noncoding region which terminates in a poly (A) tract. It was recently been shown that a single nucleotide change in the coxsackievirus B3 5' NTR is associated with determining cardiovirulence (TU et al. 1995). If nucleotide 234 is U, the virus is cardiovirulent, if nucleotide (nt) 234 is C the virus is attenuated. These studies demonstrated no role for the capsid or other regions to determine the virulence phenotype for the two strains studied. There was no difference in viral protein translation as a function of the viral phenotype but there was a tenfold decrease in replication efficiency in the noncardiovirulent strain compared to the cardiovirulent strain.

Table 1. Comparison of base pairing frequencies (GA and MFOLD) for nucleotide 234

CVB strain	CVB3/20	CVB3/20nv	CVB3/0	CVB3/0v
Nucleotide (nt) at position 234	U	C	C	U
Percent of time nt 234 was involved in pairing (genetic algorithm)	50	71	68	55
Percent of time nt 234 was involved in pairing (MFOLD)	0	35	35	6

Depicted are the nucleotides (nt) at position 234 in laboratory sequences (CVB3/20 and CVB3/0) and computer derived artificial sequences (CVB3/0v and CVB3/20nv) and the base pairing frequencies involving that nucleotide. Results from both the genetic algorithm and MFOLD are shown.

The common secondary structure map of the 5' NTR for all enteroviruses places nt 234 in a region devoid of structure between two structural elements (LE and ZUKER 1990), as does the earlier derived secondary structure model for poliovirus which utilized several enterovirus sequences (including coxsackievirus) in addition to biologic data from poliovirus to arrive at the model (SKINNER et al. 1989). The genetic algorithm (SHAPIRO and NAVETTA 1994) was used to predict the secondary structure of the CVB3 sequences as well as the more commonly used MFOLD (ZUKER 1989b). The GA was employed because of the demonstrated good agreement between the GA predicted structure for poliovirus 3 and the currently accepted model for poliovirus and the enteroviruses using the single sequence analysis described earlier.

The 5' NTR of the cardiovirulent strain CVB3/20 has U at nucleotide position 234, whereas the attenuated strain CVB3/0 has C at nucleotide position 234 in addition to two other mutations at other locations in the 5' NTR. Two artificial sequences were created, CVB3/20nv and CVB3/0v, with the substitution at nucleotide 234 as indicated in Table 1. When the (+) strand of all four sequences was folded with the genetic algorithm, nucleotide 234 was involved in pairing 50% and 55% of the time in CVB3/20 and CVB3/0v, respectively, and 68% and 71% of the time in CVB3/0 and CVB3/20 nv, respectively. The base pairings for each sequence occurred in two small, 2 and 3 base pair competing or alternate stemlets, which were found within a larger structure bounded by a 10 base pair in length stem from nucleotide 195 to 434 (Fig. 5). This 10 base pair stem was predicted at a frequency greater than 50% and had a free energy of –14.8 kcal; it is felt to be stable because of its size and free energy. When these sequences were folded with MFOLD, utilizing its suboptimal feature and applied to the individual sequences, nucleotide 234 was found in a 3 bp long stem less than 10% of the time in CVB3/20 and CVB3/0v, and 35% of the time in CVB3/0 and CVB3/20nv. The long range interaction of nucleotides 195 pairing to 434 was seen below the threshold of 50% at 17% in the MFOLD outputs, but a different stem occurred at greater than 50% from nucleotide 210 to 482, 7 base pairs in length with a free energy of

a

a

b

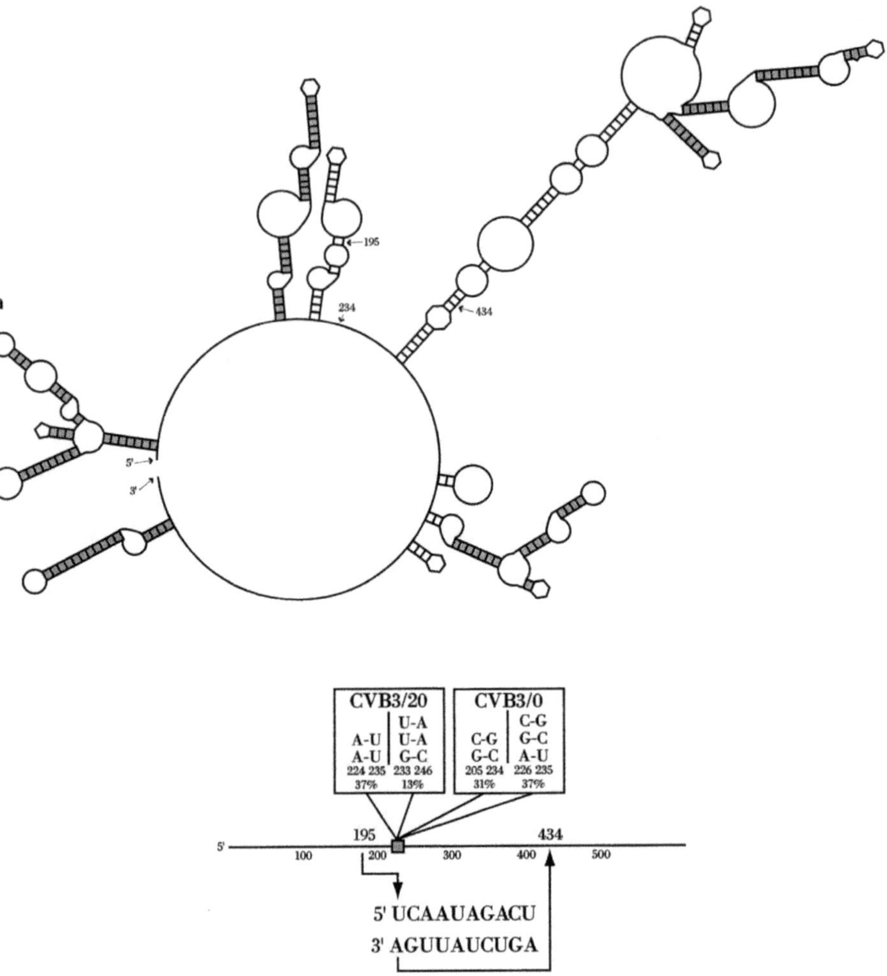

Fig. 5A, B. The global secondary structure model for enteroviruses predicted by MFOLD (Le and Zuker 1990) with the stems and structures aligned corresponding to CVB3 nucleotide positions.
A Note that nucleotide 234 is located in a position lacking secondary structure. The arrow indicates the location of nucleotide 234. The stems and structures predicted by the genetic algorithm which overlap the MFOLD model are represented with the bases superimposed over the outline.
B The10 base pair stem from nucleotide 195 to 434 with the alternate smaller stems involving nucleotide 234 shown in the boxes. The stems and structures unique to the genetic algorithm model result in an alternate structure with nucleotide 195 and 434 forming a 10 base pair in length stem as the base of this structure. Nucleotide 234 resides within this structure

−11.4 kcal. Nucleotide 234 is found within the loop of this stem but is base paired at a low frequency. In the currently accepted secondary structure model of the enteroviruses using MFOLD, nucleotide 234 is located in a region devoid of structure. This is reflected in the markedly lower percent base pairing with nucleotide 234 although the general trend is preserved. The data with the genetic algorithm however suggest that secondary structural elements are present in this region and that a change of nucleotide 234 from U to C results in stronger base pairing in the secondary structure. This suggests that the existence of the secondary structure and its increased stability in the 5' NTR may contribute to nonvirulence, perhaps related to the observed decreased transcriptional efficiency.

Table 2. Base pairing frequencies with nucleotide substitution at position 234

CVB strain	CVB3/20	CVB3/0	CVB3/20sa	CVB3/20sg
Nucleotide(nt) at position 234	U	C	A	G
Percent of time nt 234 was involved in pairing	50	68	83	79

Depicted are the nucleotides at position 234 in naturally occurring sequences CVB3/20 and CVB3/0 and in artificially created sequences CVB3/20sa and CVB3/20sg and their base pairing frequencies. Results are from the genetic algorithm.

Mutations at nucleotide position 234 with either A or G have not been found to date in wild-type CVB3s. Therefore two sequences were created using CVB3/20 as the template (Table 2). In one sequence, A replaced U at nucleotide 234 and in the other sequence G replaced U at nucleotide 234. When these sequences were folded with the genetic algorithm, the sequence with A at position 234 (CVB3/20sa), nucleotide 234 paired 83% of the time. When G was substituted at position 234 (CVB3/20sg), nucleotide 234 paired 79% of the time. Furthermore, both sequences had nucleotide 234 involved in longer (i.e., 5 and 8 base pairs in length) and thus more stable stems. Comprising the total pairing frequency, each had an individual stem above the 50% threshold thus contributing to the overall secondary structure map. CVB3/20sa had nucleotide 234 involved in a stem 5 base pairs in length occurring at a threshold of 54%, and CVB3/20sg had nucleotide 234 involved in a stem 8 base pairs in length at a threshold of 52%. The genetic algorithm predicts that G or A base substitution at position 234 contributes to enhanced stability of the secondary structure(s) in that region even more pronounced than C base substitution. Might the enhanced stability of these two variants correspond to decreased virulence and replication? Our preliminary data indicate that in a construct where G replaces U at nucleotide 234, replication is severely diminished, even more so than when C replaces U in CVB3/0 (S. Tracy, unpublished data).

The structure analysis of the CVB3 RNA has been extended by applying the genetic algorithm to fold the entire 5' NTR of the CVB3 sequences. This structure was generated by applying a 50% threshold to the predicted stems. This model has multiple stems in common with both the model proposed by Skinner et al. (1989), for poliovirus and the enteroviruses, and the global enterovirus model proposed by Le and Zuker (1990); however, there are unique features (Fig. 5). In comparing the three models, the genetic algorithm predicted 14 of the 22 stems present in Skinner's model and five stems not predicted in Skinner's model. In addition, three stems were predicted that were also present in the global enterovirus model. The stems unique to the genetic algorithm predicted model for CVB3 are primarily within the domain described by stem 195–434. Although the genetic algorithm has stems in common with the global enterovirus model in this region, the stems unique to the genetic algorithm model include stems involving nucleotide 234 (Fig. 5). These unique features of CVB3 indicate a difference between the secondary structure of this virus and the consensus secondary structure map for the enteroviruses. There is agreement between the genetic algorithm predicted model and the global enterovirus model except for the region that includes nucleotide 234. The global enterovirus model shows nucleotide 234 in an unpaired region. Data from Dr. Tracy's laboratory are not inconsistent with a secondary structure involving this nucleotide.

Several lines of evidence support the concept of a consensus structure for the enteroviruses: (1) the individual sequences can be folded in the "consensus" structure; (2) there appear to be compensatory mutations among the sequences to allow and or support these structures (Rivera et al. 1988); and (3) some of the functional domains as well as entire 5'NTRs of the enteroviruses may be substituted one for the other without sacrificing viability (Johnson and Semler 1988; Alexander et al. 1994). A consensus secondary structure map of the 5' NTR of the enteroviruses was first reported by Rivera et al. (1988). Phylogenetic comparion and computer methods were used to fold the individual sequences of different enterovises and arrive at a "consensensus" structure. Compensatory mutations among the sequences were used to support the structure. Skinner et al. (1989) compared the structural elements in related enteroviruses to arrive at their secondary structure map for poliovirus 3 and used biologic data from their laboratory to support their model. Le and Zuker (1990) predicted a consensus secondary structure of the 5' NTR for the enteroviruses by using MFOLD with phylogenetic comparison of multiple enteroviruses and rhinovirus. Their model was similar to that of Skinner. Tertiary elements in the 5' NCR have also been reported to be conserved for enteroviruses although the case made for the coxsackieviruses was weaker (Le et al. 1992).

5 Future Directions

The body of knowledge about the genetics and mechanisms of enteroviral pathogenesis is primarily derived from poliovirus studies. Likewise, the biological and biochemical data supporting the secondary structure model for the enteroviruses come from poliovirus studies and primarily from studies of poliovirus 3. Yet there are unique features among the enteroviruses, i.e., a site for the artificial attenuation determination of cardiovirulence is in a different structural and functional domain than is the site for neurovirulence in poliovirus. In the common secondary structure map for the enteroviruses, CVB3's site for determination of cardovirulence supports the concept of coxsackievirus possessing unique structural features within the enterovirus "family." It will be interesting to see what role tertiary interactions play in these various functions and how the enteroviruses differ from each other.

Numerous experimental observations have shown that different secondary structures can appear during RNA processing. This implicates that different structures are preferred at various times (MIRONOV et al. 1985; KISTER et al. 1993) and it is possible that the structure starts fulfilling its biological function before it reaches equilibrium. Moreover, the lifetime of an RNA molecule may be shorter than the time to reach equilibrium of the "final" structure. In fact, formation of the RNA structure is a stochastic process, and thus an ensemble of secondary structures exists in vivo. Rarely is the "true" structure the most energetically favorable one, rather, as Zuker has pointed out and many have found, it often exists within 10% of the most optimal energy. Within this 10% exist many structures, and at present we have no way of knowing which are the biologically viable structures in the melange of structures. In large sequences, the window may be even larger resulting in many more solutions to search (SHAPIRO and WU 1996a). The genetic algorithm performs better at predicting known structural elements which may be due in part to its different approach and stochastic nature as described above. In previous analyses, when watching 5s RNA molecules fold as one base at a time was added, we noted that structures alternated conformations implying metastable states. This probably happens with the larger molecules as well, although computational constraints prevent observing, capturing, and storing them. Another factor is the protein-RNA interactions: proteins may well stabilize interim structures. Perhaps, a kinetic approach to RNA structure prediction may provide additional insight into RNA structure-function relationships. Currently, approaches that utilize energy, but also take into account other processes, have come the closest to accurate RNA structure prediction.

As computer prediction of higher order structures becomes more sophisticated and we are not limited by requiring phylogentic comparison for confirmation, we may be able to elucidate unique structural features of individual genomes. As genomes continue to evolve, structural features may persist that may not be functionally relevant and we may be obscuring unique structure-function relationships by using phylogenetic anaysis. The biological mechanisms in the enteroviruses with their similarities, but also their subtle differences, will undoubtedly play an important role as we gain further insight into and understand RNA structure-function relationships more fully.

Acknowledgements. The authors would like to thank John Owens for his help and expertise with the design and drafting of figures. Computer prediction of RNA structure by Michael Zuker in Methods in Enzymology (ZUKER 1989a) served as an excellent review for some of the basic fundamentals. We acknowledge the National Cancer Institute for allocation of computing time and staff support at the Frederick Biomedical Supercomputing Center of the Frederick Cancer Research and Development Center.

References

Alexander L, Lu HH, Wimmer E (1994) Polioviruses containing picornavirus type 1 and/or type 2 internal ribosomal entry site elements: genetic hybrids and the expression of a foreign gene. Proc Natl Acad Sci USA 91:1406–1410

Andino R, Rieckhof GE, Baltimore D (1990) A functional ribonucleoprotein complex forms around the 5' end of poliovirus RNA. Cell 63:369–380

Blyn LB, Ruoying C, Semler B, Ehrenfeld E (1995) Host cell proteins binding to domain IV of the 5' noncoding region of poliovirus RNA. J Virology 69:4381–4389

Cech TR, Tanner NK, Tinoco I, Weir BR, Zuker M, Perlman PS (1983) Secondary structure of the tetrahymena ribosomal RNA intervening sequences: structural homology with fungal mitochondrial intervening sequences. Proc Natl Acad Sci USA 80:3903–3907

Crothers DM, Cole PE (1978) Conformational changes of tRNA. In: Altman S (ed) Transfer RNA, MIT, Cambridge, MA

Currey K, Shapiro B (1996) Secondary structure computer prediction of the poliovirus 5' non-coding region is improved with a genetic algorithm. CABIOS, in press

Dildine SL, Semler BL (1989) The deletion of 41 proximal nucleotides reverts a poliovirus mutant containing a temperature-sensitive lesion in the 5' noncoding region of genomic RNA. J Virology 63:847–862

Evans DMA, Dunn G, Minor PD, Schild GC, Cann AJ, Stanway G, Almond JW, Currey K, Maizel J (1985) A single nucleotide change in the 5' non-coding region of the genome of the sabin type 3 poliovaccine is associated with increased neurovirulence. Nature 314:548–550

Freier SM, Kierzek R, Jaeger JA, Sugimoto N, Carothers MH Neilson T, Turner DH (1986) Improved free-energy parameters for predictions of RNA duplex stability. Proc Natl Acad Sci USA 83:9373–9377

Gautheret D, Damberger SH, Gutell RR (1995) Identification of base-triples in RNA using comparative sequence analysis. J Mol Biol 248:27–43

Gouy M (1987) Nucleic acid and protein sequence analysis: a practical approach. In: Bishop MJ, Rawlings CJ (eds), IRL, Washington DC

Gultyaev AP, van Batenburg FHD, Pleij CWA (1995) The computer simulation of RNA folding pathways using a genetic algorithm. J Mol Biol 250:37–51

Haller AA, Semler BL (1992) Linker scanning mutagenesis of the internal ribosome entry site of poliovirus RNA. J Virol 66:5075–5086

Haller AA, Nguyen JHC, Semler B (1993) Minimum internal ribosome entry site required for poliovirus infectivity. J Virol 67:7461–7471

Hellen CUT, Pestova TV, Litterst M, Wimmer E (1994) The cellular polypeptide p57 (pyrimidine tract-binding protein) binds to multiple sites in the poliovirus 5' nontranslated region. J Virol 68:941–950

Jackson RJ, Hunt SL, Gibbs CL, Kaminski A (1994) Internal initiation of translation of picornavirus RNAs. Mol Biol Rep 19:147–159

Jacobson SJ, Konings DAM, Sarnow P (1993) Biochemical and genetic evidence for a pseudoknot structure at the 3' terminus of the poliovirus RNA genome and its role in viral RNA amplification. J Virol 67:2961–2971

Jaeger JA, Turner DH, Zuker M (1989) Improved predictions of secondary structures for RNA. Proc Natl Acad Sci USA 86:7706–7710

Jaeger JA, Turner DH, Zuker M (1990) Predicting optimal and suboptimal secondary structure for RNA. Methods in Enzymol 183:281–306

Jaeger JA, SantaLucia J, Tinoco I (1993) Determination of RNA structure and thermodynamics. Annu Rev Biochem 62:255–87

Johnson VH, Semler BL (1988) Defined recombinants of poliovirus and coxsackievirus: sequence-specific deletions and functional substitutions in the 5' noncoding regions of viral RNAs. Virology 162:47–57

Kanehisa MI, Goad WB (1982) Pattern recognition in nucleic acid sequences II: an efficient method for finding locally stable secondary structure. Nucleic Acids Res 10:265–278

Kim SH, Suddah FL, Quigley GJ, McPherson A, Sussman JL, Wang AHJ, Seeman NC, Rich A (1974) Three-dimensional tertiary structure of yeast phenylalanine transfer RNA. Science 185:435–440

Kister A, Magarshak Y, Malinsky J (1993) The theoretical analysis of the process of RNA molecule self-assembly. Biosystems 30:31–48

Kuge S, Nomoto A (1987) Construction of viable deletion and insertion mutants of the Sabin strain of type 1 of poliovirus: function of the 5' noncoding sequence in viral replication. J Virol 61:1478–1487

LaMonica N, Meriam C, Racaniello V (1986) Mapping of sequences required for mouse neurovirulence of poliovirus type 2 Lansing. J Virol 57:515–525

LaMonica N, Almond J, Racaniello V (1987) A mouse model for poliovirus neurovirulence identifies mutations that attenuate the virus for humans. J Virol 61:2917–2920

LaMonica N, Racaniello V (1989) Differences in replication of attenuated and neurovirulent polioviruses in human neuroblastoma cell line SH-SY5Y. J Virol 63:2357–2360

Le SY, Zuker M (1990) Common structures of the 5' noncoding RNA in enteroviruses and rhinoviruses: thermodynamical stability and statistical significance. J Mol Biol 216:729–741

Le SY, Chen JH, Sonenberg N, Maizel J (1992) Conserved tertiary structure elements in the 5' untranslated region of human enteroviruses and rhinoviruses. Virol 191:858–866

Macadam AD, Ferguson G, Burlison J, Stone D, Skuce R, Almond JW, Minor PD (1992) Correlation of RNA secondary structure and attenuation of sabin vaccine strains of poliovirus in tissue culture. Virology 189:415–422

Martinez HM (1984) An RNA folding rule. Nucleic Acids Res 12:323–334

Martinez HM (1988) An RNA secondary structure workbench. Nucleic Acids Res 16:789–1798

Martinez HM (1990) Detecting pseudoknots and other local base-pairing structures in RNA sequences. Methods in Enzymol 183:306–317

Meerovitch K, Svitkin YV, Lee HS, Lejbkowicz F, Kenan DJ, Chan EKL, Agol VI, Keene JD, Sonenberg N (1993) La autogen enhances and corrects aberrant translation of poliovirus RNA in reticulocyte lysate. J Virol 67:3798–3807

Mironov AA, Dyakonova LP, Kister AE (1985) A kinetic approach to the prediction of RNA secondary structures. J Biomol Struct Dynam 2:953–962

Ninio J (1979) Prediction of pairing schemes in RNA molecules-loop contributions and energy of wobble and non-wobble pairs. Biochimie 61:1133–1150

Omata T, Kohara M, Kuge S, Komatsu T, Abe S, Semler B, Kammeda A, Itoh H, Arita M, Wimmer E, Nomoto A (1986) Genetic analysis of the attenuation phenotype of poliovirus type 1. J Virol 58:348–358

Pelletier J, Sonenberg N (1988) Internal initiation of translation of eukaryotic mRNA directed by a sequence derived from poliovirus RNA. Nature 334:320–325

Percy N, Belsham GJ, Brangwyn JK, Sullivan M, Stone DM, Almond JW (1992) Intracellular modifications induced by poliovirus reduce the requirement for structural motifs in the 5' noncoding region of the genome involved in internal initiation of protein synthesis. J Virol 66:1695–1701

Pierangeli A, Bucci M, Pagnotti P, Degener AM, Bercoff RP (1995) Mutational analysis of the 3' terminal extra-cistronic region of poliovirus RNA: secondary structure is not the only requirement for minus strand replication. FEBS 374:327–332

Pilipenko EV, Blinov VM, Romanov LI, Dinyakov AN, Maslova SV, Agol VI (1989) Conserved structural domains in the 5'-untranslated region of picornaviral genomes: an analysis of the segment controlling translation and neurovirulence. Virology 168:201–209

Pleij CWA (1990) Pseudoknots: a new motif in the RNA game. TIBS 15 April:143–147

Pleij CWA, Rietveld K, Bosch L (1985) A new Principle of RNA folding based on pseudoknotting. Nucleic Acids Res 13:1717–1731

Pley HW, Flaherty KM, McKay DB (1994) Three-dimensional structure of a hammerhead ribozyme. Nature 372:68–74.

Riesner D, Romer R (1973) Thermodynamics and kinetics of conformational transitions in oligonucleotides and tRNA. In: Duchesne J (ed) Physicochemical properties of nucleic acids. Academic, New York, 2:237–318

Rivera VM, Welsh JD, Maizel JV (1988) Comparative sequence analysis of the 5' noncoding regions of the enteroviruses and rhinoviruses. Virology 165:42–52

Rohll JB, Moon DH, Evans DJ, Almond JW (1995) The 3' untranslated region of picornavirus RNA: features required for efficient genome replication. J Virol 69:7835–7844

Rohll JB, Percy N, Ley R, Evans DJ, Almond JW, Barclay W (1994) The 5'-untranslated regions of picornavirus RNAs contain independent functional domains essential for RNA replication and translation. J Virol 68:4384–4391

Salser W (1977) Globin mRNA sequences: analysis of basepair and evolutionary implication. Cold Spring Harbor Symp Quant Biol 42:985–1002

Shapiro BA, Navetta J (1994) A massively parallel genetic algorithm for RNA secondary structure prediction. J Supercomputing 8:195–201

Shapiro BA, Kasprzak W (1995) STRUCTURELAB: A heterogeneous computer system for the analysis of RNA structures. Association of Lisp User Meeting and Workshop Proceedings. Cambridge, MA, pp 1–19

Shapiro BA, Wu JC (1996a) An annealing mutation operator in the genetic algorithms for RNA folding. CABIOS, in press

Shapiro BA, Wu JC (1996b) Predicting RNA H-type pseudoknots with the massively parallel genetic algorithm. Submitted

Shapiro BA, Lipkin LE, Maizel J (1982) An interactive technique for the display of nucleic acid secondary structure. Nucleic Acids Res 10:7041–7052

Shapiro BA, Chen JH, Busse T, Navetta J, Kasprzak W, Maizel JV (1995) Optimization and performance of a massively parallel dynamic programming algorithm for RNA secondary structure prediction. Int J Supercomputer Appl 9:29–39

Skinner MA, Racaniello VR, Dunn G, Cooper J, Minor PD, Almond JW (1989) New model for the secondary structure of the 5' noncoding RNA of poliovirus is supported by biochemical and genetic data that also show that RNA secondary structure is important in neurovirulence. J Mol Biol 207:379–392

Steger G, Hofman H, Fortsch J, Gross HJ, Randles JW, Sanger HL, Riesner D (1984) Conformational transitions in viroids and virusoids: comparison of results from energy minimization algorithm and from experimental data. J Biomol Struct Dyn 2:543–571

Trono D, Andino R, and Baltimore D (1988) An RNA sequence of hundreds of nucleotides at the 5' end of poliovirus RNA is involved in allowing viral protein synthesis. J Virol 62:2291–2299

Trono D, Pelletier J, Sonenberg N, Baltimore D (1988) Translation in mammalian cells of a gene linked to the poliovirus 5' noncoding region. Science 241: 445–448.

Tu Z, Chapman NM, Hufnagel F, Tracy S Romero JR, Barry WH, Zhao L, Currey K, Shapiro B (1995) The cardiovirulent phenotype of coxsackievirus B3 is determined at a single site in the genomic 5' non-translated region. J Virol 69:4607–4618

Turner DH, Sugimoto N, and Freier S M (1988) RNA structure Predictions. Ann Rev Biophys Biophys Chem 17:167–192

Xiang W, Harris K, Alexander L, Wimmer E (1995) Interaction between the 5'-terminal cloverleaf and 3AB/3CD^{pro} of poliovirus is essential for RNA replication. J Virol 69:3658–3667

Zuker M (1989) Computer prediction of RNA structure. Methods in Enzymology 180:262–288

Zuker M (1989) On finding all suboptimal foldings of an RNA molecule. Science 24:48–52

Zuker M, Stiegler P (1981) Optimal computer folding of large RNA sequences using thermodynamics and auxiliary information. Nucleic Acids Res 9:133–148

Zuker M, Jaeger JA, Turner DH (1991) A comparison of optimal and suboptimal RNA secondary structures predicted by free energy minimization with structures determined by phylogenetic comparison. Nucleic Acids Res 19:2707–2714

The Structure of Coxsackievirus B3

J. K. Muckelbauer and M. G. Rossmann

Department of Biological Sciences, Purdue University, 1392 Lilly Hall of Life Science, West Lafayette, IN 47907, USA

1 Introduction

Picornaviruses are small, nonenveloped, icosahedral viruses. The mature virions are approximately 300 Å in diameter and the interior of the capsid shell is packed with single-stranded plus sense RNA (Mw ~ 8×10^6 Daltons) (RUECKERT 1990). The virus capsid is composed of 60 protomers, each consisting of one copy of the viral proteins VP1, VP2, VP3 and VP4 arranged in a $T = 1$ (*Pseudo* T=3) icosahedral lattice (CASPAR and KLUG 1962) (Fig. 1). Atomic resolution structures have been determined for all *Picornaviridae* genera (HOGLE et al. 1985; ROSSMANN et al. 1985; LUO et al. 1987; ACHARYA et al. 1989) with the exception of the hepatoviruses. Their structures demonstrate conservation of the eight-stranded antiparallel β-sandwich of VP1, VP2 and VP3 while VP4, which is completely interior, is not well defined in many of the structures. The regions of most structural dissimilarity for VP1, VP2 and VP3 occurs at the NH_2- and COOH-terminals and the loops that connect the β-strands. Many of the differences at the COOH-terminals and β-strand loops present themselves on the surface of the virion and give each virus its unique surface topology and hence define how these viruses interact with cells and mediate early steps in the viral life cycle including receptor recognition and viral uncoating.

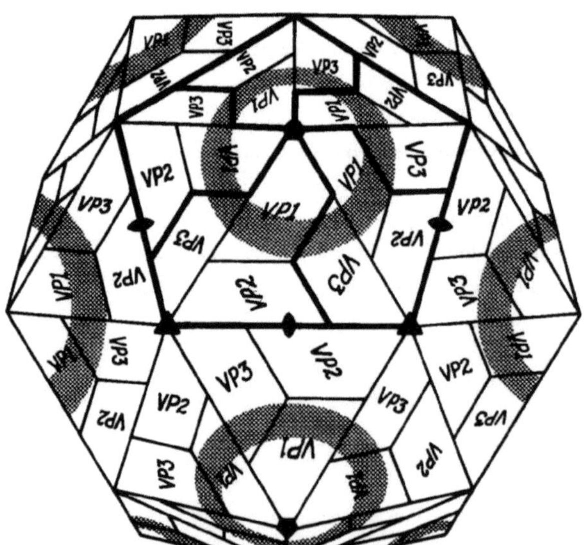

Fig. 1. The T=1 icosahedral arrangement of the picornavirus major structural proteins. The *shaded circle* around each icosahedral fivefold axes represents the canyon. The *thick black outline* about VP1, VP2 and VP3 represent one biological protomer

The attachment sites of intercellular adhesion molecule 1 (ICAM-1) on the surface of human rhinovirus 16 (HRV16) and human rhinovirus 14 (HRV14) have been shown to be a 15 Å deep depression termed the "canyon" which surrounds each icosahedral fivefold axes (OLSON et al. 1993; N. Olson, P. Kolatkar, M. Rossmann, T. Baker unpublished results). These results corroborated the earlier proposed canyon hypothesis (ROSSMANN 1989) which was based on the three-dimensional structure of HRV14 (ROSSMANN et al. 1985), the conservation of amino acid residues lining the canyon (ROSSMANN and PALMENBERG 1988; CHAPMAN and ROSSMANN 1993) and mutational studies of amino acids located inside the canyon of rhinoviruses (COLONNO et al. 1988) and polioviruses (COLSTON and RACANIELLO 1994). There is evidence for, at least, two coxsackievirus B3 (CVB3) receptors on human HeLa cells (MAPOLES et al. 1985; HSU et al. 1990). This suggests that the coxsackie B viruses (CVBs) have the ability to recognize and utilize multiple receptors, or they may be part of a receptor complex comprised of a number of different proteins that define the CVB receptor.

Sterling Winthrop Pharmaceuticals Research Division and others (ZHANG et al. 1993) have synthesized a number of antiviral agents that are active against rhino-, polio- , coxsackie- and other enteroviruses by inhibiting viral uncoating (FOX et al. 1986) and sometimes also by blocking attachment of the virus to the cell (PEVEAR et al. 1989; HEINZ et al. 1990). Crystallographic studies have shown that these antiviral compounds bind underneath the canyon floor into a hydrophobic pocket within the β-sandwich of VP1 (SMITH et al. 1986; BADGER et al. 1989; KIM et al. 1993; OLIVEIRA et al. 1993; GRANT et al. 1994). When the compounds bind into the drug binding pocket of HRV14 they cause large conformational changes, deforming the canyon floor and blocking receptor attachment (PEVEAR et al. 1989; HEINZ et al. 1990). The native structures of poliovirus 1-Mahoney (PV1/M), poliovirus 3-Sabin (PV3/S), human rhinovirus 1A (HRV1A) and HRV16 showed that the antiviral binding pocket was occupied in its native state by an unidentified cofactor called the "pocket factor" (FILMAN et al. 1989; KIM et al. 1989; OLIVEIRA et al. 1993). The pocket factor is displaced when the drugs bind and, hence, in the case of HRV1A, HRV16 and polioviruses, the drugs do not cause significant conformational changes upon binding. It has been suggested that the pocket factor plays a role in stabilizing the virion when in transit between cells, while mediating capsid instability after attachment prior to uncoating (OLIVEIRA et al. 1993; GRANT et al. 1994; ROSSMANN 1994). Although these pocket factors have been modeled as fatty acids, based on the shape of observed electron density and their hydrophobic environments, there has been no chemical identification of pocket factors.

CVBs are etiologic agents of a number of human diseases that range in severity from asymptomatic to lethal infections (MELNICK 1990). Structural studies of CVB3 were initiated in light of the information available on the cellular receptors for this virus as well as the relevance to the design of antiviral capsid binding compounds for the CVBs. The structure of CVB3 at 3.5 Å resolution (MUCKELBAUER et al. 1995, 1996) shows the conserved β-sandwich structure of VP1, VP2 and VP3 while structural differences between CVB3 and other entero- and rhinoviruses are

a

```
                                                                                                           |αZ=|    |==βB===|
         --------10--------------------20--------30--------40--------50--------60--------70--------80-------
CVB3     Gpvedaitaaigrvadtvg.............tGPtnseaiPaLTAaETGhTsqvvPSDtmqTRhvknyhsrSEstiEnFlcRsACVyfteyeNsg....
PV1/M    GlgqmlesmidntvretVgaatsrdalpnteasGPthskeiPaLTAvETGaTnplvPSDtvqTRhvvqhrsrSEssiEsFfaRgACVtimtvdNpast.t
HRV14    Glgdeleevivektkqtva.........sissGPkhtqkvPiLTAnETGaTmpvlPSDsieTRttymhfngSEtdvEcFlgRaACVhvteiqNk..dat

                 |=βC|       |==αA==| |=====βD=========|               |=βE==|            |αB|     |βF|     |βG1|
         ------------90-------100-------110-------120-------130--------140-------150-------160--------170-------180---
CVB3     ..a.....KryaeWvItprqaaQLRrKLEfFTYvRFDlElTfviTstqqpsttqnq.daqilthQiMYVPPGqPvPdkvDsYvWQtstNPSvFwteGnappRmSVPfls
PV1/M    nkd.....KlfavWkItykdtvQLRrKLEfFTYsRFDmElTfvvTanfte...tnnghalnqvyQiMYVPPGaPvPekwDdYtWQtssNPSiFytyGtapaRiSVPyvg
HRV14    gidnhreaKlfndWkInlsslvQLRkKLElFTYvRFDsEyTilaTasqpd...sanyss.nlvvQaMYVPPGaPnPkewDdYtWQsasNPSvFfkvG.dtsRfSVPyvg

            |βG2|                                   |=βH=|          |========βI=======|
         ---190-------200--------------210-------220-------230-------240-------250--------260-------270--------280-
CVB3     ignAYsnFYDGwSefsr........nqvYGintLNnmGtlyaRhVNagstgpikstIRiYfkpKhvkaWiPRpPRlcqYeka.knvnfqpsqvtttr..qsittmtnt
PV1/M    isnAYshFYDGfSkvplkdqsaalgdslYGaasLNdfGilavRvVNdhnptkvtskIRvYlkpKhirvWcPRpPRavaYygp.gvdykdgtltplst..kdltty
HRV14    lasAYncFYDGyShdda........etqYGitvLNhmGsmafRiVNehdehktlvkIRvYhraKlveaWiPRaPRalpYtsigrtnypkntepvikkrkgdiksy
```

b

```
                         |βA1|  |βA2|                                       |αZ|   |==βB==|      |βC|       |==αA===|==
         --------10--------20--------30--------40--------50--------60--------70--------80--------90-------100
CVB3     SPtvEeCGYSDRVrsiTLGNSTITTQEcANvVVgYgvWPdYLkDseAtaedqptqPDvatCRFYTLDsvqWqktSpGWwWKLPDALsnlGlFGQNMqyHy
PV1/M    SPniEaCGYSDRVlqlTLGNSTITTQEaANsVVaYgrWPeYLrDseAnpvdqptePDvaaCRFYTLDtvsWtkeSrGWwWKLPDALrdmGlFGQNMyyHy
HRV14    SPnvEaCGYSDRVqqiTLGNSTITTQEaANaVVcYaeWPeYLpDvdAsdvnktskPDtsvCRFYTLDsktWttgSkGWcWKLPDALkdmGvFGQNMffHs

         ==βD======|        |==βE====|----------------PUFF--------------------------------------|=αB=|  |=βF|
         -------110-------120-------130---------140-------150-------160-----------170-------180-------190----
CVB3     LGRtGYTiHVQCNAsKFHqGcLlVvcvPEaemgcatln.n.tpssaellgGdtakefadkpvasqs....nklvqrvvynagmGvgvGNltiFPHQwINL
PV1/M    LGRsGYTvHVQCNAsKFHqGaLgVfavPEmclagdsntttmhtsyqnanpGekggtftgtftpdnnqtsparrfcpvdyllgnGtllGNafvFPHQiINL
HRV14    LGRsGYTvHVQCNAtKFHsGcLlVvviPEhqlasheggnvs.vkytfthpGergidlssa..nev.....ggpvkdviynmnGtllGNlliFPHQfINL

            |βG1=|      |βG2|   |====βH====|      |=======βI======|
         ---200-------210-------220-------230-------240-------250-------260------
CVB3     RTNNsATivmPYtNSvpmDnMfrHNNvtlmviPfvPLdycpqsttyvPITvTIAPMCaEynGlRlaqhq
PV1/M    RTNNcATlvlPYvNSlsiDsMvkHNNwgiailPlaPLnfasesspeiPITlTIAPMCcEfnGlRnitlprlq
HRV14    RTNNtATiviPYiNSvpiDsMtrHNNvslmviPiaPLtvptgatpslPITvTIAPMCtEfsGiRsksivpq
```

```
c
                                           |=αZ=|           |---Knob---|=βB|     |=βC=|        |=
         --------10--------20--------30--------40--------50--------60--------70--------80--------90-------100
CVB3     GLPtmntPGScQfLTsDdfQSPsAmPqydvTPemrIpGeVkNlmEiaevDsvvPvgn.vqekvnsmeaYqipvrsneqsqtqvfgfplqpgyssvfsrTlL
PV1/M    GLPvmntPGSnQyLTaDnfQSPcAlPefdvTPpidIpGeVkNmmElaeiDtmiPfdl.satkkntmemYrvrlsdkphtddpilclslspasdprlshTmL
HRV14    GLPtttlPGSgQfLTtDdrQSPsAlPnyepTPrihIlGkVhNllEiiqvDtliPmnnth..tkdevnsYliplnanrqneqvfgtnl..figdgvfktTlL

         αA=|  |====βD======|       |=βE==|        |αB|  |=βF=|    |=βG1|     |βG2|    |==βH==|
         -------110-------120-------130-------140-------150-------160-------170-------180-------190-------200
CVB3     GEIlnYYTHWsGSikltfmfcGsamatgKfllaYsPPGAgaPtkRvdAMLGTHVvWDvGLQSscvlciPWiSqthyRyvasDeyTagGfitcwYQTnivv
PV1/M    GEIlnYYTHWaGSlkftflfcGfmmatgKllvsYaPPGAdpPkkRkeAMLGTHViWDiGLQSsctmvvPWiSnttyRqtidDsfTegGyisvfYQTrivv
HRV14    GEIvqYYTHWsGSlrfssmytGpalssaKltlaYtPPGArgPqdRreAMLGTHVvWDiGLQStivmtiPWtSgvqfRytdpDtyTsaGflscwYQTslil

              |==βI====|
         -------210-------220-------230---------
CVB3     PadaqsscyimcFvSACnDFsvRLlkDTpfIsQenffq
PV1/M    PlstpremdilgFvSACnDFsvRLlrDTthIeQkalaq
HRV14    PpettgqvyllsFiSACpDFklRLmkDTqtIsQtvalte

d
         1-------10--------20--------30--------40--------50--------60-------69
CVB3     MGAQVStQKtGaHEtglnasgnSiihyTnINYYkDaASnsanrQdftqDPSKFTEPvKDimiKslPaLN
PV1/M    MGAQVSsQKvGaHEnsnrayggStinyTtINYYrDsASnaaskQdfsqDPSKFTEPiKDvliKtaPmLN
HRV14    MGAQVStQKsGsHEnqniltngSnqtfTvINYYkDaAStssagQslsmDPSKFTEPvKDlmlKgaPaLN
```

Fig. 2a-d. Sequence alignments of CVB3, PV1/M and HRV14 for VP1 (**a**), VP2 (**b**), VP3 (**c**) and VP4 (**d**). The numbering shown is for the CVB3 sequence. Secondary structural elements (for a definition see ROSSMANN et al. 1985) are shown above the sequences. *Capital letters* denote identical residues for the sequences shown. *Underlined residues* indicate surface residues for CVB3. *Bold type* face residues line the β-sandwich pocket of VP1. (From MUCKELBAUER et al. 1995)

located primarily on the viral surface. The structure of CVB3 shows not only the canyon depression about the fivefold axes but a large depression at the twofold axes. Both of these depressions are possible receptor recognition sites. The hydrophobic pocket of the CVB3 VP1 β-sandwich is occupied by a pocket factor modeled as a C_{16} fatty acid. It was also shown that the pocket factor can be displaced by an antiviral compound. Myristate is observed covalently linked to the NH_2-terminal of VP4 and putative ions are observed on the icosahedral threefold and fivefold axes.

2 Coxsackievirus B3 Amino Acid Sequence and Alignments

The strain of CVB3 for which the structure was solved is CVB3/M, originally characterized by GAUNTT et al. (1979). The amino acid sequence of the capsid protein of CVB3/M (Fig. 2) was provided by Dr. Steven Tracy, University of Nebraska Medical Center at Omaha and Dr. Charles Gauntt, University of Texas Health Science Center at San Antonio. Sequence alignments of the capsid proteins for CVB3/M, HRV14 and PV1/M (Fig. 2) were based on structural superpositions and they agree well with sequence alignments (PALMENBERG 1989) with some differences in loop regions.

3 Coxsackievirus B3 Atomic Model

The quality of the electron density map at 3.5 Å resolution was very good. The main chain density was continuous at a density level of 3σ above background which allowed for the building of the CVB3 sequence into the electron density map without ambiguity. VP1, VP2 and VP3 possessed the conserved β-sandwich folding motif observed for many icosahedral viruses (Fig. 3) (CHELVANAYAGAM et al. 1992). The β-sandwich consists of eight, antiparallel β-strands, B through I, that form two sheets, BIDG and CHEF. Regions of the amino acid sequence for which there was no electron density were the NH_2-terminal residues 1001–1012 (amino acid numbering convention adds 1000, 2000, 3000 and 4000 to the residues of VP1, VP2, VP3 and VP4, respectively) plus the densities of residues 1128 and 1130 of VP1, the NH_2-terminal residues 2001–2008 of VP2 and residues 4012–4024 of VP4. Coordinates for CVB3 coat protein are deposited in the Brookhaven Protein Data Bank (1COV).

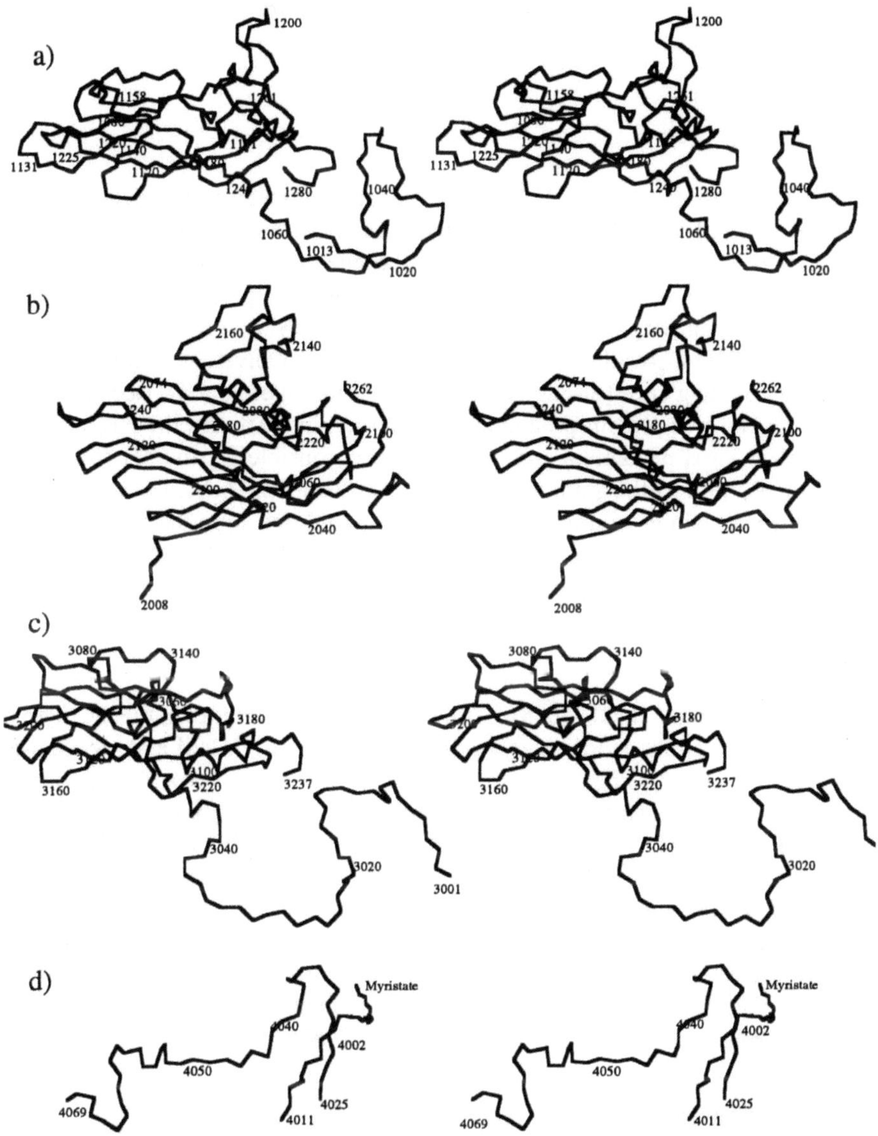

Fig. 3a–d. Alpha carbon (Cα) structure of CVB3. Stereo diagrams of the Cα tracings for (**a**) VP1, (**b**) VP2, (**c**) VP3 and (**d**) VP4 are shown with every 20th residue labeled. (From MUCKELBAUER et al. 1995)

4 The Structure of Coxsackievirus B3 Capsid Proteins

4.1 VP1

4.1.1 DE Loop

The DE loop (residues 1125–1137) is the most prominent of the exterior loops about the fivefold axis. It runs closest and parallel to the fivefold axis and, as a result, makes extensive interactions with its icosahedral fivefold symmetry related DE loops. The size and orientation of the CVB3 DE loop present greater opportunity for these symmetry related interactions to occur than do shorter DE loops of PV1/M and HRV14. Residues in the DE loop are also involved in interactions with putative ion binding sites (discussed below) along the fivefold axis. Because of the length and extensive protein-protein and protein-ion interactions, the DE loop may help stabilize virus pentamers and be partially responsible for the pH stability of CVB3. In addition to the stabilizing role of the DE loop, CAGGANA et al. (1993) have identified a single amino acid (Thr-1129) on the DE loop that is responsible for the expression of a virulent vs nonvirulent phenotype of CVB4. Therefore, it appears that the DE loop may play a role in both virion stability and pathogenicity.

4.1.2 BC Loop

The BC loop (residues 1081–1085) is nine amino acids shorter than the BC loop of HRV14 and five amino acids shorter than the BC loop of PV1/M and, as a consequence, it is not as prominent a surface loop as it is for the polio- and rhinoviruses. This loop flanks the north rim of the canyon and is an important neutralizing immunogenic (NIm) site for the polio- and rhinoviruses (SHERRY et al. 1985, 1986; MINOR et al. 1986; PAGE et al. 1988). CVB4/CVB3 chimera studies have shown that the BC loop of CVB4 contains a major NIm site (REIMANN et al. 1991). Residues located on the BC loop have been determined to be responsible for the host range specificity of poliovirus 2-Lansing (PV2/L) (MURRAY et al. 1988) and, in the case of HRV14 and HRV16, residues in the BC loop are located within the ICAM-1 receptor footprint and may play a role in receptor recognition.

4.1.3 Pocket Factor

A long, thin worm-like density was found inside the VP1 β-sandwich at the binding site of antiviral compounds. This unidentified pocket factor, modeled as a palmitate molecule (16 carbon fatty acid) (Fig. 4.a), is very similar in size and shape to that observed in PV1/M and PV3/S (FILMAN et al. 1989). The polar head group of the pocket factor is towards the pore end of the pocket while the aliphatic chain extends the length of the hydrophobic pocket towards the fivefold axis. For the two oxygens of the polar head group, the electron density suggests that one oxygen is hydrogen bonded to the side chain nitrogen of Arg-1095 while the second oxygen

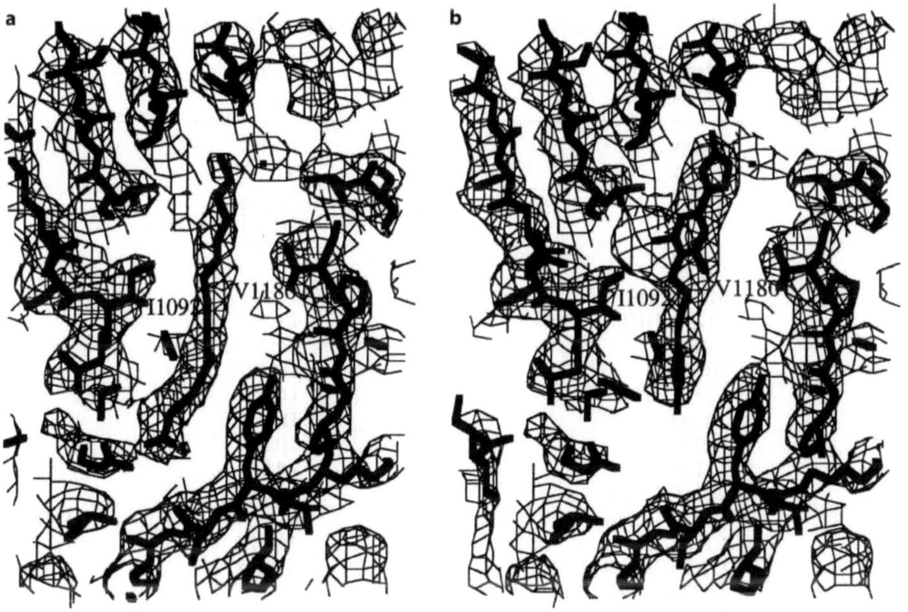

Fig. 4a, b. Averaged electron density maps showing the VP1 β-sandwich pocket. **a** CVB3 native electron density and structure showing the pocket factor in the middle of the figure modeled as palmitate. The map was computed with $|F_{native}|$(unit weights) coefficients and the final phases determined by molecular replacement electron density averaging. The model is that obtained from X-PLOR refinement. The structure of the palmitate molecule, representing the pocket factor, was obtained by graphical model building. **b** CVB3 complexed with the antiviral compound, WIN 66393. The view is the same as shown for (**a**). The iodine atom can be seen as the large lump of density off the phenyl ring of the drug. The map was computed with $(|F_{win}| - k|F_{native}|)$ (unit weights) coefficients where k=0.65 was chosen to equalize the WIN compound and protein density. The final phases and the superimposed structure are as in (**a**). The WIN compound structure is based on graphical model building. $|F_{native}|$ and $|F_{win}|$ are the structure amplitudes for the native and drug soaked crystals. (Adapted from MUCKELBAUER et al. 1995)

may indirectly interact with the side chain oxygen of Ser-1184 via a water molecule. The carbon chain of the fatty acid makes extensive interactions with the side chains of hydrophobic residues lining the β-sandwich pocket (Table 1). When comparing the position of the pocket factors for CVB3, PV1/M and HRV1A and HRV16, the polar head groups are all in similar locations at the pore end of the pocket while the aliphatic chains differ significantly in their conformations and lengths with respect to each other. This is, presumably, due to the different shapes of the pocket interiors for each of the viruses which are defined by the amino acids that line the pocket.

Table 1. CVB3 residues within 4.0 Å of palmitate and WIN 66393 (adapted from Muckelbauer et al. 1995).

WIN 66393 atom	CVB3 residue	Palmitate atom	CVB3 residue
C51	Y1189	O1	R1095
N5	P1094	O2[1]	S1184
C4	P1094	C1	R1095
C3	M1213	C2	N1211, M1213
C2	M1213	C3	P1094
N1	F1237	C4	Y1189
C1#	M1213, I1092	C5	P1094
C2#	F1237, I1092	C6	F1237
C3#	F1237	C7	M1213
O1*	I1092	C8	I1092, L1114
C1*	L1116, I1092	C9	I1092
C2*	Y1143	C10	I1092
C3*	Y1143	C11	I1092
C4*	L1116, L1216	C12	L1116
C5*	L1116, L1216	C13	L1116
C6*	L1116, L1216	C14	M1178, Y1143, L1116, L1216
C21	L1183, Y1143	C15	M1178, Y1143
I61	L1116, W1090, L1216	C16	M1178
N1'	Y1116, M1178, P1165		
C2'	Y1143, M1178, L1216		
N3'	M1178, Y1143, L1216		
C4'	M1178, P1165		
O5'	M1178, P1165, Y1143		
C41	M1178, V1167, S1166, P1165		

[1] Possible water mediated hydrogen bond.

4.2 VP2

4.2.1 EF Loop (Puff)

The largest, most variable surface loop of VP2 is the EF loop or puff (residues 2129–2180). The puff of CVB3 is six residues shorter than the puff of PV1/M and three residues longer than the puff of HRV14. The most exterior region of the puff loop contains NIm sites for the polio- and rhinoviruses (SHERRY et al. 1985, 1996; MINOR et al. 1986; PAGE et al. 1988). The puff may play a structural role in maintaining the periphery topology of the canyon and may be directly or indirectly involved in receptor binding or recognition. In the case of HRV14, residues within the ICAM-1 footprint include residues from the puff loop near the south rim of the canyon (OLSON et al. 1993). In addition, mutational studies of CVB3 have shown that the puff is fairly intolerable to mutagenesis, especially to those residues that line the rim of the canyon and the depression around the twofold axis (B. Coller, S. Tracy, unpublished results).

4.2.2 VP2 Residues Responsible for CVB3-RD Phenotype

Two amino acid residues in VP2 have been identified as being responsible for the ability of a CVB3 variant (CVB3-RD) to utilize a secondary receptor, decay-accelerating factor (DAF) (BERGELSON et al. 1995), as a functional receptor on normally unsusceptible rhabdomyosarcoma (RD) cells (REAGAN et al. 1984; LINDBERG et al. 1992). One of the residues, Asp-2108 to Val, is internal and located on βD. The other residue, Thr-2151 to Ser, is located on the VP2 puff, adjacent to the twofold depression and may be involved in receptor recognition. Both Val- 2108 and Ser-2151 have been observed in other isolates of CVB3. The isolate, CVB3/Ø, which has a Ser at 2151, does not infect RD cells (S. Tracy, personal communication), which suggests that this change alone is insufficient to permit RD cell infection.

4.3 VP3

The NH_2-terminal 25 residues of VP3 are located near the icosahedral fivefold axes around 110 Å from the viral center. The fivefold symmetry related NH_2-terminals form an intertwining β-barrel cylinder whose structure is highly conserved in all picornavirus structures determined so far. This β-cylinder structure is likely to be important in pentamer formation after the release of the VP3 terminals by post-translational cleavage (HOGLE et al. 1985; ROSSMANN et al. 1985;). Residues 3058–3069 form the knob which is the major surface protrusion of VP3 and is the location of NIm sites for rhino- and polioviruses (SHERRY et al. 1985, 1986; MINOR et al. 1986; PAGE et al. 1988).

4.4 VP4

Electron density extending from the NH_2-terminals of VP4 was identified as myristate, a 14 carbon fatty acid, covalently bound to the NH_2-terminal glycine of VP4. There was continuous electron density up to atom C_{12} of myristate while the atoms C_{13} and C_{14} were missing in the map and presumably assume multiple conformations. The fivefold related myristates cluster at the interior surface of the virion just below the NH_2-terminal VP3 β-cylinder. The myristate mediates interactions between the NH_2-terminals of VP3 and VP4 and is thought to play an important role in virus assembly by stabilizing 14S pentamers (ANSARDI et al. 1992). It has also been suggested that myristate may be involved in disassembly by mediating the association of virus to cellular membranes (CHOW et al. 1987). The position of the myristate is essentially identical to that found in PV1/M (FILMAN et al. 1989). The carboxylate oxygen of myristate hydrogen bonds to a fivefold symmetry-related side chain oxygen of Thr-4028, and it has been shown by mutagenesis that this interaction is important for normal viral assembly and infectivity for the polioviruses (Moscufo and Chow 1992). VP4 meanders from the fivefold axis along the capsid interior to the threefold axis close to the NH_2-terminal of VP2.

5 Coxsackievirus B3 Surface Characteristics

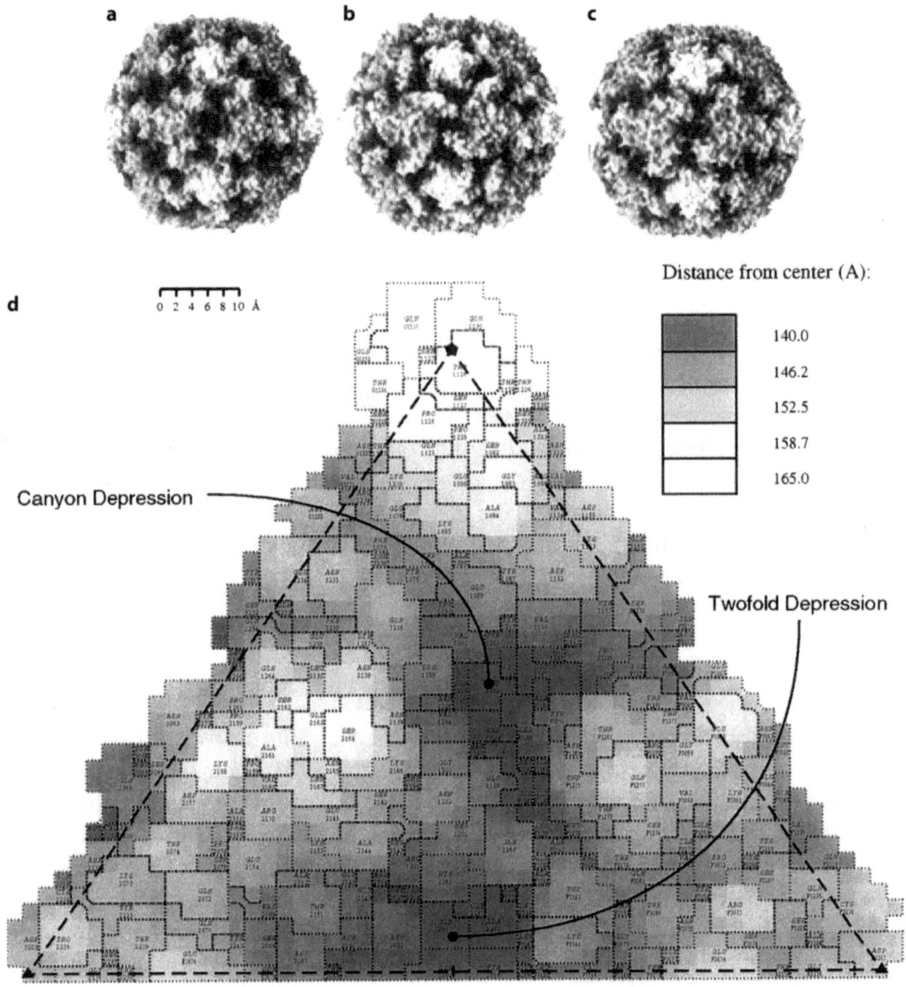

Fig. 5a–d. Radial depth-cued images viewed down the icosahedral twofold axis of (**a**) CVB3, (**b**) PV1/M and (**c**) HRV14. Regions in *dark gray* represent surface depressions while regions of *white* represent surface protrusions. Images courtesy of Jean-Yves Sgro, Institute for Molecular Virology, University of Wisconsin-Madison. (**d**) A CVB3 roadmap of the icosahedral asymmetric unit viewed down the icosahedral twofold axis showing the surface residues relative to the distance from the viral center. The depth range is from 140.0 to 165.0 Å with *dark gray* and *white* corresponding to 140.0 Å and 165.0 Å from the viral center, respectively. Positions of the canyon and twofold depression are labeled. (Adapted from Muckelbauer et al. 1995)

5.1 Surface Topology

The water-accessible surfaces of CVB3, PV1/M and HRV14 (Fig. 5a–c) demonstrate the similarities and differences between these three viruses. The canyon surrounding the icosahedral fivefold axis for CVB3 is not as pronounced as the canyon for the PV1/M and HRV14. The CVB3 canyon appears to be bridged at the most narrow region between VP1 and VP2 giving the impression of five pits approximately 15 Å deep. These pits correspond to the HRV14 and HRV16 receptor binding site (Olson et al. 1993; N. Olson, P. Kolatkar, M. Rossmann, T. Baker, unpublished results). Flanking the icosahedral twofold axes is a second 15 Å deep surface depression. The floor and walls of the twofold depression are lined, predominately, with residues from VP2 and some from VP3 (Fig. 5d). Residue 2151, one of the two amino acids responsible for the RD phenotype of CVB3-RD (discussed above), projects directly into the twofold depression (shown in Fig. 5d as a surface residue adjacent to the twofold depression) and suggests that this surface depression may serve as a possible receptor binding site.

5.2 Conservation of Amino Acids

Sequence alignments of CVB1, CVB3, CVB4 and CVB5 (A. Palmenberg, personal communication) show that the regions of greatest sequence conservation coincide with the canyon surrounding the fivefold axis and the twofold depression. Amino acids in the canyon of rhinoviruses have been shown to be the most conserved surface amino acids (Rossmann and Palmenberg 1988; Chapman and Rossmann 1993). The conservation of amino acids in the canyon and, to a lesser extent, the twofold depression for the CVBs, supports the hypothesis that the canyon and/or the twofold depression are receptor binding sites. The more extensive conservation of residues in the canyon, as opposed to the twofold depression, may distinguish the canyon as the major receptor binding site for all six CVB serotypes, whereas the twofold depression may function as an alternate or second receptor site.

6 Ions

6.1 Threefold

An unidentified density, twice the height of the surrounding protein density, was observed on the icosahedral threefold axis at the outer surface of the virion, approximately 148 Å from the viral center. A similar density to that observed for HRV14 has recently been determined, by EGTA soaking experiments, to be Ca^{2+} liganded to Glu-3200 (R. Zhao, M. Rossmann, unpublished results). Although soaking experiments with EGTA have not been carried out, we propose that the observed density on the threefold axis of CVB3 is also Ca^{2+}. Bidentate coordination (~2.5 Å distances) occurs with the six side chain oxygens of threefold related Asp-3203. An additional ligand to a water molecule occurs directly below the Ca^{2+} closer to the viral center on the threefold axis. Residue 3203 is conserved among the CVBs as aspartic or glutamic acid, suggesting that all CVBs may bind Ca^{2+} at this position. However, the trend of conservation does not hold for all rhino-and enteroviruses and, hence, the threefold Ca^{2+} site for the picornaviruses is not ubiquitous.

6.2 Fivefold

Two peaks of electron density, observed on the fivefold axis at 153 Å and 147 Å from the viral center, are also proposed to be ion binding sites. These peaks, approximately 75% the height of the surrounding protein electron density, appear to make water mediated interactions with the side chain oxygen of Asp-1133 and with the main chain oxygen of Gln-1132. Residue Asp-1133 is conserved as aspartic acid for the CVBs and both Asp-1133 and Gln-1132 are located on the DE loop (discussed above) of VP1. The density of these peaks is weaker than the threefold ion peak and is not well defined, suggesting that the sites on the fivefold axis are not as well occupied as the putative Ca^{2+} ion on the threefold. The peak at 153 Å corresponds to the putative ion binding site previously observed for HRV14, Mengo virus, HRV1A and HRV16 (Rossmann et al. 1985; Luo et al. 1987; Kim et al. 1989; Oliveira et al. 1993). The same EGTA soaking experiments that identified Ca^{2+} on the threefold of HRV14 also identified Ca^{2+} at the fivefold site (R. Zhao, M. Rossmann, unpublished results). There is evidence for HRV14 that the occupancy of the fivefold ion site, at times, is variable (Giranda et al. 1992) (M. Rossmann, unpublished results). The Ca^{2+} may bind to the rhinoviruses due to its presence in the crystallization solution or, alternatively, it may bind during the assembly process. For CVB3, the purification and crystallization solutions do not contain Ca^{2+} and crystallization is from NaCl. The favorable protein environment for ion binding and the analogous ions observed for other picornaviruses suggests that CVB3 can bind a cation at the fivefold site.

7 Coxsackievirus B3 Complexed with Antiviral Compound

7.1 WIN 66393

The antiviral compound WIN 66393, synthesized by Sterling Winthrop Pharmaceuticals Research Division, was used for CVB3 crystal soaking studies. WIN compounds do not inhibit CVB3 attachment to monkey kidney cells but are active against CVB3 by inhibiting viral uncoating (P. Fox, G. Diana, F. Dutko, J. Groarke,

D. Pevear, unpublished results). WIN 66393 is iodinated on the phenyl ring (Table 1) and was used as a potential heavy atom marker for crystallographic experiments in attempts to solve particle positional, orientational and packing ambiguities described elsewhere (MUCKELBAUER et al. 1996).

7.2 Coxsackievirus B3: WIN 66393 Complex

WIN 66393 bound into the hydrophobic β-sandwich pocket of VP1 (Fig. 4b). Little conformational change occurred upon the binding of drug because the drug displaced the pocket factor found in the native state. As WIN 66393 is considerably shorter than the native pocket factor, it was apparent by the absence of pocket factor at the ends of the pocket that the antiviral compound displaced the pocket factor completely (compare Fig. 4a with Fig. 4b). There was some indication that I-1092 moved (~1.0 Å) to accommodate the substituted phenyl ring on the WIN compound. In light of the small conformational changes in the pocket and canyon on the binding of WIN 66393, assuming that the canyon is the receptor binding site, it is not surprising that there is no inhibition of receptor attachment. Moreover, if the receptor binding site were elsewhere, the WIN compound would also not inhibit attachment. Amino acid residues in the pocket close to the drug are predominately hydrophobic (Table 1). The main chain oxygen of Pro-1165 at the interior end of the pocket and the hydroxyl group of Tyr-1143 are the only two polar atoms within 4.0 Å of WIN 66393.

7.3 Coxsackievirus B3/M vs Coxsackievirus B3-Nancy

WIN compounds with substituted phenyl rings active against the Gauntt strain (CVB3/M) were found to be inactive against the Nancy strain (CVB3-Nancy) (P. Fox, G. Diana, F. Dutko, J. Groarke, D. Pevear, T. Woods, unpublished results). There are two residues located in the binding pocket that differ between CVB3/M and CVB3-Nancy. For CVB3-Nancy, residue 1092 is leucine instead of isoleucine and residue 1180 is isoleucine instead of valine. Residue 1092 is conserved as isoleucine except for CVB3-Nancy and residue 1180 is conservatively hydrophobic (valine, leucine or isoleucine) for the rhino- and enteroviruses. These two residues are situated across the pocket from each other with their side chains extended into the pocket (Fig. 4) and both are near the substituted phenyl ring of the WIN compound. The change from isoleucine to leucine at 1092 and of valine to isoleucine at 1180 may be responsible for the inability of CVB3-Nancy to bind WIN compounds. The amino acid changes fill the pocket and may block the binding of drugs in a manner analogous to the exclusion mutants of HRV14 (HEINZ et al. 1990). In support of this conclusion is the observation that compounds with unsubstituted phenyl rings do not differentiate between CVB3/M and CVB3-Nancy (G. Diana, F. Dutko, J. Groarke, D. Pevear, unpublished results).

8 Future Directions

8.1 Coxsackievirus B3 Receptors Complexes

The interaction of viruses with their cellular receptors is an initial component in the life cycle of a virus. Identification, isolation and overexpression of the CVB receptor(s) has, however, not been an easy task. We are in the process of overexpressing a soluble form of DAF (see above) to form complexes with the virus. Such a complex can then be studied structurally using cryoelectron microscopy and X-ray crystallography (Olson et al. 1993; N. Olson, P. Kolatkar, M. Rossmann, T. Baker, unpublished results). Such work is also in progress for the homologous rhinovirus. It will be interesting to compare how different cellular molecules are utilized by similar viruses for attachment and cell entry.

8.2 Identification of Pocket Factors

Cellular attachment is only the first stage in the infectious process. Attachment must initiate subsequent steps of cell entry, perhaps by endocytosis. Displacement of the pocket factor by the receptor has been suggested as a possible mechanism for initiating the disassembly process (Rossmann 1994). Thus, determining the chemical nature of the pocket factor is important. Attempts at displacing the pocket factor using better binding antiviral compounds in HRV16 and then identifying the pocket factor with mass spectroscopy have not yet succeeded. Using CVB3, where the pocket factor is larger and more fully substituted, may be a possibility.

8.3 Antiviral Compounds for Coxsackieviruses

Coxsackie A and B viruses can cause severe and even fatal diseases. Hence, the availability of therapeutic antiviral agents would be of considerable benefit. While the further development of potential WIN compounds merely for coxsackie viruses may not be economically viable, their prophylactic use to prevent the common cold gives considerable hope. It is, therefore, necessary to further explore the relationship between low minimal inhibitory drug concentrations and the binding mode of the drugs in the coxsackie A and B viruses. Such studies require both experimental and crystallographic data and their computational analysis.

Acknowledgements. We gratefully acknowledge Steven Tracy and Charles Gauntt for the amino acid sequence of the CVB3/M isolate. We are also grateful to Sterling Winthrop Pharmaceuticals Research Division for providing the initial stock of CVB3/M and the antiviral compound WIN 66393. This work was funded by grants from the National Institute of Health and Sterling Winthrop Pharmaceuticals Research Division to MGR.

References

Acharya R, Fry E, Stuart D, Fox G, Rowlands D, Brown F (1989) The three-dimensional structure of foot-and-mouth disease virus at 2.9Å resolution. Nature 337:709–716

Ansardi DC, Porter DC, Morrow CD (1992) Myristylation of poliovirus capsid precursor P1 is required for assembly of subviral particles. J Virol 66:4556–4563

Badger J, Minor I, Oliveira MA, Smith TJ, Rossmann MG (1989) Structural analysis of antiviral agents that interact with the capsid of human rhinoviruses. Proteins 6:1–19

Bergelson JM, Mohanty JG, Crowell RL, St. John N, Lublin DM, Finberg RW (1995) Coxsackievirus B3 adapted to growth in RD cells binds to decay-accelerating factor (CD55). J Virol 69:1903–1906

Caggana M, Chan P, Ramsingh A (1993) Identification of a single amino acid residue in the capsid protein VP1 of coxsackievirus B4 that determines the virulent phenotype. J Virol 67:4797–4803

Caspar DLD, Klug A (1962) Physical principles in the construction of regular viruses. Cold Springs Harbor Symp Quant Biol 27:1–2

Chapman MS, Rossmann MG (1993) Comparison of surface properties of picornaviruses: strategies for hiding the receptor site from immune surveillance. Virology 195:745–756

Chelvanayagam G, Heringa J, Argos P (1992) Anatomy and evolution of proteins displaying the viral capsid jellyroll topology. J Mol Biol 228:220–242

Chow M, Newman JFE, Filman D, Hogle JM, Rowlands DJ, Brown F (1987) Myristylation of picornavirus capsid protein VP4 and its structural significance. Nature 327:482–486

Colonno RJ, Condra JH, Mizutani S, Callahan PL, Davies M, Murcko MA (1988) Evidence for the direct involvement of the rhinovirus canyon in receptor binding. Proc Natl Acad Sci 85:5449–5453

Colston E, Racaniello VR (1994) Soluble receptor-resistant poliovirus mutants identify surface and internal capsid residues that control interaction with the cell receptor. EMBO J 13:5855–5862

Filman DJ, Syed R, Chow M, Macadam AJ, Minor PD, Hogle J (1989) Structural factors that control conformational transitions and serotype specificity in type 3 poliovirus. Embo J 8:1567–1579

Fox MP, Otto MJ, McKinlay MA (1986) The prevention of rhinovirus and poliovirus uncoating by WIN 51711: a new antiviral drug. Antimicrob Agents Chemother 30:110–116

Gauntt CJ, Trousdale MD, LaBadie DRL, Paque RE, Nealon T (1979) Properties of coxsackievirus B3 variants which are amyocarditic or myocarditic for mice. J Med Virol 3:207–220

Giranda VL, Heinz BA, Oliveira MA, Minor I, Kim KH, Kolatkar PR, Rossmann MG, Rueckert, RR (1992) Acid-induced structural changes in human rhinovirus 14: possible role in uncoating. Proc Natl Acad Sci USA 89:10213–10217

Grant RA, Hiremath CN, Filman DF, Syed R, Andries K, Hogle JM (1994) Structures of poliovirus complexes with anti-viral drugs: Implications for viral stability and drug design. Curr Biol 4:784–797

Heinz BA, Shepard DA, Rueckert RR (1990) Escape mutant analysis of a drug-binding site can be used to map functions in the rhinovirus capsid. In: Laver WG, Air GM (eds) Use of x-ray crystallography in the design of antiviral agents. Academic, New York, pp173–186

Hogle JM, Chow M, Filman DJ (1985) Three-dimensional structure of poliovirus at 2.9Å resolution. Science 229:1358–1365

Hsu KL, Paglini S, Alstein B, Crowell RL (1990) Identification of a second cellular receptor for a coxsackievirus B3 variant, CVB3-RD. In: Brinton MA, Heinz FX (eds) New aspects of positive strand RNA viruses. American Society for Microbiology, Washington, DC, pp 271–277

Kim KH, Willingmann P, Gong ZX, Kremer MJ, Chapman MS, Minor I, Oliveira MA, Rossmann MG, Andries K, Diana GD, Dutko FJ, McKinlay MA, Pevear DC (1993) A comparison of the anti-rhinoviral drug binding pocket in HRV14 and HRV1A. J Mol Biol 230:206–227

Kim S, Smith TJ, Chapman MS, Rossmann MG, Pevear DC, Dutko FJ, Felock PJ, Diana GD, McKinlay MA (1989) The crystal structure of human rhinovirus serotype 1A (HRV1A). J Mol Biol 210:91–111

Lindberg MA, Crowell RL, Zell R, Kandolf R, Pettersson U (1992). Mapping of the RD phenotype of the Nancy strain of coxsackievirus B3. Virus Research 24:187–196

Luo M, Vriend G, Kamer G, Minor I, Arnold E, Rossmann MG, Boege U, Scraba DG, Duke GM, Palmenberg A (1987) The atomic structure of Mengo virus at 3.0Å resolution. Science 235:182–191

Mapoles JE, Krah DL, Crowell RL (1985) Purification of a HeLa cell receptor protein for the group B coxsackieviruses. J Virol 55:560–566

Melnick JL (1990) Enteroviruses: polioviruses, coxsackieviruses, echoviruses and newer enteroviruses.

In: Fields B, Knipe D (eds) Virology, 2nd edn. Raven, New York, 1:549–605

Minor PD, Ferguson M, Evans DM, Almond JW, Icenogle JP (1986) Antigenic structure of polioviruses of serotypes 1, 2 and 3. J Gen Virol 67:1283–1291

Moscufo N, Chow M (1992) Myristate-protein interactions in poliovirus: Interactions of VP4 threonine 28 contribute to the structural conformation of assembly intermediates and the stability of assembled virions. J Virol 66:6849–6857

Muckelbauer JK, Kremer M, Minor I, Diana G, Dutko FJ, Groarke J, Pevear DC, Rossmann MG (1995) The structure of coxsackievirus B3 at 3.5Å resolution. Structure 15:653–667

Muckelbauer JK, Kremer M, Minor I, Tong L, Zlotnick A, Johnson JE, Rossmann MG (1996) The structure determination of coxsackievirus B3 to 3.5Å resolution. Acta Crystallogr D51:871–887

Murray MG, Bradley J, Yang XF, Wimmer E, Moss EG, Racaniello VR (1988) Poliovirus host range is determined by a short amino acid sequence in neutralizing antigenic site 1. Science 241:213–215

Oliveira MA, Zhao R, Lee W-M, Kremer MJ, Minor I, Rueckert RR, Diana GD, Pevear DC, Dutko FJ, McKinlay MA, Rossmann MG (1993). The structure of human rhinovirus 16. Structure 1:51–68

Olson NH, Kolatkar PR, Oliveira MA, Cheng RH, Greve JM, McClelland A, Baker TS, Rossmann MG (1993) Structure of a human rhinovirus complexed with its receptor molecule. Proc Natl Acad Sci USA 90:507–511

Page GS, Mosser AG, Hogle JM, Filman DJ, Rueckert RR, Chow M (1988) Three-dimensional structure of poliovirus serotype 1 neutralizing determinants. J Virol 63:1781–1794

Palmenberg AC (1989) Sequence alignments of picornaviral capsid proteins. In: Semler BL, Ehrenfeld E (eds) Molecular aspects of picornavirus infection and detection. American Society for Microbiology, Washington, DC, pp 211–241

Pevear DC, Fancher MJ, Felock PJ, Rossmann MG, Miller MS, Diana G, Treasurywala AM, McKinlay MA, Dutko FJ (1989) Conformational changes in the floor of the human rhinovirus canyon blocks adsorption to HeLa cell receptors. J Virol 63:2202–2207

Reagan KJ, Goldberg G, Crowell RL (1984). Altered receptor specificity of coxsackievirus B3 after growth in rhabdomyosarcoma cells. J Virol 49:635–640

Reimann BY, Zell R, Kandolf R (1991) Mapping of a neutralizing antigenic site of coxsackievirus B4 by construction of an antigen chimera. J Virol 65:3475–3480

Rossmann MG (1989) The canyon hypothesis. J Biol Chem 264:14587–14590

Rossmann MG (1994) Viral cell recognition and entry. Prot Sci 3:1712–1725

Rossmann MG, Palmenberg AC (1988) Conservation of the putative receptor attachment site in picornaviruses. Virology 164:373–382

Rossmann MG, Arnold E, Erickson JW, Frankenberger EA, Griffith JP, Hecht H-J, Johnson JE, Kamer G, Luo M, Mosser AG, Rueckert RR, Sherry B, Vriend G (1985) Structure of a human common cold virus and functional relationship to other picornaviruses. Nature 317: 145–156

Rueckert RR (1990) Picornaviridae and their replication. In: Fields B, Knipe D (eds) Virology, 2nd edn, Raven, New York, 1:507–548

Sherry B, Rueckert RR (1985) Evidence for at least two dominant neutralization antigens on human rhinovirus 14. J Virol 53:137–143

Sherry B, Mosser AG, Colonno RJ, Rueckert RR (1986) Use of monoclonal antibodies to identify four neutralization immunogens on a common cold picornavirus, human rhinovirus 14. JVirol 57:246–257

Smith TJ, Kremer MJ, Luo M, Vriend G, Arnold E, Kamer G, Rossmann MG, McKinlay MA, Diana GD, Otto MJ (1986) The site of attachment in human rhinovirus 14 for antiviral agents that inhibit uncoating. Science 233:1286–1293

Zhang A, Nanni RG, Li T, Ferstandig Arnold G, Oren DA, Jacobo-Molina A, Williams RL, Kamer G, Rubenstein DA, Li Y, Rozhon E, Cox S, Buontempo P, O'Connell J, Schwartz J, Miller G, Bauer B, Versace R, Pinto P, Ganguly A, Girijavallabhan V, Arnold E (1993) Structure determination of antiviral compound SCH 38057 complexed with human rhinovirus 14. J Mol Biol 230:857–867

Identification and Biology of Cellular Receptors for the Coxsackie B Viruses Group

R. J. KUHN

1 Introduction

The picornaviruses represent a diverse and well studied family of RNA-containing animal viruses. In addition to having a simple structural organization and a limited size plus-strand RNA genome, members of the family constitute a significant group of human and animal pathogens. The coxsackie B viruses (CVB), which belong to the enterovirus genus of *Picornaviridae*, have long been recognized as the causative agents for a number of human ailments frequently involving the heart. Infection of neonates with CVB commonly occurs and can sometimes result in severe consequences. The tissue tropism of these and other viruses has long been thought to be controlled at least in part by the interaction of surface residues on the virus with specific cellular proteins found on the membrane of the target cells. The binding of the virus to these cellular proteins, which act as receptors, results ultimately in the entry of viral nucleic acid into the target cell through processes that are poorly understood (ROSSMANN 1994).

Department of Biological Sciences, Purdue University, 1392 Lilly Hall of Life Sciences, West Lafayette, Indiana 47907, USA

This attachment step in the virus life cycle has been intensively investigated since the introduction of cell culture techniques for the propagation of viruses. Early biochemical experiments were designed to measure the binding ability of a virus to a host cell. These studies were closely followed by competition assays which examined viruses based on their ability to compete with one another for receptors on a given cell type. The use of polyacrylamide gels to separate proteins and blotting techniques permitted the identification of specific polypeptides that could bind to viruses. These experiments were instrumental in establishing the nature and quantity of cellular receptors for viruses. With the introduction of molecular genetics it became possible to precisely identify cellular receptors involved in attachment and entry using functional assays. The ability to clone cDNAs that encoded the receptor permitted an analysis of the interactions on a molecular level using structural, genetic and biochemical techniques. Consequently, the identification of cellular receptors for viruses has proceeded rapidly over the past several years using these technologies.

The decade of the 1980s saw rapid progress in the understanding of the molecular biology of the picornaviruses. The construction of cDNA clones from which infectious virus could be rescued allowed for the first time molecular genetic analyses on these viruses (Racaniello and Baltimore 1981). The high resolution crystal structures of human rhinovirus 14 (Rossmann et al. 1985) and of poliovirus type 1 (Hogle et al. 1985) provided detailed information about the structure of the virus particles and showed that rhinovirus and poliovirus were remarkably similar in many structural features despite their significant biological differences. On the basis of structural and antibody studies, Rossmann et al. (1985) proposed the "canyon hypothesis." They suggested that the cellular receptor for rhinovirus would bind into the depression or canyon that surrounds the fivefold axes of the virion and would be largely impenetrable by antibodies (Rossmann et al. 1985). The first cellular receptors identified for the picornaviruses, intercellular adhesion molecule-I for the major group of rhinoviruses, and the poliovirus receptor for polioviruses types 1–3, were members of the immunoglobulin (Ig) superfamily of proteins (Mendelsohn et al. 1986; Greve et al. 1989; Staunton et al. 1989). These initial identifications suggested that all of the picornaviruses might utilize receptors from this family of proteins. However, it is now evident that the picornaviruses utilize a wide variety of distinct cell surface receptors, and the concept of a single family of cellular receptors for the picornaviridae has been abandoned. Table 1 lists the cellular receptors that have been identified to date that picornaviruses use to enter cells.

Table 1. Cellular receptors utilized by picornaviruses

Virus	Cellular receptor family	Reference
Coxsackie A9	Vitronectin receptor ($\alpha_v\beta_3$)integrin	ROIVAINEN et al., 1994
Coxsackie A13, 18, 21	Intercellular adhesion molecule 1 (ICAM-1)/immunoglobulin	COLONNO et al., 1986
Coxsackie B 1–6	Nucleolin (?)	DE VERDUGO et al., 1995
Coxsackie B3-RD	Decay accelerating factor (DAF)/ regulators of complement activation (RCA)	BERGELSON et al., 1995
Coxsackie B1, 3, 5	Decay accelerating factor (DAF)/ regulators of complement activation (RCA)	SHAFREN et al., 1995
Echo 1, 8	VLA-2 ($\alpha_2\beta_1$)/integrin	BERGELSON et al., 1992; BERGLESON et al., 1993
Echo 6, 7, 11, 12, 13, 20, 21, 29, 33	Decay accelerating factor (DAF)/ regulators of complement activation (RCA)	BERGELSON et al., 1994 WARD et al., 1994
Encephalomyocarditis	Vascular cell adhesion molecule 1 (VCAM-1)/immunoglobulin	HUBER 1994
Foot-and-mouth-disease A_{12}	Vitronectin receptor ($\alpha_2\beta_1$)/integrin	BERINSTEIN et al., 1995
Human rhino (major group)	Intercellular adhesion molecule 1 (ICAM-1)/immunoglobulin	GREVE et al., 1989 STAUNTON et al., 1989 TOMASSINI et al., 1989
Human rhino (major group)	Low density lipoprotein receptor (LDLR)/LDLR	HOFER et al., 1994
Polio 1–3	Poliovirus receptor (PVR)/immunoglobulin	MENDELSOHN et al., 1989

2 Early Studies of Cellular Receptors for the Coxsackie B Viruses

The concept of a cellular receptor controlling the initial step in virus pathogenesis dates to the early work by Holland and colleagues, who demonstrated that the introduction of the viral RNA of poliovirus into nonsusceptible cells resulted in replication and production of progeny virions (HOLLAND et al. 1959a, b). These studies suggested that there was a specific recognition event between the virus and the host cell that was required for virus infection. Virus attachment interference assays were used to examine the ability of viruses to protect cells from superinfection by homologous and heterologous viruses. These studies demonstrated that infection with one serotype of CVB rendered a cell refractory to another CVB serotype infection. In contrast, attachment and infection were unaffected for poliovirus and at least three serotypes of coxsackie A virus (13, 15, 18) (CROWELL 1963, 1966). This virus blockade occurred at the cell surface and was observed to affect both virus adsorption and cell penetration. Treatment of intact HeLa cells with chymotrypsin rendered the cells insensitive to CVB infection, demonstrating that receptor activity was present on the extracellular surface of the plasma membrane (ZAJAC and CROWELL 1965a). In addition, within 4 h of this protease treatment, cells recovered 50% of their receptor activity and this activity was not found on intracellular membranes (ZAJAC and CROWELL 1965b).

In addition to the sensitivity of the CVB receptor to chymotrypsin, it was shown that the CVB receptor on HeLa cells could be incubated for 30 min at 56°C and retain activity. This differed from the poliovirus receptor which lost the ability to attach virus under these conditions (ZAJAC and CROWELL 1969), confirming that these viruses use different cellular receptors. Furthermore, binding of coxsackievirus B3 (CVB3) was able to saturate the receptors on HeLa cells, and the rates of attachment of the six CVB serotypes to the cells were found to differ (CROWELL 1963; CROWELL and LANDAU 1983). This suggested that each virus might bind to the same receptor with a different charged surface. This was further supported by the observation that each of the six serotypes had different pH optimums for attachment.

Enteroviruses that attach to cells under physiological conditions undergo an irreversible alteration that renders the virions noninfectious (HOLLAND 1962). This "eclipse" of virus infectivity suggests that association of the virus with the cellular receptor initiates the process of particle disassembly. CROWELL and PHILIPSON (1971) showed that at 0°C CVB3 bound and remained attached to HeLa cells with no subsequent penetration. As was previously shown with poliovirus (HOLLAND 1962), shift of the cell-associated CBV3 to 37°C, resulted in the elution of greater than 50% of the virus from the HeLa cells. Analysis of the eluted virus showed that it sedimented slower than native virus in sucrose gradients and that it had lost the internal VP4 capsid protein. The eluted virus, termed the "A particle," was no longer able to attach to HeLa cells and was more labile than native virions; however, the RNA genome was still resistant to RNase. As will be described later (Sect. 6), the A particle is believed to be an intermediate in the entry and uncoating of the virus. Therefore, the binding of virus to the receptor can be functionally divided into at least two steps: (1) attachment, which is a reversible step that can occur at temperatures ranging from 0°C through physiological temperatures; and (2) virus eclipse, which is an irreversible step that occurs at 37°C and converts bound infectious virions into A particles. Although this multistep process was first demonstrated using intact cells, subsequent experiments have demonstrated that a similar process can occur in the presence of solubilized receptor proteins or under certain artificial conditions such as high pH or temperature (ZAJAC and CROWELL 1969; ROESING et al. 1975).

Competition experiments using radiolabeled CVB3 or adenovirus type 2 revealed the unexpected result of competition of both viruses for the same receptor in suspension HeLa cells (LONBERG-HOLM et al. 1976). It was also shown that the adenovirus 2 fiber protein alone could compete with CVB3 for attachment to the cells. More recent experiments with adenovirus have shown that this virus requires at least two cellular proteins for entry: the yet unidentified attachment protein that recognizes the fiber proteins and competes with CVB3, and the integrin $\alpha_v\beta_3$ which recognizes the penton base proteins and is required for the penetration step across the plasma membrane (WICKHAM et al. 1993).

Previous work on the cellular receptor for adenovirus type 5 had shown that the receptor protein could be solubilized using sodium deoxycholate (DOC) (BUTTERS and HUGHES 1974). Since adenovirus types 2 and 5 share the same attachment receptor (PHILIPSON et al. 1968), it was quite likely that the CVB receptor could also be isolated using DOC-solubilized membrane preparations. CROWELL and SIAK (1978) were able to isolate a soluble CVB receptor using DOC and could not only demonstrate the ability of the soluble receptor to bind CVB3, but could also show that the virus could be converted to A particles or uncoated after incubation at 37°C. Whether these two processes, A particle conversion and uncoating, were mediated by the same protein responsible for attachment of CVB3 was not established.

Additional studies using DOC solubilized HeLa cell plasma membranes revealed that the CVB receptor could be isolated in a semipurified form from a Sepharose 4B size exclusion column (KRAH and CROWELL 1985). The predicted mass of the receptor based on its mobility through the matrix was approximately 275 kDa and the virus-binding activity was sensitive to chymotrypsin, periodate and several types of glycosidases. Work from the same laboratory had demonstrated that the virus-receptor complex was stable in the presence of DOC and Triton X-100. Thus, the investigators were able to use a differential and sucrose density purification scheme to isolate the virus-receptor complex in the presence of these detergents (MAPOLES et al. 1985). This strategy revealed a protein with a molecular mass of 49.5 kDa and was designated RP-α. It was proposed that RP-α was a subunit in the 275 kDa multimeric CVB receptor complex. A protein of similar mass was identified on the mouse T lymphoma cell line YAC-1 and was shown to bind to all six CVB serotypes (HSU and CROWELL 1989).

3 Coxsackie B Viruses Can Utilize Multiple Receptors for Cell Entry

The isolation of virus host-range mutants has been an extremely useful approach in the analysis of virus-host interactions. Crowell and colleagues used such a strategy and isolated a series of host-range mutants of CVB1, 3, 5 and 6 (REAGAN et al. 1984). The six CVB prototype viruses were blindly passaged in human rhabdomyosarcoma (RD) cells, a cell line that is refractory for CVB attachment and subsequent infection. Four variants (CVB1-RD, CVB3-RD, CVB5-RD, CVB6-RD) out of the six prototype CVB serotypes were isolated. Each produced cytopathic effects, were competent for plaque formation on RD cells and retained the ability to infect HeLa cells. Based on evidence showing that the cell line chosen for propagation of enteroviruses could influence their ability to agglutinate human erthrocytes (COVA and AYMARD 1980), Crowell and colleagues examined the parental CVB viruses and the RD-variants for hemagglutination activity. Interestingly they observed that the parental viruses were HA⁻ but all of the RD-variants were HA⁺. This suggested, along with the previous data, that the RD viruses had an expanded host range. This was the result of altered receptor specificity and could be explained by either recognition of a new domain on the multicomponent receptor complex or recognition of a novel receptor.

The ability of the RD viruses to infect HeLa cells was further supported by virus attachment interference assays. The CVB3-RD virus was competent for attachment to both RD and HeLa cells. If the HeLa cells were saturated with CVB1, and the attachment of CVB3 and CVB3-RD was measured, it was found the CVB1 blocked the attachment of CVB3 but not the attachment of CVB3-RD, which bound normally (REAGAN et al. 1984). Binding of saturating amounts of CVB3-RD to the HeLa cells blocked binding of the parental CVB3. These results suggested that, in addition to the RD viruses acquiring an expanded receptor specificity, they retained the ability to bind to their original HeLa cell receptor and that the affinity for that receptor was not signifcantly altered. It is likely that the RD viruses had altered their capsid proteins in such a manner as to create a second receptor binding site on the surface of the virion. This suggestion will be discussed in Sect. 4.3.

A completely different line of experiments also suggested that the CVB group was capable of utilizing multiple receptors for attachment to permissive cells. Crowell and colleagues used HeLa cells to immunize mice and produced monoclonal antibodies that were screened for their ability to block the cytopathic effects of CVB3 and CVB3-RD (CROWELL et al. 1986). Three types of monoclonals were isolated: (1) those that blocked HeLa cells from CVB3 infection; (2) those that blocked RD cells from CVB3-RD infection; and (3) those that blocked both viruses on their respective cells. Unfortunately, only one antibody with the latter activity survived (RmcA) and was analyzed in greater detail. Surprisingly RmcA did not protect HeLa cells from CVB3-RD but did protect RD cells. However, the antibody was able to block a substantial percentage of CVB3-RD attachment to HeLa cells suggesting an alternative receptor for CVB3-RD infection in HeLa cells (HSU et al. 1988). The RmcA antibody was also effective at protecting HeLa cells from echovirus 6 and coxsackievirus A9, although it was shown that echovirus 6 did not share a receptor on HeLa cells with CVB3 (REAGAN et al. 1984). Indirect immunofluorescence assays confirmed the binding of RmcA to the surface of HeLa and RD cells and the antibody was also used to determine that there were 3.5×10^6 binding sites per HeLa cell. A similar strategy had been previously used to isolate several monoclonal antireceptor antibodies of CVB1 (CAMPBELL and CORDS 1983). At least one of these antibodies had a specificity similar to those that blocked HeLa cells from CVB3 infection, in that it blocked CVB3 attachment to HeLa cells but had no effect on CVB3-RD attachment to RD cells (CROWELL et al. 1986).

The 49.5 kDa Rp-α protein, described in the previous section, was also used as an immunogen to generate monoclonal antibodies and the resulting hybridomas were screened for their ability to protect HeLa cells from infection with CVB3 (HSU et al. 1988). A single monoclonal antibody was isolated (RmcB) which protected HeLa cells from all six CVB serotypes but did not protect against echovirus 6 infection and did not compete for the same epitope as the RmcA monoclonal antibody. It also did not recognize any sites on the surface of RD cells. The RmcB antibody had substantially fewer binding sites on HeLa cells than did the RmcA antibody, with only 5.0×10^4 binding sites per HeLa cell. This number correlated fairly well with the number of binding sites for CVB3 on HeLa cells (1.8×10^4) but suggested that binding may require multiple receptor proteins for each virion. The RmcB antibody blocked only a small percentage of CVB3-RD binding on HeLa cells but was able to protect the cells from infection with this virus, suggesting that although CVB3-RD can bind to an alternative site on HeLa cells, this binding cannot produce a functional infection. Using both RmcA and RmcB, the attachment of CVB3-RD to HeLa cells could be completely blocked, therby providing evidence that RmcA and RmcB recognized distinct molecules (HR2 and HR1, respectively) on the surface of HeLa cells that provided sites for CVB3-RD and CVB3 attachment. Based on these results, it seemed likely that HR1 and HR2 were closely associated with one another on the surface of HeLa cells.

The protein (HR2) that was recognized by the RmcA antibody was identified as a 60 kDa polypeptide (HSU et al. 1990). Using octylthioglucoside solubilzed HeLa cells the protein was purified by two approaches. The first used isoelectric focusing and preparative SDS-polyacrylamide gel electrophoresis; the second used RmcA affinity chromatography (MOHANTY and CROWELL 1993). In dot blot experiments, CVB3-RD was found to bind to both HR1 and HR2 receptor preparations whereas CVB3 only bound to preparations containing HR1.

4 Identification of a Coxsackie B Virus Receptor

4.1 Decay-Accelerating Factor

In an attempt to identify the cellular receptor for echovirus 7, Bergelson and colleagues immunized mice with intact HeLa cells to obtain monoclonal antibodies with anti-receptor activity (BERGELSON et al. 1994). One antibody with protective activity was isolated which immunoprecipitated a HeLa cell protein with a molecular mass of 70 kDa. HeLa cells incubated with phosphatidylinositol-specific phospholipase C showed a significant decrease in the surface expression level of the 70 kDa protein suggesting that it might be a glycosylphosphatidylinositol (GPI)-anchored protein such as decay-accelerating factor (DAF; CD55). Anti-DAF antibodies precipitated a protein with an identical mobility as the 70 kDa protein precipitated by the anti-receptor antibody and confirmed that the receptor for echovirus 7 was DAF. The data also showed that echovirus 6, along with several other echovirus serotypes (see Table 1), utilized DAF as an attachment receptor.

The observation that the RmcA monoclonal antibody could protect HeLa cells from echovirus 6 infection prompted BERGELSON et al. (1995) to investigate whether the 60 kDa protein recognized by RmcA was DAF. HeLa cells were radioiodinated and RmcA and a DAF-specific monoclonal antibody were used to immunoprecipitate a detergent-extracted fraction from HeLa cells. Both antibodies recognized proteins that had identical mobilities in an SDS-polyacrylamide gel. It was further shown that CVB3-RD could bind to chinese hamster ovary (CHO) cells transfected with DAF whereas the parental CVB3 virus was unable to attach to the transfected cells.

In contrast to the results of BERGELSON et al. (1995), work reported by SHAFREN et al. (1995) showed that DAF could function as an attachment receptor for the CVB1, B3 and B5 prototype strains. Again, a monoclonal antibody was used to inhibit binding and plaque formation of human HEp2 cells and immunoprecipitation using this antibody identified a protein having the same mobility as DAF. CVB1, B3 and B5 could bind to WOP cells (polyoma-transformed mouse 3T3 fibroblasts) transfected with DAF but could not establish lytic infections. Therefore, expression of DAF itself is not the only requirement for the establishment of a CVB infection. Presumably other polypeptides, either as part of a multicomponent receptor complex or as separate functional molecules, are involved in the multistep process of virus attachment, eclipse and uncoating.

The discrepancy between the results of SHAFREN et al. (1995) and BERGELSON et al. (1995) relating to the binding of the parental CVB3 to DAF expressing cells are difficult to reconcile but can be analyzed in light of the earlier antibody protection experiments (CROWELL et al. 1986; HSU et al. 1988). As was mentioned, these protection experiments showed that the RmcA antibody, which has been shown to bind to DAF, was able to protect HeLa cells from CVB1, B3 and B5 even though these viruses may not be directly binding to DAF. This close association between DAF and the major receptor HR1 suggests that the two proteins could function together in a multicomponent receptor complex. Mutation of a few capsid residues in CVB may favor association of the virion with the DAF component of the receptor complex.

4.2 Physical and Biochemical Characterization of Decay-Accelerating Factor

Decay-accelerating factor, or CD55, is a member of the RCA (regulators of complement activation) family of proteins (for reviews see LUBLIN and ATKINSON 1989; NICHOLSON-WELLER 1994) and is involved in modulating the complement cascade on the surface of cells. The complement system is controlled by two pathways: (1) the classical pathway is activated by the binding of antibody to antigen, which initiates a series of cleavages starting with C1; (2) the alternative pathway is constituitively active at a low rate but is stimulated by foreign surfaces. These two pathways converge at the C3 to C3b conversion step and it is here that DAF inter-

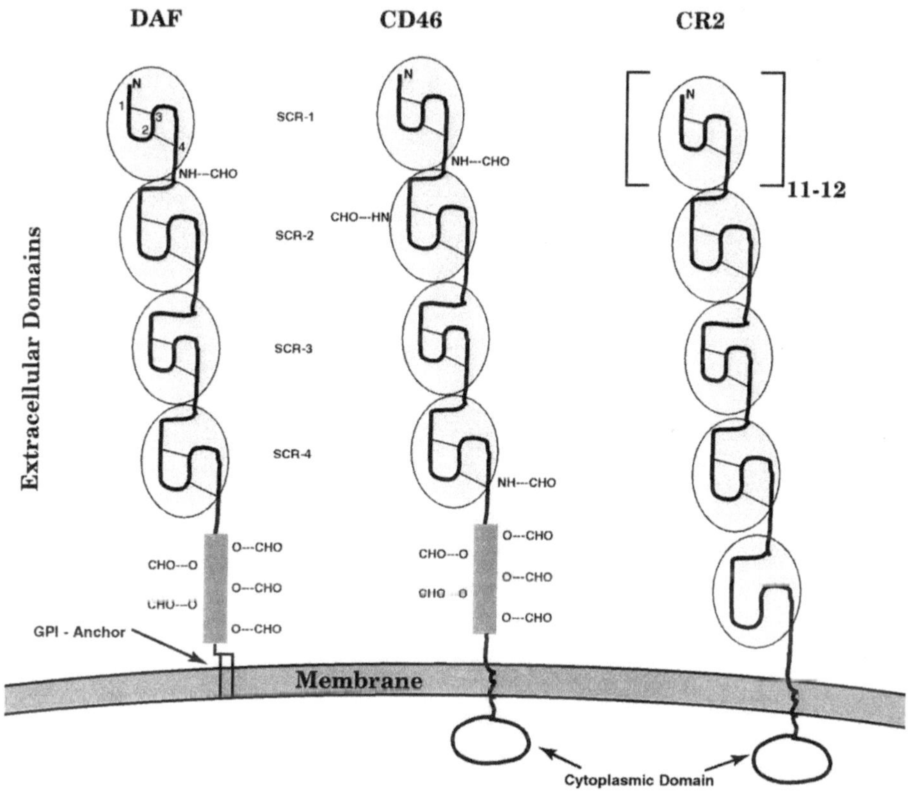

Fig. 1. Short consensus repeat (SCR) domain proteins that serve as viral receptors. The diagram shows the structures of DAF, CD46 and CR2, receptors for CVB-RD, measles and Epstein-Barr viruses, respectively. All three proteins are involved in complement regulation and share a common structural domain, the SCR (see text for detail). The SCR domains are shown as *shaded elipses* with the *heavy line* representing the polypeptide chain and the *thin lines* representing disulfide bonds. Numbers within the elipse for SCR-1 of DAF indicate the order of cysteine residues from the NH_2-terminus (*N*). Both DAF and CD46 have four SCR domains while CR2 has either 15 or 16 domains. The Ser/Thr-rich region of DAF and the Ser/Thr/Pro-rich region of CD46 are indicated by the *shaded rectangles*. Glycosylation sites, either O-linked or N-linked, are indicated for both DAF and CD46

acts with the membrane-bound C3/C5 convertase to modulate inhibition of the complement pathway. DAF acts by binding to the C3/C5 convertase on the same cell to accelerate the decay dissociation of the convertase to the components C3b and Bb and it can also function to accelerate the dissociation of C2a from C4b. Therefore, DAF acts to control the self-induced lysis of the complement system to prevent host cell destruction.

DAF is a single chain GPI-linked membrane protein having an apparent molecular weight of 70 kDa on an SDS-polyacrylamide gel under reducing conditions (Lublin et al. 1986). Two mRNA forms of DAF have been described which result from a single splicing event (Caras et al. 1987; Medof et al. 1987). The major form is GPI-linked and is derived from the spliced DAF mRNA; it is a 381 amino acid polypeptide with a 34 amino acid signal sequence at its NH_2-terminus. The minor (10%) unspliced form could encode a 440 amino acid protein that is predic-

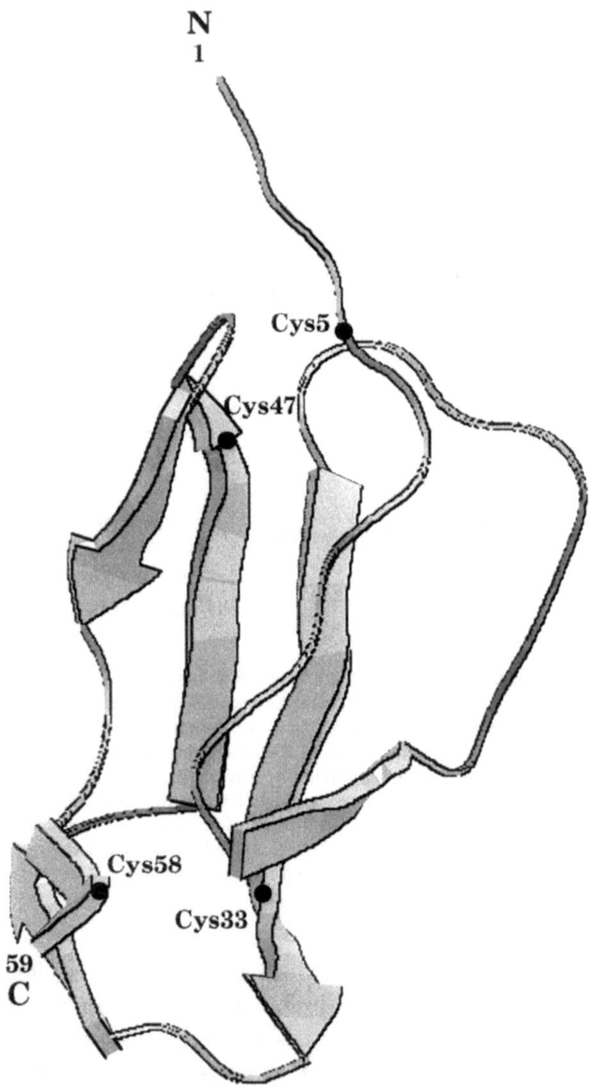

Fig. 2. Three-dimensional structure of a short consensus repeat (SCR) domain. The three-dimensional structure of the sixteenth SCR domain from the human complement factor H protein as solved by NMR is depicted (NORMAN et al. 1991). This 59 amino acid protein has 5 β-stranded sheets (*arrows*) folded into a compact core. The NH_2- and COOH-terminals are indicated by the letters *N* and *C*, respectively. Cysteine residues involved in forming intramolecular disulfides are listed and indicated by the *black circle*. The image was prepared from the PDB coordinates (1HCC) and drawn using the Macintosh program Molview (SMITH 1992)

ted to be a secreted form of DAF, although the functional relevance of this form of the protein has yet to be established (LUBLIN and ATKINSON 1989). At its COOH-terminus the protein has a 24 amino acid hydrophobic domain that probably acts as a signal for linkage to the GPI anchor (Fig. 1). This is followed by a serine/threonine-rich region that is extensively O-linked glycosylated and serves as a spacer for the four short consensus repeat (SCR) domains that are responsible for DAF activity. The SCR domains are approximately 60 amino acids in size and are related to domains found on other complement regulatory proteins. A single N-linked glycosylation site is found on the connecting region between SCR-1 and SCR-2. Experiments have shown that DAF can function equally well having either a normal transmembrane anchor or the GPI anchor. In addition, the O-linked glycosylation domain can be substituted by a nonspecific spacer region (LUBLIN and COYNE 1991).

The SCR domain is a common structural motif found in many complement-related proteins and several noncomplement related proteins. Proteins may have as many as 30 SCR domains linked in tandem, as is found in the complement receptor type 1 (CR1) protein, or as few as two domains as is found in the interleukin 2 receptor. The 60 amino acids that comprise the SCR domain are folded into a β-sandwich arrangement composed of five β-strands (Fig. 2) (NORMAN et al. 1991). There are four conserved cysteine residues that are involved in disulfide bond formation, required for the correct conformation with the first cysteine bonded to the third cysteine and the second cysteine bonded to the fourth cysteine. Both faces of the molecule contribute conserved hydrophobic amino acid side chains that project into the compact core of the SCR, whereas the loop regions contain residues that vary widely among the different SCR domains. Removal of SCR-1 in DAF has no effect on DAF activity whereas deletion of either SCR-2, SCR-3 or SCR-4 eliminates activity (COYNE et al. 1992). A monoclonal antibody directed against SCR-3 completely abolished DAF activity suggesting that this domain might be involved in the protein-protein interactions thought to be required for DAF function (COYNE et al. 1992).

4.3 Virus-Receptor Interactions

The use of an RCA family member for virus attachment and entry is not unique for the CVB group. Epstein-Barr virus has been shown to use the CR2 molecule as its cellular receptor (Fig. 1) (COOPER et al. 1988). CR2, also known as CD21, binds the C3dg fragment of the complement component C3, and interestingly the virus glycoprotein gp350/220 contains a C3dg-like sequence. The virus, therefore, binds to CR2 via SCR-1 and SCR-2 domains which are at the NH_2-terminus of the 16 SCR domain protein, in a manner analagous to the normal substrate for CR2. Measles virus, an enveloped virus of the paramyxovirus family, uses the human membrane cofactor protein (CD46) as its cellular receptor (DÖRIG et al. 1993; NANICHE et al. 1993). CD46 is involved in degradation of the complement proteins

C3b and C4b and therefore acts to protect cells from autolysis by the complement pathway (GERLIER et al. 1995). CD46 is structurally and functionally similar to DAF (Fig. 1), and is its closest known relative. It has four SCR domains, with domains 1 and 2 likely to be involved in measles virus binding. An O-linked series of glycosylation sites are found on a serine/threonine/proline-rich spacer domain that separates the SCRs from the membrane anchor, which in this case is a transmembrane domain connected to one of two possible cytoplasmic domains. Like the immunoglobulin family and integrin family, the RCA family represents a series of proteins related by conserved structural motifs that serve as cellular receptors for a diverse group of animal viruses.

The binding of echovirus 7 to DAF has been suggested to involve the interaction of the virus particle with domains SCR-2, -3, and -4 but not SCR-1 (CLARKSON et al. 1995). Recombinant deletion mutants of DAF were stably expressed in CHO cells and the requirement for each SCR domain was assessed by sequential deletion of each domain. The resulting, partially deleted, DAF molecules were examined for their ability to bind to echovirus 7. In addition to the deleted DAF molecules, the binding of echo 7 to CD46 was also examined, along with binding to chimeric molecules that exchanged the GPI anchor and the O-linked glycosylation domain. There was no measurable binding to the related RCA protein CD46, or DAF/CD46 chimeras in which the CD46 SCR-2, SCR-3 or SCR-4 were individually exchanged with the DAF SCR domains. However, there was no effect on binding when either the GPI anchor was exchanged for the CD46 transmembrane domain or when the serine/threonine-rich domain was exchanged for another spacer domain.

A related set of experiments that examined the SCR domain requirements for CVB binding were performed with CVB3-RD and various DAF SCR-deletion constructs (BERGELSON et al. 1995). Deletions of SCR-1, -3, -4 individually resulted in approximately 50% lower virus binding than with intact DAF. However, deletion of SCR-2 completely abolished virus binding indicating that CVB3-RD was attaching to DAF through interaction with SCR-2. SHAFREN et al. (1995) used a monoclonal antibody that was specific for SCR-3 to demonstrate inhibition of CVB5 binding. It is likely that the antibody sterically hindered the attachment of virus; therefore the virus binding site might be on SCR-3 or more likely in the vicinity of SCR-3. The exact location of the virus binding site on DAF will require a more detailed analysis of the interaction. Experiments are currently in progress to examine the binding of DAF to CVB using both cryoelectron microscopy and X-ray crystallography of virus-receptor complexes (JK Muckelbauer, MG Rossmann, TS Baker, RJ Kuhn, unpublished data).

In addition to the site of virus binding on DAF, it is intriguing to speculate on the site of DAF binding to virus, the receptor attachment site. The expanded host range seen in the CVB-RD variants was the result of a series of blind passages of the prototype viruses in RD cells (REAGAN et al. 1984). The resulting RD viruses accumulated a sufficient number of nucleotide substitutions to allow the viruses to attach to RD cells using DAF. Lindberg et al. carried out nucleotide sequencing of CVB3-RD and generated chimeric CVB3-RD/CVB3 viruses to show that two amino acid substitutions in VP2 were responsible for the RD phenotype (LINDBERG et al. 1992). Of the two amino acids, Thr151, which was changed to Ser, is located on the surface of the virus in close proximity to the large twofold depression (see Chap. 9, Sect. 4.2.2; MUCKELBAUER et al. 1995). The other amino acid substitution, Asp108 to Val, is located internally in the virus and would not be expected to directly interact with DAF. However, the serine at 151 does not appear by itself to be sufficient for RD attachment since the CVB3/0 strain which has Ser151 is unable to infect RD cells (S Tracy, personal communication). These observations raise the possibility that the binding site for DAF might be in the twofold depression and not in the canyon that surrounds the fivefold axis. In light of this, it is noteworthy that early studies of CVB neutralizing antibodies suggested that the VP2 capsid protein contained the major sites for virus neutralization (CROWELL and LANDAU 1983). Additional sequencing of the CVB1-RD, CVB5-RD and CVB6-RD genomes would be useful to compare the sequence changes required for each virus to expand its receptor specificity.

5 Additional Receptors for Coxsackie B Viruses

With the establishment of DAF as the minor CVB receptor (HR2), the search for the major receptor (HR1) has continued. This receptor, although found in lower amounts on the surface of HeLa cells, can bind to all six serotypes of CVB. Recently, DE VERDUGO et al. (1995) have provided evidence that a 100 kDa protein related to nucleolin functions as a binding protein for all six serotypes of CVB. A virus overlay protein-binding assay (VOPBA) was used to identify a prominent band that exhibited binding affinity with CVB3. Permissive cell lines were shown to express this 100 kDa protein whereas nonpermissive cells did not exhibit expression. It was not shown if adenovirus 2, which shares its receptor with the CVB group, could also attach to the 100 kDa protein as would be predicted from the virus competition assays (LONBERG-HOLM et al. 1976). Purification of the protein using two-dimensional gel electrophoresis permitted peptide sequencing which revealed that the 100 kilodalton protein had sequences identical to those found within the COOH-terminus of human nucleolin.

Nucleolin is an abundantly expressed phosphoprotein found predominantly in the nucleolus in exponentially growing cells (SRIVASTAVA et al. 1989). It is believed to function in the binding of preribosomal RNA and to regulate transcription of polymerase I genes. Nucleolin has been shown to shuttle between the nucleolus and the cytoplasm (BORER et al. 1989) and nucleolin-related proteins have been shown to reside on the surface of cells (SEMENKOVICH et al. 1990; JORDAN et al. 1994). Interestingly, CVB3 did not bind to a partially purified nuclear nucleolin suggesting that the cytoplasmic protein is different from the nuclear-localized version of nucleolin (DE VERDUGO et al. 1995). Cloning of the cDNA encoding this putative CVB receptor and demonstration that its expression in receptor minus cells confers on them the ability to bind all six serotypes of CVB will be required to positively identify this nucleolin-related protein as a receptor for CVB.

The biochemical experiments of Crowell and coworkers had previously suggested that the major receptor for CVB was a 49.5 kDa protein, although it was believed that this polypeptide was a component of a larger 275 kDa receptor complex (MAPOLES et al. 1985). How this observation might relate with a nucleolin-like molecule is unclear. The possibility remains that the major receptor, as originally defined by Crowell, has yet to be identified at the molecular level. An alternative explanation is that there are multiple binding proteins for the CVB group. These binding proteins might have different affinities for viral attachment based on the cell type and on the surrounding proteins that might interact to assemble into a functional viral receptor. In all of the experiments described in the above sections, cDNA, when transfected into nonpermissive cells, only permitted virus attachment not virus infection, therefore suggesting a missing component in a functional virus-receptor complex.

6 Role of the Cellular Receptor in Virus Entry

Despite many of the similarities among members of the picornavirus family, and in particular viruses within the enterovirus genus, the selection of cellular proteins that serve to permit virus entry appears to be diverse and to lack a central structural or functional theme. Perhaps this is not surprising if the receptor plays only a passive role in virus entry and is only responsible for attachment of the virus to the cell surface. Indeed, many proteins have been identified as viral receptors based only on their ability to attach virus to their cell surface. However, as demonstrated by the DAF transfection experiments, expression of an attachment protein on the cell surface may not be sufficient to initiate infection. It appears likely, based on the biochemical data described above and more recent experiments using poliovirus and rhinovirus together with their respective receptors, that the picornaviruses must undergo several steps following attachment to the cell in order for a productive infection to initiate (KAPLAN et al. 1990; GREVE et al. 1991; RACANIELLO 1992).

The tight, stable picornavirus particle must undergo a substantial conformational change in order to uncoat the genome RNA. This is likely to be closely linked to the attachment of the virus particle to the cellular receptor. This receptor, either by itself or in conjunction with additional cellular proteins, engages the virus particle in a tight binding that affects the conformation of the virus and maybe also the receptor. This conformational change would be sufficient to induce the movement of the hydrophobic VP1 NH_2-terminus of one fivefold vertex toward the cellular membrane with the concomitant release of the VP4 internal capsid proteins resulting in the production of

A particles. The next defined step, uncoating of the genome RNA, may occur at either the plasma membrane or in endocytic vesicles depending upon the virus and/or the receptor. The details of such an entry process await future experiments that will resolve the molecular details of interaction of the virus with its cellular receptor.

7 Future Directions and Conclusions

Understanding the role of cellular receptors in the entry of the CVB group into susceptible cells is far from complete. Identification of all of the cellular participants in the entry process will be a prerequisite for a complete understanding. However, the pieces of this process are coming together rapidly. An exciting possibility is that the CVB particle may be capable of binding multiple cellular proteins to multiple sites on the virus surface. In particular, it is possible to imagine that DAF binds to the twofold depression on the virus surface whereas the major receptor protein (nucleolin?) binds in the canyon that surrounds the fivefold axis. A related possibility, but one that has not been observed in other picornaviruses, is that two cellular proteins might be required to bind sequentially or simultaneously to two distinct sites on the virus particle to affect the entry of CVB into cells. It should be emphasized that DAF normally functions by binding to other cellular, usually complement-related, proteins and this may be functionally related to CVB group receptor interactions. Molecular genetics and structural biology will certainly be required to investigate these fundamental questions concerning the function of cellular receptors. Ultimately, viral pathogenesis will be explained by a series of molecular interactions that are initiated when a virus binds to its cellular receptor.

References

Bergelson JM, Shepley MP, Chan BMC, Hemler ME, Finberg RW (1992) Identification of the integrin VLA-2 as a receptor for echovirus 1. Science 255:1718–1720

Bergelson JM, Chen M, Solomon KR, St John NF, Lin H, Finberg RW (1994) Decay-accelerating factor (CD55), a glycosylphosphatidylinositol-anchored complement regulatory protein, is a receptor for several echoviruses. Proc Natl Acad Sci US 91:6245–6248

Bergelson M, Mohanty JG, Crowell RL, St John NF, Lublin DM, Finberg RW (1995) Coxsackievirus B3 adapted to growth in RD cells binds to decay-accelerating factor (CD55). J Virol 69:1903–1906

Berinstein A, Roivainen M, Hovi T, Mason PW, Baxt B (1995) Antibodies to the vitronectin receptor (integrin $\alpha_v\beta_3$) inhibit binding and infection of foot-and-mouth disease virus to cultured cells. J Virol 69:2664–2666

Borer RA, Lehner CF, Eppenberger HM (1989) Major nucleolar proteins shuttle between nucleus and cytoplasm. Cell 56:379–390

Butters TD, Hughes RC (1974) Solubilization and fractionation of glycoproteins and glycolipids of KB cell membranes. Biochem J 140:469–478

Campbell BA, Cords, CE (1983) Monoclonal antibodies that inhibit attachment of group B coxsackieviruses. J Virol 48:561–564

Caras IW, Davitz MA, Rhee L, Weddell G, DW, Martin J, Nussenzweig V (1987) Cloning of decay-accelerating factor suggests novel use of splicing to generate two proteins. Nature 325: 545–549
Clarkson NA, Kaufman R, Lublin DM, Ward T, Pipkin PA, Minor PD, Evans DJ, Almond, JW (1995) Characterization of the echovirus 7 receptor: Domains of CD55 critical for virus binding. J Virol 69:5497–5501
Colonno RJ, Callahan PL, Long WL (1986) Isolation of a monoclonal antibody that blocks attachment of the major group of human rhinoviruses. J Virol 57:7–12
Cooper NR, Moore MD, Nemerow GR (1988) Immunobiology of CR2, the lymphocyte receptor for Epstein-Barr virus and the C3d complement fragment. Annu Rev Immunol 6:85–113
Cova L, Aymard M (1980) Isolation and characterization of non-hemagglutinating echovirus 11. J Gen Virol 51:219–222
Coyne KE, Hall,SE, Thompson ES, Arce MA, Kinoshita T, Fujita T, Anstee DJ, Rosse W, Lublin DM (1992) Mapping of epitopes, glycosylation sites, and complement regulatory domains in human decay accelerating factor. J Immunol 149:2906–2913
Crowell RL (1963) Specific viral interference in HeLa cell cultures chronically infected with coxsackie B5 virus. J Bacteriol 86:517–526
Crowell RL (1966) Specific cell-surface alteration by enteroviruses as reflected by viral-attachment interference. J Bacteriol 91:198–204
Crowell RL, Field AK, Schleif WA, Long WL, Colonno RJ, Mapoles JE, Emini EA (1986) Monoclonal antibody that inhibits infection of the HeLa and rhabdomyosarcoma cells by selected enteroviruses through receptor blockade. J Virol 57:438–445
Crowell RL, Landau BJ (1983) Receptors in the initiation of picornavirus infections. In: Fraenkel-Conrat H, Wagner RR (eds) Comprehensive virology, Plenum, New York, 18:1–42
Crowell RL, Philipson L (1971) Specific alterations of coxsackievirus B3 eluted from HeLa cells. J Virol 8:509–515
Crowell RL, Siak J-S (1978) Receptor for the group B coxsackieviruses: characterization and extraction from HeLa cell plasma membrane. In: Pollard M (ed) Perspectives in Virology. Raven, New York, pp 39–53
de Verdugo UR, Selinka,HC, Huber M, Kramer B, Kellermann J, Hofschneider PH, Kandolf R (1995) Characterization of a 100-kilodalton binding protein for the six serotypes of coxsackie B viruses. J Virol 69:6751–6757
Dörig RE, Marcil A, Chopra A, Richardson CD (1993) The human CD46 molecule is a receptor for measles virus (Edmonston strain). Cell 75:295–305
Gerlier D, Varior-Krishnan G, Devaux, P (1995) CD46-mediated measles virus entry: a first key to host-range specificity. Trends in Microbiol 3:338–345
Greve JM, Davis G, Meyer AM, Forte CP, Yost SC, Marlor CW, Karmarck ME, McClelland, A (1989) The major human rhinovirus receptor is ICAM-1. Cell 56:839–847
Greve JM, Forte CP, Marlor CW, Meyer AM, Hoover-Litty H, Wunderlich D, McClelland A (1991) Mechanisms of receptor-mediated rhinovirus neutralization defined by two soluble forms of ICAM-1. J Virol 65:6015–6023
Hofer F, Gruenberger M, Kowalski H, Machat H, Huettinger M, Kuechler E, Blaas D (1994) Members of the low density lipoprotein receptor family mediate cell entry of a minor-group common cold virus. Proc Natl Acad Sci USA 91: 1839–1842
Hogle JM, Chow M, Filman DJ (1985) Three-dimensional structure of poliovirus at 2.9 Å resolution. Science 229:1358–1365
Holland JJ (1962) Irreversible eclipse of poliovirus by HeLa cells. Virology 16:163–176
Holland JJ, McLaren LC, Syverton JT (1959a) Mammalian cell-virus relationship III. Poliovirus production by nonprimate cells exposed to poliovirus ribonucleic acid. Proc Soc Exp Med 100: 843–845
Holland JJ, McLaren LC, Syverton JT (1959b) The mammalian cell-virus relationship IV. Infection of naturally insusceptible cells with enterovirus ribonucleic acid. J Exp Med 110:65–80
Hsu KHL, Paglini S, Alstein B, Crowell RL (1990) Identification of a second cellular receptor for a coxsackievirus B3 variant, CB3-RD. In: M A Brinton and F X Heinz (ed) New aspects of positive-strand RNA viruses, American Society for Microbiology, Washington, DC pp 271–277
Hsu KL, Crowell RL (1989) Characterization of a YAC-1 mouse cell receptor for Group B caxsackieviruses. J Virol 63:3105–3108
Hsu KL, Longberg-Holm K, Alstein B, Crowell RL (1988) A monoclonal antibody specific for the cellular receptor for the group B coxsackieviruses. J Virol 62:1647–1652

Huber SA (1994) VCAM-1 is a receptor for encephalomyocarditis virus on murine vascular endothelial cells. J Virol 68:3453–3458

Jordan P, Heid H, Kinzel V, Kübler D (1994) Major cell surface-located protein substrates of an ecto-protein kinase are homologs of known nuclear proteins. Biochemistry 33:14696–14706

Kaplan G, Freistadt MS, Racaniello VR (1990) Neutralization of poliovirus by cell receptors expressed in insect cells. J Virol 64:4697–4702

Krah DL, Crowell RL (1985) Properties of the deoxycholate-solubilized HeLa cell plasma membrane receptor for binding group B coxsackieviruses. J Virol 53:867–870

Lindberg AM, Crowell RL, Zell R, Kandolf R, Pettersson U (1992) Mapping of the RD phenotype of the Nancy strain of coxsackievirus B3. Virus Res 24:187–196

Lonberg-Holm K, Crowell RL, Philipson L (1976) Unrelated animal viruses share receptors. Nature 259:679–681

Lublin DM, Atkinson JP (1989) Decay-accelerating factor: biochemistry, molecular biology, and function. Annu Rev Immunol 7:35–58

Lublin,DM, Coyne KE (1991) Phospholipid-anchored and transmembrane versions of either decay-accelerating factor or membrane cofactor protein show equal efficiency in protection from complement-mediated cell damage. J Exp Med 174:35–44

Lublin DM, Krsek-Staples J, Pangburn MK, Atkinson JP (1986) Biosynthesis and glycosylation of the human complement regulatory protein decay-accelerating factor. J Immunol 137:1629–1635

Mapoles JE, Krah DL, Crowell RL (1985) Purification of a HeLa cell receptor protein for group B coxsackieviruses. J Virol 55:560–566

Medof ME, Lublin DM, Holers VM, Ayers DJ, Getty RR, Leykan JF, Atkinson JP, Tykocinski ML (1987) Cloning and characterization of cDNAs encoding the complete sequence of decay-accelerating factor of human complement. Proc Natl Acad Sci USA 84:2007–2011

Mendelsohn C, Johnson B, Lionetti KA, Nobis P, Wimmer E, Racaniello VR (1986) Transformation of a human poliovirus receptor gene into mouse cells. Proc Natl Acad Sci USA 83:7845–7849

Mohanty JG, Crowell RL (1993) Attempts to purify a second cellular receptor for a coxsackievirus B3 variant, CB3-RD from HeLa cells. Virus Res 29:305–320

Muckelbauer JK, Kremer M, Minor I, Rossmann MG, Diana G, Dutko FJ, Groarke J, Pevear DC (1995) The structure of coxsackievirus B3 at 3.5 Å resolution. Structure 3:653–667

Naniche D, Varior-Krishnan G, Cervoni F, Wild TF, Rossi B, Rabourdin-Combe C, Gerlier D (1993) Human membrane cofactor protein (CD46) acts a cellular receptor for measles virus. J Virol 67:6025–6032

Nicholson-Weller (1994) Structure and function of decay accelerating factor CD55. J Lab Clin Med 123:485–491

Norman DG, Barlow PN, Baron M, Day AJ, Sim RB, Campbell ID (1991) Three-dimensional structure of a complement control protein module in solution. J Mol Biol 219:717–725

Philipson L, Lonberg-Holm K, Pettersson U (1968) Virus-receptor interaction in an adenovirus system. J Virol 2:1064–1075

Racaniello VR (1992) Interaction of poliovirus with its cell receptor. Sem Virol 3:473–482

Racaniello VR, Baltimore D (1981) Cloned poliovirus complementary DNA is infectious in mammalian cells. Science 214:916–919

Reagan KJ, Goldberg B, Crowell RL (1984) Altered receptor specificity of coxsackievirus B3 after growth in rhabdomyosarcoma cells. J Virol 49:635–640

Roesing TG, Toselli PA, Crowell RL (1975) Elution and uncoating of coxsackievirus B3 by isolated HeLa cell plasma membranes. J Virol 15:654–667

Roivainen M, Piirainen L, Hovi T, Virtanen I, Riikonen T, Heino J, Hyypiä T (1994) Entry of coxsackievirus A9 into host cells: specific interactions with $\alpha_v\beta_3$ integrin, the vitronectin receptor. Virology 203:357–365

Rossmann MG (1994) Viral cell recognition and entry. Prot Sci 3:1712–1725

Rossmann MG, Arnold E, Erickson JW, Frankenberger EA, Griffith JP, Hecht HJ, Johnson JE, Kamer G, Luo M, Mosser AG, Rueckert RR, Sherry B, Vriend G (1985) Structure of a human common cold virus and functional relationship to other picornaviruses. Nature 317:145–153

Semenkovich CF, Ostlund RE, Olson MO, Yang JW (1990) A protein partially expressed on the surface of HepG2 cells that binds lipoproteins specifically in nucleolin. Biochemistry 29:9708–9713

Shafren DR, Bates RC Agrez MV, Herd RL, Burns GF, Barry RD (1995) Coxsackieviruses B1, B3, and B5 use decay accelerating factor as a receptor for cell attachment. J Virol 69:3873–3877

Smith TJ (1992) MacInPlot II-an updated program to display electron density and atomic models on the MacIntosh personal computer. J Appl Cryst 26:496–498
Srivastava M, Fleming P, Pillard HB, Burns AL (1989) Cloning and sequencing of the human nucleolin cDNA. FEBS Letters 250:99–105
Staunton ED, Merluzzi VJ, Rothlein R, Barton R, Marlin SD, Springer, TA (1989) A cell adhesion molecule, ICAM-1, is the major surface receptor for rhinoviruses. Cell 56:849–853
Tomassini JE, Graham D, DeWitt CM, Lineberger DW, Rodkey JA, Colonno RJ (1989) cDNA cloning reveals that the major group rhinovirus receptor on HeLa cells is intercellular adhesion molecule 1. Proc Natl Acad Sci USA 86:4907–4911
Ward T, Pipkin PA, Clarkson NA, Stone DM, Minor PD, Almond JW (1994) Decay-accelerating factor CD55 is identified as the receptor for echovirus 7 using CELICS, a rapid immuno-focal cloning method. EMBO J 13:5070–5074
Wickham TJ, Mathias P, Cheresh DA, Nemerow GR (1993) Integrins $\alpha_v\beta_3$ and $\alpha_v\beta_5$ promote adenovirus internalization but not virus attachment. Cell 73:309–319
Zajac I, Crowell RL (1965a) Effect of enzymes on the interaction of enteroviruses with living HeLa cells. J Bacteriol 89:574–582
Zajac I, Crowell RL (1965b) Location and regeneration of enterovirus receptors of HeLa cells. J Bacteriol 89:1097–1100
Zajac I, Crowell RL (1969) Differential inhibition of attachment and eclipse activities of HeLa cells for enteroviruses. J Virol 3:422–428

Genetics of Coxsackievirus Virulence

N. M. CHAPMAN[1], A. I. RAMSINGH[2], AND S. TRACY[1]

1 Introduction

The first reported isolate of the coxsackieviruses occurred in the upstate New York town of Coxsackie (DALLDORF and SICKLES 1948). By the mid-late 1950s (DALLDORF 1955), the coxsackie B viruses (CVBs) had been shown to be agents of severe human inflammatory heart muscle disease (myocarditis) (MONTGOMERY et al. 1955; GEAR et al. 1956; JAVETT et al. 1956; VAN CREVELD and DE JAGER 1956; VERLINDE et al. 1956; FLETCHER and BRENNAN 1957; KIBRICK and BENIRSCHKE 1958; SIMENHOFF and ULYS 1958). In the intervening years, the CVBs have been

[1]Department of Pathology and Microbiology, University of Nebraska Medical Center, 600 S 42 Street, Omaha NE 68198–6495, USA

[2]David Axelrod Institute, Wadsworth Center, New York State Department of Health, 120 New Scotland Avenue, Albany NY 12201–2002, USA

confirmed as etiologic agents of, or etiologically linked to, a host of human diseases ranging from flu-like illness through severe inflammatory diseases of the heart and central nervous system (MELNICK 1990; Chaps. 2–4). Based on molecular detection data, it is estimated that enteroviruses (likely in most cases to be CVBs) are etiologic agents in perhaps 20%–25% of the cases of inflammatory heart disease (or myocarditis) and dilated cardiomyopathy (DCM), although individual reports range from 0%–53% (BOWLES et al. 1986; EASTON and EGLIN 1988; TRACY et al. 1990; KANDOLF and HOFSCHNEIDER 1991; GRASSO et al. 1992; reviewed in MARTINO et al. 1995). DCM alone afflicts approximately five to eight per 100 000 per year worldwide (WILLIAMS and OLSEN 1985; MANOLIO et al. 1992). In the aggregate, these data indicate that there are at least 5000 or more occurrences of enterovirus-induced DCM alone per year in the United States.

With the exception of the polioviruses (PVs), vaccines are not available at present against any of the human enteroviruses. The success of the PV vaccines in eradicating PV disease in the Western Hemisphere strongly suggests that vaccines against other enteroviruses would be equally effective at eradicating the plethora of CVB-caused diseases. Partly for this reason, the study of the genetics of viral virulence has assumed increasing importance. Detailed understanding of the viral determinants of virulence and interactions of the virus and host in virulence determination is valuable for effective and economic vaccine design. Additionally, and nontrivially, the CVBs represent the only other highly characterized, non-PV enterovirus system in which detailed molecular, genetic and pathogenic studies can be pursued at present.

We have learned much more about the molecular genetics of enteroviruses than the pathogenesis from the numerous studies on PVs. In terms of mechanisms underlying the virulence phenotype of human enteroviruses, the fact remains that the greatest focus has been upon understanding the molecular mechanisms underlying the artificially induced attenuation of the wild-type neurovirulent PV strains. It remains to be seen whether the characterization of the genetics of naturally occurring CVB virulence will follow the PV paradigm.

2 Genetics and Mechanisms of Coxsackievirus B3 Virulence

Although coxsackievirus B3 (CVB3) is also etiologic in serious pediatric central nervous system diseases (GRIST et al. 1978; Chap. 4), it is as an agent of inflammatory heart muscle disease that this virus has become widely studied (MARTINO et al. 1995; Chapter 3). Not long after its discovery as an agent of human myocarditis (GEAR et al. 1956), CVB3 was exploited by Woodruff and coworkers in studies of viral pathogenesis in a relevant murine model of enteroviral myocarditis (WOODRUFF and KILBOURNE 1970; WOODRUFF and WOODRUFF 1974; HUBER et al. 1980; WOODRUFF 1980). Other murine models of CVB3-induced murine myocarditis have since been developed as well (SCHNURR et al. 1984; GAUNTT et al. 1989; LESLIE et al. 1989; BECK et al. 1990; CHOW et al. 1992; MATSUMORI 1992), providing a variety of sound experimental systems in which to study specific aspects of the disease.

2.1 Mapping the Determinants of Cardiovirulence in the Coxsackievirus B3 Genome

There are presently three infectious cDNA copies of the CVB3 genome that have been characterized (KLUMP et al. 1990; TRACY et al. 1992; CHAPMAN et al. 1994). Only one of these clones is of a noncardiovirulent CVB3 genome (CHAPMAN et al. 1994), while the other two represent cardiovirulent CVB3 phenotypes (KLUMP et al. 1990; TRACY et al. 1992). In addition, Gauntt and coworkers (C Gauntt, unpublished data) and Huber and Knowles (Huber, unpublished data) have constructed infectious cDNA copies of cardiovirulent CVB3 genomes (WOODRUFF and KILBOURNE 1970; TROUSDALE et al. 1976; VAN HOUTEN et al. 1991). Sequence analysis demonstrates that all of these CVB3 strains are closely related to each other (Chap. 6). At this writing, the identity and location of the site(s) in the CVB3 genome that naturally determines the cardiovirulence phenotype remain unclear.

2.2 Is There a Role for the CVB3 5' Nontranslated Region in Determining the Cardiovirulence Phenotype?

The 5' nontranslated region (NTR) clearly plays a significant role in contributing to the artificial attenuation of the naturally occurring neurovirulent phenotypes of the three PV serotypes (Sect. 5.1 below). The 5' NTRs of CVBs and PVs share significant sequence and structural identity; for example, CVB3 shares approximately 70% nucleotide identity with that of PV type 1 (PV1), strain Mahoney (TRACY et al. 1985). Computer modeled RNA folding studies (CURREY et al. 1986; PILIPENKO et al. 1989; SKINNER et al. 1989; LE and ZUKER 1990; LE et al. 1992) of the 5' NTR sequences from a variety of enteroviruses have produced consensus higher order structures, many aspects of which have been confirmed by biochemical probing (SKINNER et al. 1989; Chap. 8). Although a comparative sequence analysis study of enteroviral 5' NTRs places the CVBs in a separate cluster of enteroviruses from the PVs (POYRY et al. 1994), the secondary structures of the CVBs and PVs are in general conserved (PILIPENKO et al. 1989; SKINNER et al. 1989; LE and ZUKER 1990; AGOL 1991). Perhaps not surprisingly then, the 5' NTRs of CVBs and PVs are functionally related as well. Semler and colleagues (SEMLER et al. 1986; JOHNSON and SEMLER 1988) were the first to demonstrate that sequences in the 5' NTR of PV1 could be substituted by the homologous sequences from CVB3 to produce viable, intertypic chimeric viruses. More recently, the entire 5' NTR of CVB4 was substituted for that of PV3 in a PV3 genome to produce a viable virus (ROHLL et al. whether the CVB3 5' NTR plays a significant role in determining the cardioviru-

```
             -   -- --      ----->     <-----        --  - -----]▼      [----- - -----    --       - ------   ->
CVB3/20    CTGAGTATCA ATAGACTGCT CACGCGGTTG AAGGAGAAAG CGTTCGTTAT CCGGCCAACT ACTTCGAAAA ACCTAGTAAC ACCGTGGAAG cardiovirulent CVB3
CVB3/0     CTGAGTATCA ATAGACTGCT CACGCGGTTG AAGGAGAAAG CGTTCGCTAT CCGGCCAACT ACTTCGAAAA ACCTAGTAAC ACCGTGGAAG nonvirulent CVB3
M33854     CTGAGTATCA ATAGACTGCT CACGCGGTTG AAGGAGAAAG CGTTCGTTAT CCGGCCAACT ACTTCGAAAA ACCTAGTAAC ACCGTGGAAG cardiovirulent CVB3
CVB3/AS    CTGAGTATCA ATAGACTGCT AGCGCGGTTG AAGGAGAAAA CGTTCGTTAT CCGGCCAACT ACTTCGAGAA ACCTAGTAAC ACCATGGAAG cardiovirulent CVB3
CVB3/DO    CTGAGTATCA ATAGGCTGCT CGCGCGGCTG AGGGAGAAAC CGTCCGTTAC CCGGCCAACT ACTTCGAGAA ACCCAGTAAC ATCATGTACG cardiovirulent CVB3
CVB3/CO    CTGAGTATCA ATAGGCTGCT CGCGCGGCTG AAGGAGAAAC CGTCCGTTAT CCGGCCAACT ACTTCGAGAA ACCCAGTAAC ATCATGGACG nonvirulent CVB3
CVB3/ZU    CTGAGTATCA ATAAACTGCT CGCGCGGTTG AGGGAGAAAC CGTCCGTTAT CCGGCCAACT ACTTCGAGAA ACCCAGTAAC ATCATGGACG cardiovirulent CVB3
CVB3/GU    CTGAGTATCA ATAGGCTGCT CGCGCGGCTG AGGGAGAAAC CGTCCGTTAT CCGGCCAACT ACTTCGAGAA ACCCAGTAAT ATCATGGACG nonvirulent CVB3
CVB3/RE    CTGAGTATCA ATAGGCTGCT CGCGCGGCTG AGGGAGAAAC CGTCCGTTAT CCGACCAACT ACTTCGAGAA ACCTAGTAAC GCCATGAACA nonvirulent CVB3
CVB3/OL    CCGAGTATCA ATAAGCTGCT TGCGCGGCTG AAGGAGAAAC CGTCCGTTAC CCGGCTAACT ACTTCGAAAA ACCCAGTAAC ACCATGGACG nonvirulent CVB3
CVB3/GA    CTGAGTATCA ATAAGCTGCT CACGTGGCTG AAAGAGAAAA CGTTCGTTAC CCGGCCAGTT ACTTCGGGAA GCCTAGTACC ACCATGAAAG nonvirulent CVB3
CXB4S      CTGAGTATCA ATAAGCTGCT TGCGCGGCTG AAGGAGAAAC CGTTCGTTAC CCGGCCAACT ACTTCGAGAA GCCTAGTAAC GCCATGAACG CVB4
S39291     CTGAGTATCA ATAAGCTGCT TGCGCGGCTG AAGGAGAAAC CGTTCGTTAC CCGGCCAACT ACTTCGAGAA GCCTAGTAAC GCCATGAACG CVB4
CXA1G      CTGAGTATCA ATAGACCGCT AACGCGGTTG AAGGAGAAAA CGTTCGTTAC CCGGCCAACT ACTTCGAAAA ACCTAGTAAC ACCATGGAAG CVB1
CXB5CGA    CCAAGTATCA ATAGACTGCT CACGCGGTTG AAGGAGAAAG CGTTCGTTAC CCGACTAACT ACTTCGAGAA ACCCAGTAGC ACTATGAAAG CVB5
CXA9CG     CTGAGTATCA ATAGGCTGCT CGCGCGGCTG AAGGAGAAAG CGTTCGTTAC CCGGCCAGCT ACTTCGAGAA GCCTAGTAAC ACCATGAAGG CVA9
EC12TCGWT  CTGAGTATCA ATAAGCTGTT GACACGGCTG AAGGAGAAAA CGCCCGTTAC CCGACCAGCT ACTTCGGAGA ACCTAGTATC ACCATAGAGG Echovirus 12
E6U16283   CCGAGTATCA ATAAGCTGCT CACGCGGCTG AAGGAGAAAG CGTTCGTTAC CCGGCTAGTT ACTTCGAGAA ACCCAGTACC ACCATGAAGG Echovirus 6
EV9GENOME  CCGAGTATCA ATAGACTGCT TGCGCGGTTG AAGGAGAAAA CGTTCGTTAT CCGGCTAACT ACTTCGAGAA ACCCAGTAAC ACCATGAAGG Echovirus 9
EV11VPCD   CCGAGTATCA ATAGACTGCT CACGCGGTTG AAGGAGAAAA CGTCCGTTAC CCGACCAACT ACTTCGAGAA ACCTAGTAAC ATCATGAATG Echovirus 11
CAU05876   CCGAGTATCA ATAGACTGCT CACGCGGTTG AGGGAGAAAA CGTCCGTTAC CCGGCTAACT ACTTCGAGAA GCCTAGTAGC ACCATGAAAG CVA16
CXA21CG    .TGAAATCAT ATAGACTGTA CCCACGGTCA AAAGTGATTG .ATCCGTTAT CCGCTTGAGT ACTTCGAGAA GCCTAGTATC GCCTTGGAAT CVA21
POLIO1A    .TGATGTCGT ATAGACTGCT TGCGTGGTTG AAAGCGACGG .ATCCGTTAT CCGCTTATGT ACTTCGAGAA GCCCAGTACC ACCTCGGAAT PV1
POLIOS1    .TGATGTTGT ATAGACTGCT TGCGTGGTTG AAAGCGACGG .ATCCGTTAT CCGCTTATGT ACTTCGAGAA GCCCAGTACC ACCTCGGAAT PV1-SABIN
PIPOLS2    .TGACATTGC ATAGACTGCT CACGCGGTTG AAAGTGATCG .ATCCGTTAC CCGCTTGTGT ACTTCGAAAA GCCTAGTATC GCCTTGGAAT PV2-SABIN
POL2CG1    .TGAGGCTGT ATAGGCTGTT TCCACGGCTA AAAGCGGCTG .ATCCGTTAT CCGCTCATGT ACTTCGAGAA GCCTAGTATC ACCTTGGAAT PV2
POL2LAN    .TGAGGCTGT ATAGGCTGTT TCCACGGCTA AAAGCGGCTG .ATCCGTTAT CCGCTCATGT ACTTCGAGAA GCCTAGTATC ACCTTGGAAT PV2-LANSING
PIPOX3     .TGAGGCCGC ATAGACTGTT CCCACGGTTG AAAGTGTCCG .ATCCGTTAT CCGCTCATGT ACTTCGAGAA GCCTAGTATC GCTCTGGAAT PV3-SABIN
PIPO3119   .TGAGGCCGC ATAGACTGTT CCCACGGTTG AAAGTGTCCG .ATCCGTTAT CCGCTCATGT ACTTCGAGAA GCCTAGTATC GCTCTGGAAT PV3
POL3L37    .TGAGGCCGC ATAGACTGTT CCCACGGTTG AAAGTGGCCG .ATCCGTTAT CCGCTCATGT ACTTCGAGAA GCCTAGTATC GCTCTGGAAT PV3
PIPOLS3    .TGAGGCCGC ATAGACTGTT CCCACGGTTG AAAGTGTCCG .ATCCGTTAT CCGACCATGT ACTTCGAGAA GCCTAGTACC GCTCTGGAAT PV3-SABIN
PIPO3XX    .TGAAGTTGC ATAGATTGTT CCCACGGTCG AAAGCGACTG .ATCCGTTAC CCGCTCATGT ACTTCGGGAA GCCTAGTATC ACTTCGGACT PV3
EV70CG     .TGAAGTTGC ATAGACTGTA CCCACGGTTG AAAGCGATGA .ATCCGTTAC CCGCTTAGGT ACTTCGAGAA GCCTAGTATC ATCTTGGAAT Enterovirus 70
PIHRV2G    .TCAATGTTG ATATGCTCCA .ACAGGGCAA AAACA.ACTG CGATCGTTAA CCGCAAAGCG CCTACGCAAA GCTTAGTAGC ATCTTTGAAA Rhinovirus 2
HRVPP      .TCAAAGTTG ATATGCTCCA .CAAGGGCAA AAACA.ACTG AGATCGTTAT CCGCAAGATG CCTACGCAAA GCCTAGTAAT ACATTGAAAG Rhinovirus 16
HRV89      .TCAATGAGG ATATGCTTTA CCCAAGGCAA AAACC.TTAG AGATCGTTAT CCCCACACTG CCTACACAGA GCCCAGTACC ATTTTTGATA Rhinovirus 89
HRVACG     .TTGACGTTG ATATGCTCTA .ACAGGGCAA AAACA.ACTG ATATCGTTAC CCGCAAAGTG CCTACACAGA GCTTAGTAGG ATTCTGAAAG Rhinovirus 1B
HRV        .AGCGAGGT. ATAGGCTGTA CCCACTGCCA AAAGC.CTTT A.ACCGTTAT CCGCCAACCA ACTACGTAAC AGTTAGTACC ATCTTGTTCT Rhinovirus 14
           197*                                           ▲                                         277*
                                                          C is attenuating in CVB3-induced myocarditis
```

whether the CVB3 5' NTR plays a significant role in determining the cardiovirulence phenotype of the virus.

With the elucidation of the nucleotide sequence of the infectious copy of the noncardiovirulent CVB3/0 genome (CHAPMAN et al. 1994), the genomes of cardiovirulent and noncardiovirulent CVB3 strains were at once compared. The noncardiovirulent strain CVB3/0 had been chosen for analysis due to its well-characterized biology (GAUNTT et al. 1979; ROESING et al. 1979; HUBER and JOB 1983; CHAPMAN et al. 1994). Comparative sequence analysis demonstrated ten differences among the genomes of the three infectious CVB3 clones that correlated with the cardiovirulence phenotypes of the viruses. TU et al. (1995) then mapped which of these sites affected the virulence phenotype using intratypic chimeric CVB3 strains, constructed by replacing sequences in the cDNA clone of the noncardiovirulent CVB3/0 strain with those from the cardiovirulent CVB3/20 strain. Progeny viruses obtained by transfection were tested in mice for cardiovirulence. Of the ten genetic sites originally identified as potential determinants of the phenotype, only the site in the 5' NTR appeared to affect the cardiovirulence phenotype. This site, nucleotide (nt) 234 (where nucleotide 1 is the 5' terminal), is cytidine in the noncardiovirulent virus and uridine in the virulent strain. It was also observed that CVB3/0 consistently became cardiovirulent when passed in *scid* (severe combined immunodeficiency) mice; sequence analyses of these cardiovirulent variants demonstrated that nt234C (in the parental CVB3/0) had mutated to U. None of the other ten sites were affected as assessed by sequence analysis of the variant strains.

TU et al. (1995) preliminarily investigated the mechanism by which the cardiovirulence phenotype was controlled by the U/C transition at nt234. Previous work with PV in neuroblastoma cell lines had demonstrated that the effect of a similar transition (albeit at a different site in the 5' NTR) in the PV1 5' NTR affected viral protein translation, such that the attenuated Sabin strains translated less efficiently than the neurovirulent, disease-causing PV strains (SVITKIN et al. 1985; LA MONICA and RACANIELLO 1989; reviewed in MINOR et al. 1993). However, no difference was discerned between cardiovirulent and noncardiovirulent CVB3 strains when viral translation was assayed by SDS-PAGE analysis in murine fetal heart fibroblast (MFHF) cultures. More recently, western blot analysis of such gels also demonstrated few if any significant differences in viral capsid protein intensities (S. Tracy, unpublished). However, examination of viral RNA transcription in both CVB3-inoculated MFHF cultures and adult murine cardiomyocyte cultures demonstrated a significant difference between the two viral phenotypes: the noncardiovirulent CVB3/0 RNA was synthesized about

←

Fig. 1. Alignment of enterovirus sequences in the region of nucleotide 234. Position of nucleotide 234 (C in noncardiovirulent CVB3-0) is indicated by *arrows*. Genbank accession numbers or locus names are listed at *left*, and virus serotype is listed at *right*. Sequence for CVB3/20 is from TRACY et al. (1990), for CVB3/0 from CHAPMAN et al. (1994), and for CVB3/AS, /DO, /CO, /ZU, /GU, /RE, /OL, /GA from unpublished data (J. Romero, S. Tracy; Chap. 6). Numbering is from CVB3/20 (TRACY et al. 1992). Secondary structures are indicated above the sequences as predicted for CVB3 (LE and ZUCKER 1990; Chap. 8)

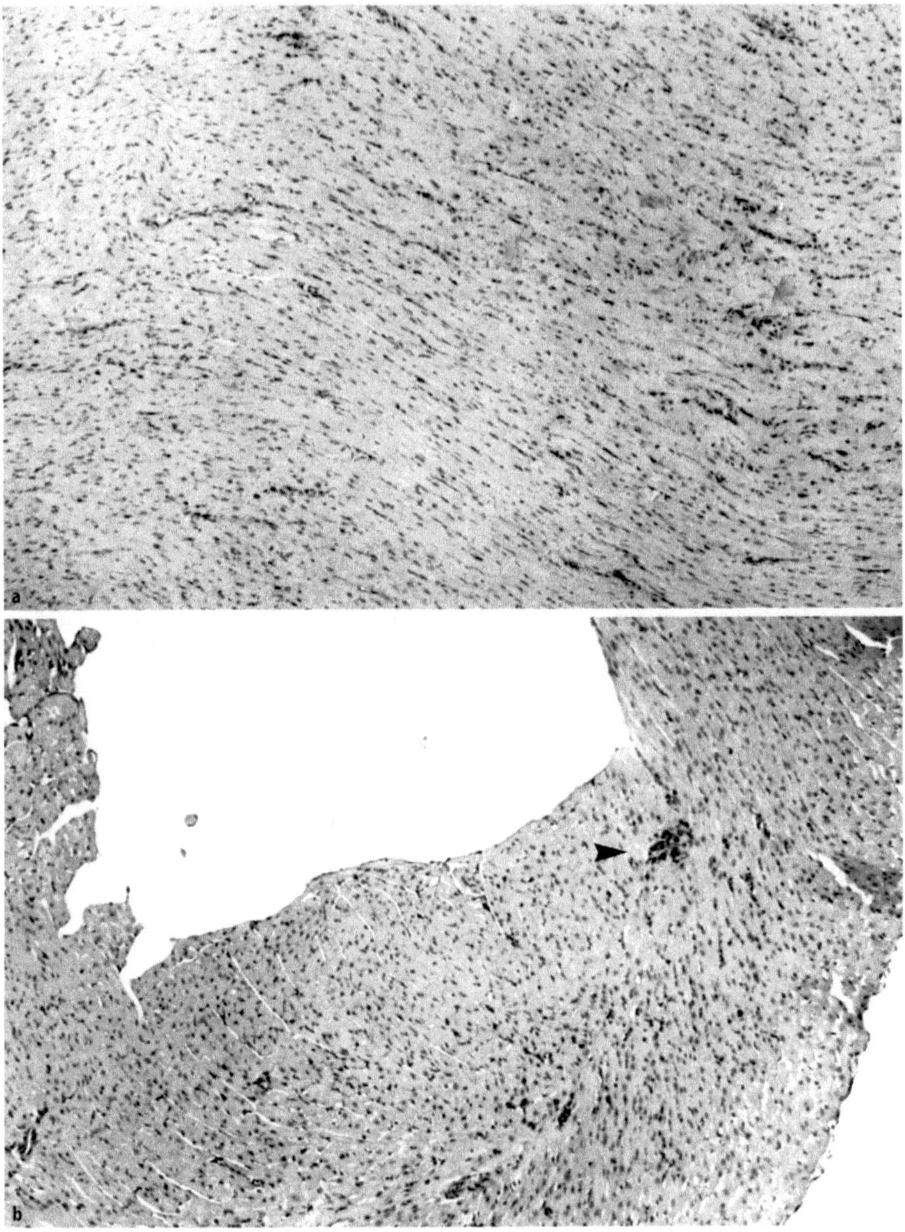

Fig. 2A–D. Pathologic changes in murine hearts as a function of wild-type CVB3 infection. C3H/HeJ male 4 week old mice were injected intraperitoneally with 10^5 $TCID_{50}$ units of (**A**) CVB3/RE: example of no detectable lesions in a typical section from an avirulent virus strain; (**B**) CVB3/CO: example of a small uncommon lesion (*arrow*) sometimes observed in sections of hearts infected with an avirulent strain; (**C**) CVB3/AS: example of typical widespread large lesions in heart of mouse infected with

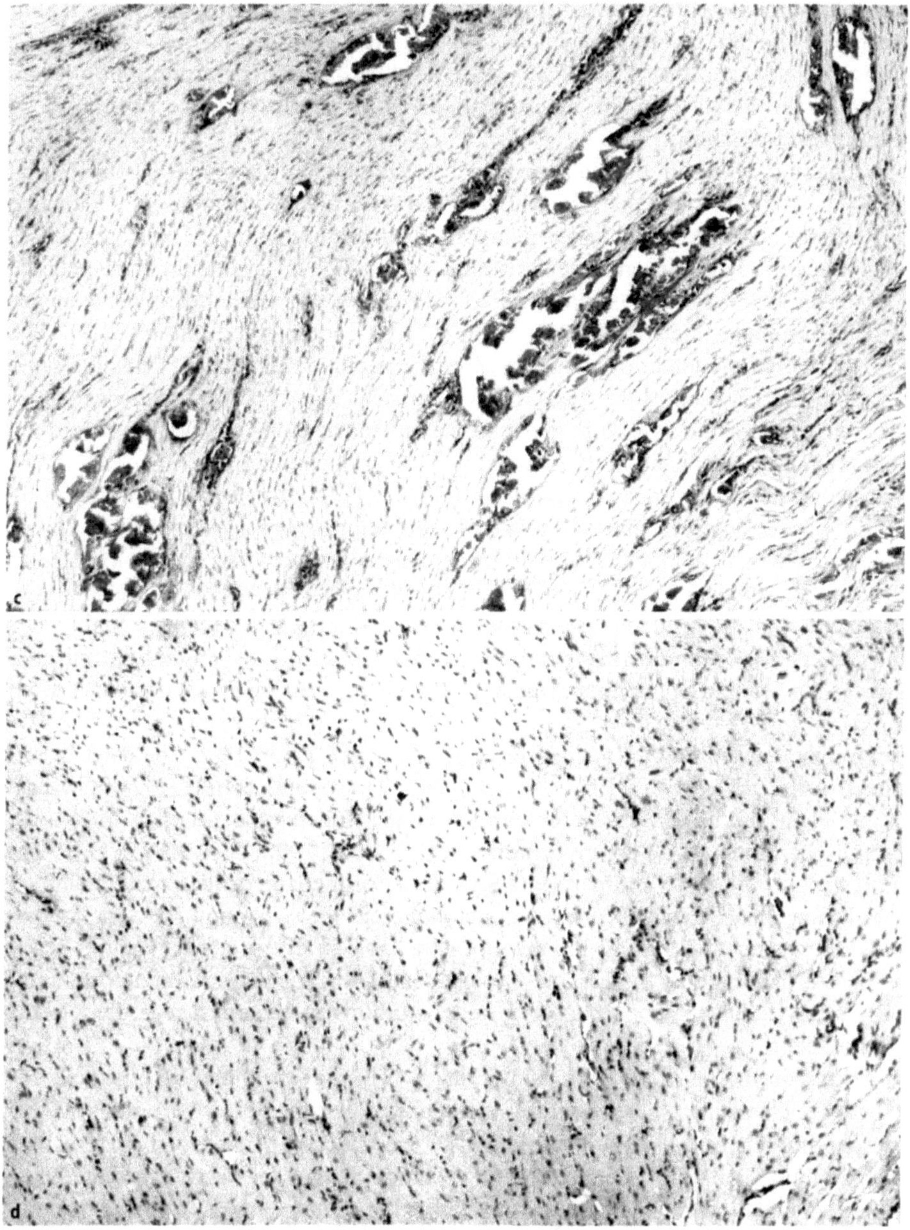

a cardiovirulent strain; (**D**) mock-infection with 0.1 ml of unsupplemented culture medium. Interestingly, of these, only the CVB3/AS strain also caused significant pancreatitis in mice. Hearts from mice at day 10 postinoculation were sectioned and stained with hematoxylin and eosin; original magnification, ×100

ten-fold more slowly than that of the cardiovirulent strain. This observation contrasted to the PV system, in which no effect was observed on the rate of RNA transcription (La Monica and Racaniello 1989). These data strongly implicated nt234 in the 5' NTR in the determination of the cardiovirulence phenotype of these two viral strains.

Ongoing work in our laboratories has convincingly shown, however, that nt234 cannot be the site that affects CVB3 cardiovirulence in nature (N. Chapman, J. Romero, S. Tracy unpublished data). An examination of the 5' NTR sequences of numerous CVB3 isolates demonstrates that nt234 is uridine in each genome (Tracy et al. 1996), and perhaps more significantly, the five nucleotide sequence which surrounds nucleotide 234 (5'-CG*U*UA) is completely conserved in each virus. Finally, alignment of available enterovirus sequences demonstrates the same absolute conservation of the 5'-CGUUA sequence (Fig. 1). Direct sequence analysis from the viral stock used to generate the molecular clone of CVB3/0 confirmed that the master sequence in the population is C at nucleotide 234 (Tu et al. 1995). Therefore, the occurrence of nt234C in CVB3/0 (Chapman et al. 1994; Tu et al. 1995) must be interpreted as the result of a mutational event, the biological consequence of which is that CVB3/0 is completely attenuated for cardiovirulence in juvenile and older mice (Gauntt et al. 1979; Chapman et al. 1994). Reversion to uridine at this site in the CVB3/0 genome causes CVB3/0 to become a highly cardiovirulent CVB3 strain: this was demonstrated not only by chimeric viral constructs but by natural reversion during passage in *scid* mice as well (Tu et al. 1995).

While determining the sequences of a variety of CVB3 strains (the dates of origin of which range from 1954 through 1987; Tracy and Gauntt 1987), we assayed eight of these viruses for cardiovirulence phenotype in mice in a well-established model of CVB3 myocarditis (Tracy et al. 1992; Chapman et al. 1994). Two strains were very cardiovirulent (defined as causing large, widespread inflammatory lesions in all murine hearts examined) whereas the remaining six were noncardiovirulent (very few, small lesions, not detectable in every heart) (Fig. 2). These results further support the lack of a role for nt234 in the natural determination of the cardiovirulence phenotype, for this site is uridine in all strains regardless of the phenotype.

We therefore assayed for a role of the 5' NTR in determination of the phenotype by constructing chimeric viruses containing either the 5' NTR from CVB3/DO (cardiovirulent) or CVB3/CO (avirulent) in the CVB3/0 background (S. Tracy, J. Romero, N. Chapman unpublished). The chimeric viruses were inoculated into C3H/HeJ male mice; at 10 days postinfection the hearts were assayed for pathologic changes characteristic of CVB3-induced inflammatory heart disease. Both chimeras were shown to be avirulent. Thus, while conversion of nt234C to U in the CVB3/0 genome converts the phenotype from avirulent to virulent, replacement of the entire CVB3/0 5' NTR by that from a cardiovirulent virus does not produce a cardiovirulent progeny viral strain.

These unexpected results have led us to suggest the following testable hypothesis. In our earlier work (Tu et al. 1995), replacing of the avirulent CVB3/0 5' NTR with that from CVB3/20 caused the chimeric virus to be virulent, a phenotype mapped specifically to the single transition at nt234. However, replacement of the CVB3/0 5' NTRs by the homologous sequences of either a cardiovirulent or avirulent strain resulted in an avirulent phenotype. In each case, as discussed above, nt234 is U. Based upon these data (and data discussed below) our working hypothesis at this time is that a CVB3 strain (all CVBs?) may have sites that affect virulence either in the 5' NTR, in the capsid protein coding region, or possibly in both sites. Consistent with data available to us at present, we suggest that such sites either contribute directly to virulence or are neutral, i.e., have no effect on the phenotype. Thus, a virus may be virulent if the site in the 5' NTR or in the capsid proteins, or in both are "on." An avirulent virus would require both sites to be "off." The CVB3 strains DO and CO are predicted by this hypothesis to have the site(s) in the 5' NTR off. Also, by this hypothesis, site(s) of cardiovirulence outside the 5' NTR would be on in the cardiovirulent CVB3/DO strain and off in the avirulent CVB3/0. Experiments to test this hypothesis are underway.

Archard and colleagues reported that a cardiovirulent CVB3 strain (Klump et al. 1990) became significantly attenuated following repeated passages in skin fibroblasts (Zhang et al. 1993). These workers sequenced the 5' NTR of the attenuated strain and found a single difference between the parental cardiovirulent strain and the attenuated strain at nt590: a transversion from A in the parental to U in the attenuated strain. The 5' NTR of the cardiovirulent parental clone was replaced with that from the attenuated strain containing the single base change, progeny virus were obtained by transfection, and the virus tested in mice for virulence. The chimeric virus was as cardiovirulent as the parental strain, indicating that the change at nt590 played no role in determining the attenuated viral phenotype. The implication of these studies was that one or more (unidentified) changes downstream of the 5' NTR must then be responsible for attenuation. In related work, Zhang et al. 1995 characterized a single amino acid change (Lys to Ser) in the NH_2-terminal sequence of the B-C loop of the CVB3 capsid protein VP1. Although this change was related to the plaque phenotype of the virus in Vero cells, it did not affect the cardiovirulence phenotype measured in mice. These two studies were performed on tissue culture-derived viral strains and may suggest that either successive passaging in cell culture is not a highly productive route toward elucidation of CVB3 genetics underlying the attenuation of cardiovirulence or, more likely, that another cell culture system might be more productive.

Recently, Gauntt and Pallansch (1996) reported the partial sequence analysis of the 5' NTRs of numerous wild-type strains of CVB3, including several of those characterized in our laboratory (above), spanning the region of nucleotides 300–599 in the 5' NTRs. In addition, these strains were assayed for the cardiovirulence phenotype in CD-1 mice, a partially outbred strain (Gauntt et al. 1979) and generally more resistant to myocardial disease than C3H/HeJ mice. Of the 15 CVB3 strains examined by Gauntt and Pallansch, two strains were shown to be cardiovirulent while the remaining 13 were not virulent. Sequence analysis showed nucleotide 565 was A in three of the four cardiovirulent strains, with U or C at this position in the avirulent strains. Although these data are suggestive, no mutational studies were performed to test the assumptions. Clearly, as nt565 corresponds to a site within the internal ribosomal entry recognition site (IRES; MEEROVITCH et al. 1991; NICHOLSON et al. 1991; PESTOVA et al. 1991) of PVs, by reasonable inference based upon genomic similarities it most likely also lies within the putative CVB3 IRES. As pointed out by these authors, nt565 also lies in a sequence that is involved in a pseudoknot structure in consensus RNA folds for this region (LE et al. 1992). It is clear that mutations at nt234 can affect viral biology (TU et al. 1995) and changes at nt565 may also affect viral biology. Are such changes involved with cellular proteins binding at these sites, or do these sites associate transiently in higher order RNA structures as well as bind cellular cofactors? These experiments are underway in several laboratories. Finding proteins that bind these sites in the CVB3 5' NTR would seem highly probable from numerous findings that cellular proteins associate with PV 5' NTR (DEL ANGEL et al. 1989; PESTOVA et al. 1991; GEBHARD and EHRENFELD 1992; HELLEN et al. 1994; SVITKIN et al. 1994; BLYN et al. 1995).

2.3 Coxsackievirus B3 Capsid Proteins and Cardiovirulence

At this writing, there are two lines of evidence, one indirect and one direct, that support a role for the CVB3 capsid proteins in determination of the viral cardiovirulence phenotype. Furthermore, a series of studies on CVB4 (discussed below, Sect. 3) also support a role for the capsid proteins in determining the CVB4 virulence phenotype.

In the study (discussed above) ZHANG et al. (1993) derived an attenuated strain of CVB3 from a highly cardiovirulent parental strain in cell culture. Characterization of the attenuated strain using an intratypic chimeric virus demonstrated no role for attenuation of the virus by the 5' NTR. The inference drawn from this study is that sites outside of the 5' NTR must be responsible for modifying CVB3 cardiovirulence.

In another series of studies, an antibody escape mutant of a highly cardiovirulent strain of CVB3 (LOUDON et al. 1991; VAN HOUTEN et al. 1991; HUBER et al. 1994) was isolated and characterized in mice and cell cultures. When assayed in established murine monocyte cultures, the cardiovirulent parental and attenuated strains induced either a Th1 or Th2 cytokine expression profile, respectively, suggesting the basis for, and extending previous observations of, a potential immunopathogenic mechanism (WEAVER et al. 1988; STREET and MOSMANN 1991; HUBER and PFAEFFLE 1994; HUBER et al. 1994). Recently, Huber and Knowlton (S. Huber, K. Knowlton, unpublished data) established that the basis for this CVB3 strain's attenuation is a single amino acid change in the exterior E-F loop of capsid protein 1B. Thus, one implication from these studies is that the E-F loop, a significant exterior structure on the surface of the CVB3 virion (MUCKLEBAUER et al. 1995), is recognized as antigenic by cells in the murine immunologic surveillance system and that individual amino acid changes in the loop's peptide sequence can help determine both viral antigenicity and, by directing which types of T cells respond, viral virulence as well. It is of course well established that single amino acid changes within an antigenic epitope recognized by T cells can alter the ability of T cells to recognize the epitope (FINNEGAN and AMBURGEY 1989: PIRCHER et al. 1990; BECK et al. 1992; TRACY et al. 1995). It can be expected that this finding will be further tested using synthetic peptides as both in vivo and in vitro antigens for immunization and testing of T cell functionality.

It should be noted that no input from any area other than the 5' NTR was observed that contributed to the cardiovirulence phenotype in the studies of TU et al. (1995). In retrospect, we realize now that CVB3/0 used in this study (CHAPMAN et al. 1994) can be in fact cardiovirulent when nt234C mutates to U. Thus, by letting nt234 be U for both strains, it is perhaps not surprising that the intratypic chimeric viruses constructed in this study failed to identify sites elsewhere in the genome, if it is assumed that these viruses are functionally identical in key sites downstream of the 5' NTR.

Beck and colleagues determined that the cardiovirulent CVB3/20 (TRACY et al. 1992), when inoculated into mice that had been nutritionally deprived for selenium (BECK et al. 1994c), became more cardiovirulent and replicated in hearts to higher titer than its parental strain that had not been passaged in selenium-deprived mice (Chap. 5). The viral genetics of this observation were not studied. Subsequent studies employed the avirulent CVB3/0 (CHAPMAN et al. 1994) in selenium deprived mice (BECK et al. 1994b; 1995). Similar disease patterns in mice were also noted by these workers in mice deprived of vitamin E (BECK et al. 1994a). These studies demonstrated that the avirulent CVB3/0 became cardiovirulent during passage in mice deprived of selenium and furthermore, not only nt234 changed from C to U but five other sites as well mutated to the base found in the related cardiovirulent strain, CVB3/20 (TRACY et al. 1992). Three of these changes were in the capsid proteins of the virus. As TU et al. (1995) had shown nt234 U to be necessary and sufficient to convert the phenotype of CVB3/0 from avirulent to cardiovirulent and that during passage in *scid* mice, no other changes occurred in the ten sites previously identified as variable with the phenotype (CHAPMAN et al. 1994), it is at present unclear what role the other five changes may play in viral biology, or whether this may be due to some aspect of the nutritional deficiencies. It is extremely interesting that the five changes were found only at the sites that differed between the cardiovirulent strain CVB3/20 (TRACY et al. 1992) and the avirulent CVB3/0 (CHAPMAN et al. 1994) used in the studies. It would be of interest to study the roles of these sites by the use of site-specific mutation in viral biology and in virus-host interactions. It also remains to be determined if virulent, wild-type CVB strains also become virulent during replication in nutritionally deprived mice.

Mutations in the 2B coding region of CVB3 which change amino acid residues in a predicted amphipathic helix have been shown to change the viability of CVB3 in cell culture and the extent of viral RNA replication (VAN KUPPEVELD et al. 1995, 1996). Protein 2B in PV1 is present in viral RNA replication complexes (BIENZ et al. 1990) and mutations in the P2-B coding region decrease viral RNA synthesis (JOHNSON and SARNOW 1991). However, because no sites within this region have been identified as determinants of virulence in natural isolates of CVBs, the role that such mutations might have in variation of virulence is at present unknown.

Because of the single mutation at nt234 that converts the noncardiovirulent CVB3/0 to a cardiovirulent strain (TU et al. 1995), and because of Tu and coworkers' inability to locate any of the other identified sites as affecting cardiovirulence in mice, it may be concluded that CVB3/0 with nt234U is phenotypically indistinguishable from other characterized cardiovirulent strains (TROUSDALE et al. 1976; KLUMP et al. 1990; VAN HOUTEN et al. 1991; TRACY et al. 1992) (although as discussed directly above, the data of Beck may suggest other possibilities). Because of this, the effects of other sites in the viral genome on the cardiovirulence phenotype could not be ruled out. To help address this problem, the genome of a wild-type isolate of CVB3 (CVB3/CO; TRACY and GAUNTT 1987; Chap. 6) that has been demonstrated to be noncardiovirulent in mice is being cloned as an infectious cDNA copy in our laboratory. Initial evaluation of the nucleotide identity between this wild-type isolate and cloned, infectious cardiovirulent CVB3 strains demonstrates significant divergence.

3 Genetics and Mechanisms of Coxsackievirus B4 Virulence

Coxsackievirus B4 (CVB4) has also been demonstrated to be an agent of human heart disease (MELNICK 1990; MAISCH et al. 1993; SATOH et al. 1994). In addition, growing circumstantial evidence suggests that CVB4 is associated with human pancreatic diseases; perhaps playing a role in acute and chronic pancreatitis (ZSOFIA et al. 1992) and in insulin-dependent diabetes (YOON et al. 1979; KING et al. 1983; BANATVALA et al. 1985; CLEMENTS et al. 1995). Mouse models of CVB4-induced pancreatic and cardiac diseases, mimicking the human situation, have been established and further support the proposition that CVB4 is an etiological agent in various pancreatic and cardiac diseases in humans.

→

Fig. 3A–J. Histopathology of pancreatic tissue from CB4-P and CB4-V-infected BIO.Q mice. Mice were injected intraperitoneally with 10^4 pfu of virus or mock-infected with 0.2 ml of PBS. Pancreata were harvested at various times postinfection and processed for routine histology. Sections were stained with hematoxylin and eosin. **A** and **F**, mock-infected; **B–E**, CB4-V; **G–J**, CB4-P. Panels **B** and **G**, 2 days postinfection; **C** and **H**, 4 days postinfection; **D** and **I**, 7 days postinfection; **E** and **J**, 14 days postinfection. *Arrows* denote ductular structures; *n*, normal acinus; *f*, fat cell replacement

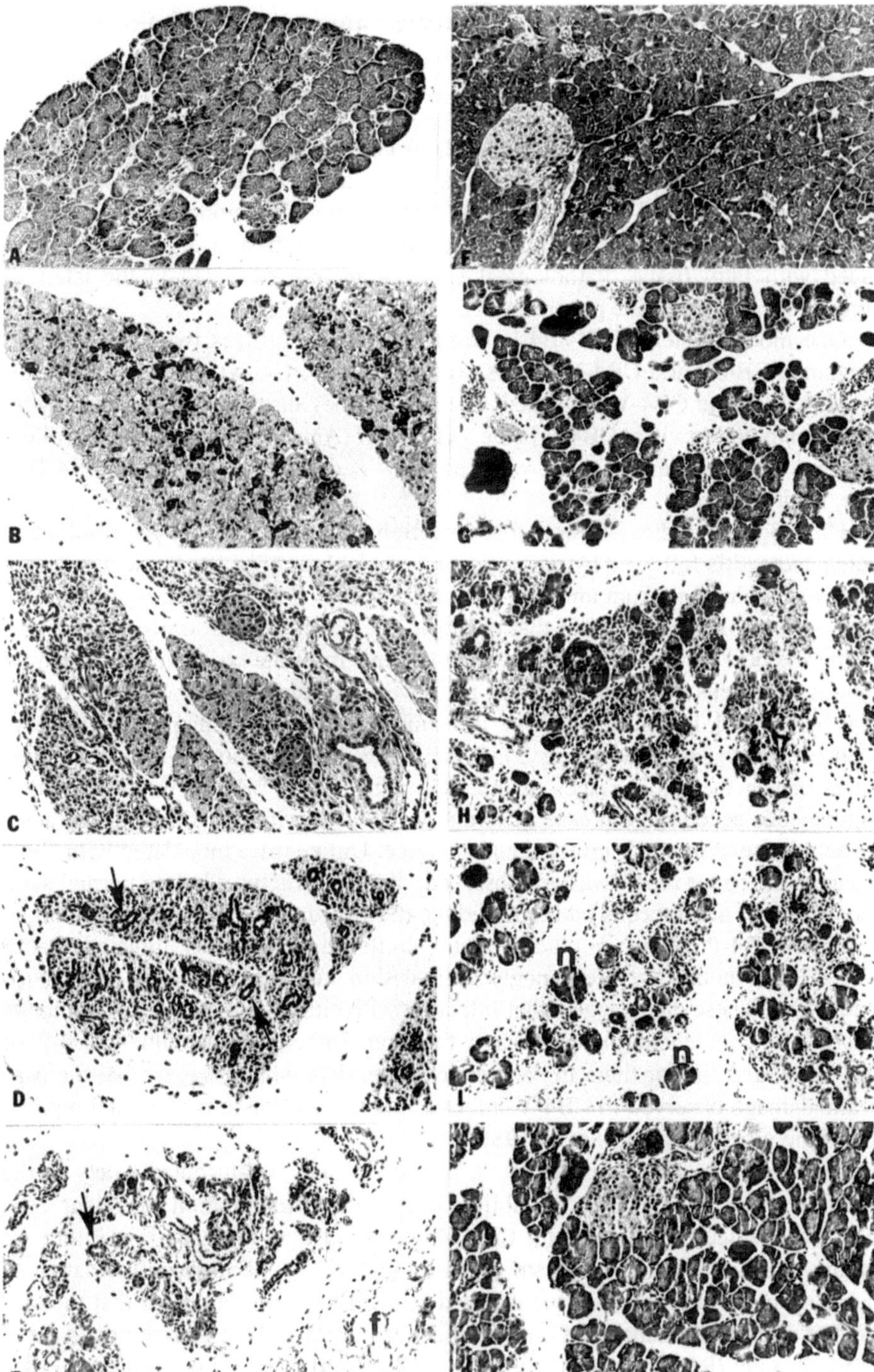
A
B
C
D
E
f
F
G
H
I
n
n

3.1 Coxsackievirus B4-Induced Murine Pancreatitis: Pathologic Aspects

Coxsackievirus B4 can induce a severe acinar pancreatitis in mice (Ross et al. 1974; Lansdown 1976; Vella et al. 1992). A common observation in these murine systems is the degeneration and necrosis of the acinar cells within the exocrine pancreas with a marked inflammatory infiltrate comprising mononuclear cells evident during acute viral infection. Destroyed exocrine tissues are eventually replaced with fatty tissue. Pathological changes are not observed in the islets of Langerhans, the endocrine pancreas.

One model system of CVB4-induced pancreatitis utilizes two CVB4 strains designated CB4-P and CB4-V (Ramsingh et al. 1989; Caggana et al. 1993). Although CB4-P and CB4-V are genetically similar, they display different phenotypes both in vivo and in vitro. CB4-P is the JVB strain (American Type Culture Collection) and is nonvirulent in mice, whereas CB4-V, derived by passaging CB4-P in mice, is virulent.

Histological studies showed that infection of mice with CB4-P resulted in acute pancreatitis followed by tissue repair 2 weeks later (Fig. 3). These data correlated with a peak in serum amylase activity at 3 days postinfection, representing a three- to fourfold increase over that of mock-infected mice (Caggana et al. 1993). During the peak of tissue injury, an inflammatory infiltrate consisting primarily of mononuclear cells was observed, but by 2 weeks postinfection the exocrine pancreas had healed. In contrast, CB4-V infection resulted in chronic pancreatitis with extensive tissue destruction (Fig. 3). Widespread degranulation of the acinar cells occurs during the initial stages of infection with CB4-V which correlated with a simultaneous sevenfold increase in serum amylase activity above that of the average activity seen in mock-infected control mice. Unlike mice inoculated with CB4-P, pancreatic tissue repair was not observed; tubular structures but no normal acini were observed and fat cells had replaced the destroyed acini. Although mice inoculated with CB4-P had serum glucose levels similar to that of mock-infected control mice, CB4-V induced severe hypoglycemia within 4 days postinfection (Ramsingh et al. 1989). These data suggest that infection with either CB4-P or CB4-V impairs exocrine but not endocrine pancreatic function. Infection with either CB4-P or CB4-V yielded similar titers in pancreatic tissue, data which suggest that the pancreatic damage observed in CB4-V-infected mice is not due to preferential replication of the virus (Ramsingh et al. 1989).

These data suggest that pancreatic damage cannot be attributed solely to the act of viral replication. One possibility is that the degranulation of acinar cells observed in mice inoculated with CB4-V may contribute to tissue injury through autodigestion. Another mechanism of tissue injury may be immune-mediated. In analyzing mortality rates in virus-infected mice, it has been observed that the outcome of infection is dependent on both the virus strain and the murine host's genetic background (Ramsingh et al. 1989). Inoculation of B10 H-2 congenic strains of mice with CB4-V resulted in significant morbidity in mice having the H-2Kqq allele while mice inoculated with avirulent CB4-P survived the infection. The correlation of a class I gene of the major histocompatibility complex (MHC) with the outcome of infection argues for an as yet unknown role of T cell-mediated destruction of the exocrine pancreas.

3.2 Genetic Mapping of the Virulent Phenotype for Coxsackievirus B4

Within the CVB4 serotype, strains exist that induce different pancreatic diseases in murine models. A β-cell tropic variant can induce a diabetes-like syndrome in mice (YOON et al. 1978; CHATTERJEE et al. 1988) while another variant induces an acinar pancreatitis (ROSS et al. 1974; LANSDOWN 1976; RAMSINGH et al. 1989; VELLA et al. 1992; CAGGANA et al. 1993).

3.2.1 CVB4 and the Murine Model of Diabetes

The viral genetic sequences that determine the diabetogenic phenotype of CVB4 in mice are at present unknown. It is clear that there are far fewer reports of CVB4 strains capable of inducing diabetes in mice than there are strains of CVB4 capable of inducing acinar pancreatitis in mice and strains of CVB3 (or CVB4) capable of inducing inflammatory heart muscle disease. This observation raises the question of the role of CVB4 infection in diabetes.

There are two strains of CVB4 that have been cloned as infectious cDNA copies: an avirulent strain and a pancreovirulent strain (CAGGANA et al. 1993). Two other groups (TITCHENER et al. 1994; KANG et al. 1994) reported the complete nucleotide sequence of two murine β-cell tropic CVB4 strains. TITCHENER et al. (1994) reported the sequence of a mouse pancreas-adapted variant (P-CB4) that caused an excessive release of insulin from pancreatic β-cells cultured in vitro. Comparison of this genomic sequence with that of the prototype JVB strain (JENKINS et al. 1987) revealed six nucleotide substitutions in the 5' NTR and seven point mutations that would result in amino acid substitutions (one in each of the capsid proteins VP4 and VP1, four in 3A and one in 3D). KANG et al. (1994) reported the sequence of the E2 strain of CVB4 which can induce a diabetes-like syndrome in mice. Comparison of this sequence with that of the JVB strain (JENKINS et al. 1987) revealed numerous genetic differences: within the two noncoding regions, 107 nucleotide substitutions were identified while 111 amino acid changes were noted scattered throughout the structural and nonstructural proteins. Based on these genomic sequence data, the P-CB4 (TITCHENER et al. 1994) and the CB4-V (CAGGANA et al. 1993) variants are more closely related to the JVB strain than is the E2 strain (KANG et al. 1994). Interestingly, these three closely related CVB4 strains each display different phenotypes in mice, consistent with relatively minor genetic alterations affecting characteristics such as host cell tropism and virulence. Identification of the determinants that govern the β-cell tropism of P-CB4 and the diabetogenic nature of E2 will require additional mapping studies.

3.2.2 Mapping Genetic Sites in the CVB4 Genome Affecting Murine Pancreatitis

Pancreatitis (inflammatory disease of the pancreas) of mice, as opposed to diabetes, is readily induced by CVB4 (Ross et al. 1974; Lansdown 1976; Vella et al. 1992) and CVB3 (S. Tracy, A. Ramsingh unpublished). To elucidate the genetic basis of virulence of CVB4-induced murine pancreatitis, a panel of recombinant, chimeric viruses (Fig. 4) was derived from the avirulent CB4-P and the virulent CB4-V parental variants (Ramsingh et al. 1990; Caggana et al. 1993; Ramsingh and Collins 1995). Each of the recombinant, chimeric cDNAs was constructed by exchanging the appropriate homologous sequence on either a CB4-P or a CB4-V genetic background. Virulence was initially scored as the percentage of B10.T(6R) mice that became moribund within 2 weeks postinfection. Mice inoculated with the uncloned parental CB4-P or the progeny virus derived from the infectious clone, vCB4-P, appeared healthy and did not succumb to the infection. In contrast, all of the mice inoculated with CB4-V or the progeny virus derived from the infectious clone, vCB4-V, became moribund within 2 weeks. Mice inoculated with the chimera vCB403 were healthy and did not succumb to infection, suggesting that the 5' NTR does not affect pancreovirulence in our model system. However, the 5' NTR influenced viral replication in vitro as evidenced by plaque size (Ramsingh et al. 1995).

Mice inoculated either with the vCB405 or vCB406 chimera also appeared healthy, suggesting that the virulent phenotype maps to the region between the 5' NTR and the P2 region. Within the P1 region, five amino acid differences are predicted from sequence data, four of which represent nonconservative changes (Ramsingh et al. 1992). To date, three chimeric viruses that contain single nucleotide changes at position 791 (VP4), 1353 (VP2) or 2833 (VP1) on the avirulent CB4-P genetic background have been constructed. Mice inoculated with these strains were monitored for the development of pancreatitis, hyperamylasemia and hypoglycemia. Interestingly, the recombinant virus, vCB420, reflected the virulent phenotype of CB4-V, suggesting that Thr-129 of VP1 is a major determinant of virulence in this model system.

In addition, the chimera vCB424 yielded a phenotype that was intermediate between that of the two parental viruses (Ramsingh and Collins 1995). Infected mice developed hypoglycemia, similar to that observed in CB4-V-inoculated mice, resulting in a mortality rate of 12%–15%. However, serum amylase activity was comparable to that of mice inoculated with avirulent CB4-P. Histological assessment of pancreatic tissues also supported the observation that vCB424 gave an intermediate phenotype.

These data suggest that both Thr-129 of VP1 and Arg-16 of VP4 influence the virulent phenotype. Arg-16 of VP4 was also shown to influence viral replication in vitro (Ramsingh et al. 1995). Thr-129 of VP1 is compatible with the development of both hyperamylasemia and hypoglycemia while Arg-16 of VP4 correlates with

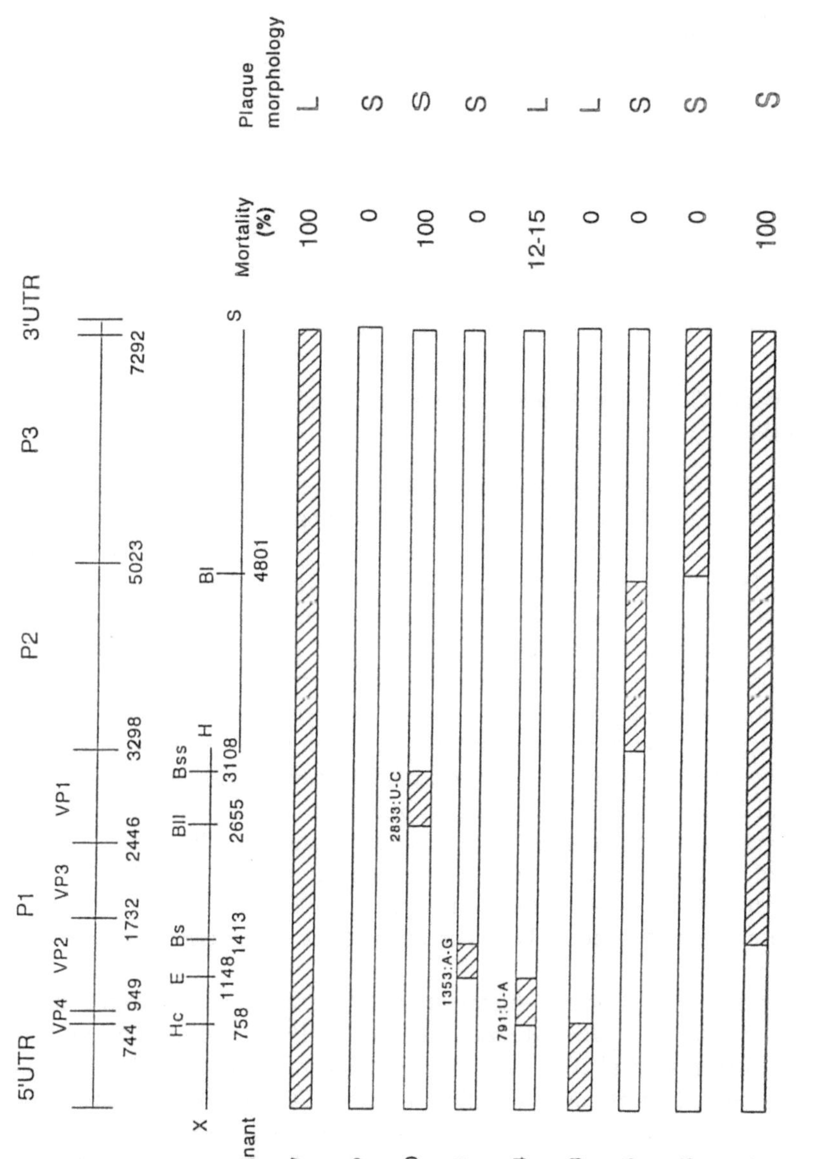

Fig. 4. Genotypes and phenotypes of recombinant, coxsackie B4 viruses. The *top line* depicts the structural organization of the coxsackievirus B4 (JVB strain) genome (JENKINS et al. 1987). A partial restriction map of the subclones used for the construction of the chimeric cDNAs is shown. One pair of subclones contained the 5' nontranslated region (NTR) and the P1 regions of CB4-P and CB4-V as *Xba*I-*Hin*dIII fragments. Another pair contained the P2 and P3 regions and the 3' NTRs of CB4-P and CB4-V as *Hin*dIII-*Sac*I inserts. Plaque phenotype is designated large (*L*) or small (*S*). *X*, *Xba*I; *Hc*, *Hinc*II; *E*, *Eco*RI; *Bs*, *Bst*BI; *BII*, *Bgl*II; *Bss*, *Bss*HII; *H*, *Hin*dIII; *BI*, *Bgl*I, *S*, *Sac*I

the development of hypoglycemia (of unknown mechanism). Hyperamylasemia is possibly the result of infection of the pancreatic acinar cells suggesting that Thr-129 of VP1 is a determinant of cellular tropism in vivo. If, as is likely, immune-mediated mechanisms play a role in the destruction of the exocrine pancreas during infection with CB4-V, then it seems likely that Thr-129 of VP1 must play a pivotal role. This amino acid substitution may alter the immunogenicity of the viral peptide that is presented to effector T cells. Ongoing studies (Ramsingh, unpublished) focus on constructing two additional chimeric viruses containing substitutions at nucleotide positions 1552. (VP2) and 2505 (VP1) on an avirulent, genetic background to determine the effects of these sites on virulence.

3.2.3 How Might Capsid Protein Sites Alter the Virulence Phenotype of CVB4?

Although the three-dimensional structure of CVB4 has not yet been solved, the location of specific amino acids may be approximated based on the structure of the recently reported CVB3 virion (Muckelbauer et al. 1995) and the PV capsids (Hogle et al. 1985; Filman et al. 1989; Yeates et al. 1991). The VP4 capsid protein of picornaviruses is found on the inner surface of the virion and interacts with the flexible NH_2-terminal extension of the other capsid proteins, forming an intertwined network within the virion that contributes to virion stability (reviewed in Hellen and Wimmer 1992). Studies with PV have shown that mutations within the VP4 coding sequence can affect cell entry (Moscufo et al. 1993), host range, and pathogenicity (Couderc et al. 1994). In one model system (Caggana et al. 1993; Ramsingh and Collins 1995), there is a gain of a positive charge at VP4-16 in the vCB424 recombinant which may affect: (1) the interaction of VP4 with the negatively charged viral genomic RNA, (2) the interaction with the NH_2-terminal extensions of the other capsid proteins, or (3) an early step in binding to the cell receptor. These interactions could be expected to affect virion assembly, stability, and/or tropism.

By using a sequence-structure alignment with PV and recent data on the three-dimensional structure of CVB3 (Muckelbauer et al. 1995), the location of Thr-129 of VP1 (Ramsingh et al. 1992) can be approximated. Thr-129 of VP1 is predicted to be on the surface loop connecting β-strands D and E, on the rim of the canyon. The D-E loop of PV is thought to play a role in both host range (Murray et al. 1988) and immunogenicity (Wiegers et al. 1989). Furthermore, the D-E loop of VP1 of PV2 also contains a major determinant of attenuation (Ren et al. 1991). As discussed above (Sect. 2.3), Huber and Knowlton suggest that a single amino acid change in the E-F loop of capsid protein VP2 differentiates between two strains of CVB3 that differ in murine cardiovirulence. In this light, it is interesting that the chimeric CVB4 strains (Caggana et al. 1993) that are virulent for the murine pancreas also demonstrated a myocarditic phenotype when tested in C3H/HeJ mice (S. Tracy, A. Ramsingh, N. Chapman, unpublished data). A salient feature of the determinants of virulence or attenuation of the picornaviruses is the diversity of nucleotide sites that govern this phenotype. The sites that significantly affect the virulence phenotype map in general to the 5' half of the viral genome, to the 5' NTR and the P1 (capsid protein coding) region. The data from two CVB serotypes and a PV serotype that show capsid sites to be significant determinants of virulence suggest a common mechanism in the pathogenesis of enterovirus infections.

4 Natural Isolates and Other Coxsackievirus B Serotypes

To date, the great majority of detailed characterization of the CVBs has been performed with laboratory strains. We feel that it will be necessary to confirm and extend what we are learning by testing the results against other, much less well-characterized, strains (so-called wild-type strains). To date, the primary paradigm for mechanisms of enterovirus attenuation derives from studies based on understanding the mechanism of the artificially induced attenuating mutations in the Sabin PV vaccine strains. As discussed above and below, the majority of CVB3 strains, when tested in mice sensitive for the development of CVB-induced myocarditis, are not cardiovirulent, suggesting that the norm for CVBs may well be an avirulent phenotype. Therefore, it will be both necessary and highly interesting to determine the genetic mechanism underlying CVB virulence phenotypes as determined in nature.

In one study, the cardiovirulence of a variety of CVB3 strains isolated in Berkeley between 1956 and 1983 was examined in semi-outbred CD-1 mice (TRACY and GAUNTT 1987). This study showed that the majority of strains (five of seven) were avirulent in mice, with only two causing detectable inflammatory heart muscle disease. The avirulent strains also replicated to lower titers in the murine hearts than did the virulent strains. Genetic similarity of these strains was assessed by probing Southern blots of a cloned CVB3 genome (TRACY et al. 1985) using full-length, representative cDNA generated from these viral RNAs as probes: clear differences were observed, demonstrating significant genetic change had occurred in the viral genome over nearly 30 years. (Many of these strains are discussed at greater length in Chap. 6.) Continuing use and characterization of wild-type CVB3 and CVB4 strains in our laboratory in the more disease-prone strain of mouse C3H/HeJ confirms that the great majority of CVB3 strains are noncardiovirulent. These data suggest that we must now learn more about the genomes of noncardiovirulent viruses, as all CVB3 cloned genomes to date have been cardiovirulent (with the exception of the mutant, CVB3/0, discussed above).

In a limited outbreak of CVB2 in the Omaha (NE) area in the Fall of 1994, one child died with severe myocarditis, another child required a heart transplant, and five other children endured severe meningitis and, in one case, hepatitis. The CVB2 strains from the child that died have been preliminarily studied in cell culture and mice. Only the isolate from the autopsied heart was cardiovirulent in

C3H/HeJ mice; the pre-mortem throat and stool isolates were avirulent in mice. Sequence comparisons of the 5' NTRs as well as coding regions of numerous CVB2 isolates from this outbreak (J. Romero, S. Tracy, N. Chapman, unpublished data) suggest, not surprisingly, that all the isolates were closely related and are consistent with one master sequence circulating at that time. The strains isolated from the fatal case of myocarditis represent a potentially useful group to study, as it seems that the initial infection was with an avirulent CVB2 strain that became cardiovirulent during replication in the child.

These preliminary studies of naturally circulating viruses suggest that most CVB strains are likely to be avirulent, assuming that the murine model represents a relevant reflection of infection in humans. These data suggest interesting questions, the answers to which will impact both our knowledge of enteroviral biology and health care. What causes a highly virulent virus population to arise? Why do a-virulent CVB strains, not virulent strains, represent the dominant quasispecies?

5 Lessons from Other Picornavirus Systems

The mapping of virus-encoded determinants of virulence with other picornaviruses has been and remains heavily dependent upon the study of sites of attenuation of virulent viruses. For example, sites in the 5' NTR have been altered to result in attenuation of the disease induced in animals, with mechanisms linked to replicative vigor. Capsid protein sites that have been mapped are involved in virulence determination through mechanisms involving a variety of interactions: capsid stability and assembly, virus-receptor interactions, or (when there is a component of immune-mediated disease) via epitopes present on capsid proteins. Because bioengineered recombinant chimeric or mutant viruses have to be tested in an animal model of the human disease to test the connection of the viral genetic change to disease pathogenesis, verification of such sites of virulence has been largely limited to the PVs, CVBs, Theiler's murine encephalomyelitis virus (TMEV), and encephalomyocarditis virus (EMCV).

5.1 Polioviruses

Polioviruses have provided an excellent model for studies of the molecular biology of picornaviruses. The impetus of the poliomyelitis epidemics of the 1940s and 1950s and the later cases of vaccine-associated poliomyelitis stimulated extensive studies of the molecular biology of the three poliovirus serotypes (PV1–3) and of the mutations present in the attenuated Sabin vaccine strains. These attenuated virus strains of each serotype were isolated after repeated passage in nonhuman primate cells (Sabin and Boulger 1973).

Use of the Sabin vaccines carries an extremely low risk of vaccine-associated paralytic poliomyelitis caused by reversion of the vaccine strain to a neurovirulent phenotype (NKOWANE et al. 1987). When this occurs, disease is most often associated with the type 2 and 3 strains, and rarely with the type 1 strain (ASSAAD and COCKBURN 1982; NKOWANE et al. 1987). The variation in the extent of expression of neurovirulence of wild-type PV strains and the genotypes underlying such wild-type variation has not been well studied (NATHANSON and MARTIN 1979). Thus, the dominant model for picornavirus disease, PV infection, relies heavily on the knowledge obtained from studies of reversion from artificially attenuated PV strains (SABIN and BOULGER 1973).

The Sabin PV strains each contain attenuating mutations within the 5' NTR and in the capsid protein coding region. Each of the three vaccine strains contains a similar attenuating mutation within a short stretch of the IRES. The picornavirus IRES enables internal initiation of translation at a site distant from the 5' end of the viral RNA in a cap-independent manner (MEEROVITCH et al. 1991; NICHOLSON et al. 1991; PESTOVA et al. 1991). The attenuating mutation in the 5' NTR of each Sabin strain (POLLARD et al. 1989; WESTROP et al. 1989, KAWAMURA et al. 1989; REN et al. 1991; MARTIN et al. 1991 MACADAM et al. 1993) is located in a secondary structure, termed domain V (MINOR 1992). The mutations in this complex stem-loop structure region have been shown to decrease both PV translational efficiency and rates of replication of the Sabin vaccine strains in cells of neuronal origin and to decrease viral translational efficiency in vitro in systems supplemented with cell extracts from these cells (SVITKIN et al. 1985, 1988; LA MONICA and RACANIELLO 1989). Other mutations in this region of domain V which alter the predicted base pairing also decrease translation in cells of neural origin (HALLER et al. 1996). Although the attenuating mutation at nt234 in the CVB3/0 genome is also in the 5' NTR, it is in a quite different region of the 5' NTR and the data do not support attenuation through a translational defect (TU et al. 1995).

Changes in the coding regions of the PV genome also affect the neurovirulence phenotype. Temperature sensitivity and attenuation is conferred upon the Sabin strain of PV3 (PV3S) by a mutation at nt2034 (C to U) that changes Ser to Phe at amino acid (aa) 91 of VP3 (MACADAM et al. 1989). Temperature resistant revertants have been shown to have changes not only at aa91 but also at second sites (Filman et al. 1989). Analysis of the structure of the capsid of PV3S and the impact of these second site mutations suggests that the attenuation of aa91 Phe is due to effects upon pentamer stability (FILMAN et al. 1989). An additional attenuating mutation at nt2493 (TATEM et al. 1992) (aa6 of VP1; I to T) contributes a small plaque phenotype to this virus. Analysis of sites differing between neurovirulent revertants from cases of vaccine-associated poliomyelitis and PV2S in primates (MACADAM et al. 1993) and in a nontransgenic mouse model of mouse-adapted PV2 (REN et al. 1991) has demonstrated that the major site of attenuation of PV2S outside of the 5'NTR is at nt2908 (aa143 in VP1). The Sabin strain of PV1 is mutated, relative to the progenitor neurovirulent strain (PV1 Mahoney, or PV1 M), at many more sites than is PV3S relative to P3/Leon/37 (NOMOTO et al. 1982). Use of recombinant viruses generated from cDNA clones of PV1 M and PV1S with tests of neurovirulence in the human poliovirus receptor transgenic mouse model identified four sites of attenuation outside the 5' NTR: nt935 (aa 65 of VP4; A to S), nt2438 (aa 225 of VP3; L to M), nt2795 (aa106 of VP1; A to T), nt2879 (aa134 of VP1; L to F) (BOUCHARD et al. 1995). The mutations at nt935 and nt2438 contribute to the temperature sensitivity of the Sabin strain, although other sites are also important. The positions in the capsid proteins of the changes encoded by capsid region attenuating mutations suggest that attenuation from these mutations might be effected through alterations in virus assembly, stability, and/or receptor interactions.

The majority of PV isolates from human beings appear to be neurovirulent to some degree based upon intraspinal inoculation in outbred primates (SABIN 1955). Certainly, studies of fecal isolates indicate that normal replication of vaccine PV strains in the gut produces neurovirulent revertants (EVANS et al. 1985; MINOR et al. 1986; GEORGESCU et al. 1994) and virus isolates from the central nervous system and feces of cases of PV vaccine-associated paralytic poliomyelitis have been shown to be neurovirulent (GEORGESCU et al. 1994). Selection for replication in the gut may, in the PVs, select for neurovirulence. Neurovirulence of PV strains appears to correlate more with damage to the motor neurons by viral multiplication in those cells than by inflammatory processes (COUDERC et al. 1989). Unlike the PV paradigm, most human isolates of CVB that have been assayed have little or no cardiovirulence or pancreovirulence as measured in disease-sensitive (and inbred) mice. For CVB3 and CVB4 infections, inflammatory processes are involved in disease and thus sites that ultimately are shown to correlate with virulence may reflect this difference in mechanism of disease.

5.2 Cardioviruses

Theiler's murine encephalomyelitis virus (TMEV) is a murine picornavirus of the cardiovirus group. Two groups of TMEV strains induce different types of disease in mice upon intracerebral inoculation. The TO strains induce a biphasic disease that consists of an acute encephalomyelitis followed by viral spread to the spinal cord where virus persists. Demyelination in the spinal cord, triggered by the murine immune response to the expression of viral antigens, is observed (LIPTON 1975). The GDVII strains induce a fatal encephalitis dependent on replication in neurons (LORCH et al. 1981; STROOP et al. 1981). The determinants of the type of disease have been mapped in large part to the capsid proteins of these strains through the use of recombinants of TO and GDVII strains (CALENOFF et al. 1990; FU et al. 1990; MCALLISTER et al. 1990; PRITCHARD et al. 1993). Mutations at aa101 (ZURBRIGGEN et al. 1991; WADA et al. 1994) and aa268 (ROOS et al. 1989) in VP1 and at aa 141 in VP2 (JAROUSSE et al. 1994) can alter the ability of TMEV to persist. All three of these sites are at or near the putative receptor binding site (GRANT et al. 1992; LUO et al. 1992) and thus have been postulated to modulate receptor binding interactions (WADA et al. 1994; JAROUSSE et al. 1994). Strains of both subgroups encode the genetic determinants of demyelination (FU et al. 1990). TMEV genomes contain an alternate open reading frame (ORF) in the leader region: the initiation codon of this ORF is missing in two GDVII viruses (MICHIELS et al. 1995) and mutations to eliminate expression of this ORF in a TO strain resulted in loss of demyelination without loss of persistence of infection (CHEN et al. 1995). As in CVB4 pancreovirulence, determinants of disease induction of TMEV strains map to the capsid coding region, and as in CVB3 cardiovirulence, viral epitopes recognized by the immune system appear to influence the immune-mediated disease.

Encephalomyocarditis virus (EMCV) and mengovirus are two, closely related and well-studied cardioviruses. Attenuation of EMCV virulence in mice (BAE and YOON 1993) and mengovirus (DUKE et al. 1990; MARTIN et al. 1996) can be accomplished by reduction of the length of a homopolymeric cytidylate (poly C) tract of variable size, located in the 5' NTR. Sequences elsewhere in the 5' NTR also affect whether the reductions in length of the poly C tract are attenuating (HAHN and PALMENBERG 1995). Deletions in the region adjoining the poly C tract can decrease replication efficiency of the virus in HeLa cell culture (MARTIN et al. 1996). As has been seen with the PVs and with CVB3, alterations in the 5' NTR in these picornaviruses can attenuate virulence.

Two strains of EMCV exist which differ in induction of diabetes in susceptible mice: the diabetogenic EMC-D and the nondiabetogenic EMC-B (YOON et al. 1980). Analysis of isolates from cultures of EMC-D and EMC-B have demonstrated that the diabetic virulence difference in these two strains maps to a single site (nt3155) in the capsid protein VP1. Diabetogenic strains express alanine at aa776 (VP1 aa152), while nondiabetogenic strains express threonine at this position (BAE and YOON 1993). By comparison to the crystalline structure of mengovirus (LUO et al. 1987), this site is within the putative receptor binding region, the so-called pit, and may correlate with the observed differences in attachment of EMC-D and EMC-B to primary cultures of murine β-cells (KAPTUR et al. 1989). A difference in cell tropism based on receptor interaction is also suggested for a determinant of CVB4 pancreovirulence (see above).

6 Future Directions

The primary question at present in terms of CVB genetics is which region(s) of the CVB genome will be mapped to the virulence phenotype. It must be stressed, however, that solving this puzzle will represent only definition of the trigger for viral disease. In addition to the specific sites which have been and will yet be identified in the viral genome, our understanding of virulence must also address the significant input from the infected host animal. Immune status, age, sex and genetic background have been established as four primary determinants that complicate an understanding of the role of one or two sites in the viral genome in pathogenesis. In addition, there is the key role of specific cells within a tissue that are infected as the virus infection spreads. Clearly, as the initial insult that triggers the disease, understanding the viral genetics of virulence is crucial. In the well-studied and related PVs, a primary site of attenuation is found in the 5' NTR for all three serotypes but, as discussed above, there are a variety of other mutations in the viral genome that also play roles in the determination of the virulence phenotype (recently reviewed by HELLEN and WIMMER 1995). It is important to recall that studies of how naturally occurring virulence is determined in CVBs differ from those pioneering studies that mapped virulence in the PVs: the PV studies mapped artificially induced sites of attenuation in proven vaccine strains but did not establish how extents of neurovirulence are determined in the wild.

Attenuating mutations in virulent CVB strains derived by passage in cell culture may not identify the sites responsible for natural variations in cardiovirulence of human isolates of CVBs. For example, it can be argued that cell culture-derived attenuating changes may be mutations that only slightly diminish the viability of the virus within a defined environment, and thus would have a lesser likelihood of being found in isolates which had to succeed in infecting hosts naturally and outracing the host immune system sufficiently to be excreted. Indeed, and if as studies in mice suggest, naturally occurring CVB cardiovirulence and pancreovirulence are the exception and not the rule, then identifying attenuating sites derived in cell culture may be the wrong approach. If most CVBs are avirulent, then the question becomes: what is the nature of the mutation that prompts a virus to become virulent from a naturally avirulent state? Mutations in the 5' NTR might increase in the degree of replication in secondary sites of infection (heart or pancreas) or changes in T cell epitopes in the capsid might generate pathologic changes in the heart or pancreas without changing viability in the gastrointestinal tract.

Several laboratories are at present involved in elucidating or confirming the CVB genetic sites that determine the virulence phenotypes. The question of which genetic site(s) affect CVB virulence phenotypes is being approached in two classical ways: (1) by the study of a parent virus and its more/less virulent derivative and (2) by investigating naturally occurring CVB strains and engineered chimeric viruses as assayed in murine models of disease. It is important to note as well that this latter approach was not practical with the PVs until the advent of human PV receptor transgenic mice (REN et al. 1990; HORIE et al. 1994), due to limited availability and the high costs of primates. It is reasonable to expect that we may soon have a well-tested molecular model for how CVBs naturally determine both pancreovirulent and cardiovirulent phenotypes. As CVBs cause inflammatory diseases of brain, heart and pancreas in humans, it will be intriguing to determine whether the same or different genetic sites in the viral genome are interpreted equivalently or differently by the host cell and immune responses depending upon the tissue in which the disease is evident. As it is clear that CVBs replicate differently in different cell types, what are the host cell cofactors that a CVB uses during replication and how do these cofactors differ between tissues? These will be among the exciting studies of the next few years: there are now well-characterized viral clones, cell culture and murine model systems which will permit rapid advancement toward a complete understanding of the molecular mechanisms underlying CVB pathogenesis.

Disease caused by the CVBs will continue to plague the human population until effective anti-CVB (anti-enteroviral) drugs and/or vaccines are developed and marketed. It is indeed ironic that at a time when funding for health care and basic research are both plummeting, we stand very close to understanding how to create entirely new generations of both vaccines and antivirals. Because the CVBs are closely related to PVs, the creation of vaccine strains of CVBs should present little difficulty. Indeed, various reports have demonstrated that CVBs can be used to vaccinate mice against subsequent CVB-induced disease (GAUNTT et al. 1983; GODNEY et al. 1987; FOHLMAN et al. 1993; TILLES 1994). It would be surprising if anti-CVB vaccines were not as effective as the Sabin and Salk anti-PV vaccines. Recombinant DNA technology now permits evaluation (at least in animal models) of genetically modified viruses. In the cardioviruses, a highly attenuated strain of mengovirus has been developed by elimination of the poly C tract in the 5' NTR (DUKE et al. 1990), permitting the use of this virus not only as a vaccine against mengo- and related cardioviruses (OSORIO et al. 1996), but for use as a vector to present foreign antigens as well (ALTMEYER et al. 1994, 1995). The 5' NTR from CVB3 can be substituted for that of PV to produce viable chimeric viruses (SEMLER et al. 1986; JOHNSON and SEMLER 1988). External loops on the PV, HRV, mengovirus and CVB3 capsid surfaces have been used to express foreign peptides (BURKE et al. 1988; EVANS et al. 1989; REIMANN et al. 1991; ALTMEYER et al. 1994; ARNOLD et al. 1994), and, recently, Andino and others expressed foreign proteins within the ORF of a PV, exploiting the mechanism of the PV 3C proteinase to cleave the foreign peptide from the PV capsid proteins (ANDINO et al. 1994; MATTION et al. 1994; YIM et al. 1996). The technology of insertion of foreign sequences into the viral ORF for processing during replication is extremely exciting, suggesting almost myriad possibilities for the expression of specific antigens, immunomodulatory peptides, etc. It seems clear then that a new generation of potentially very useful and stable attenuated enterovirus vaccines may be near to hand, providing that the pharmaceutical and health care industries take an interest in their development and marketing.

Acknowledgements. This work was supported in part by funding from the National Institutes of Health (S.T., N.M.C., A.R.) and the American Heart Association (S.T., A.R.).

References

Agol VI (1991) The 5'-untranslated region of picornaviral genomes. Adv Virus Res 40:103–180

Altmeyer R, Escriou N, Girard M, Palmenberg A, van der Werf S (1994) Attenuated Mengo virus as a vector for immunogenic human immunodeficiency virus type 1 glycoprotein 120. Proc Natl Acad Sci USA 91:9775–9779

Altmeyer R, Girard M, Van der Werf S, Mimic V, Seigneur L, Saron M (1995) Attenuated Mengo virus: a new vector for live recombinant vaccines. J Virol 69:3193–3196

Andino R, Silvera D, Suggett SD, Achacoso PL, Miller CJ, Baltimore D, Feinberg MB (1994) Engineering poliovirus as a vaccine vector for the expression of diverse antigens. Science 265:1448–1451

Arnold GF, Resnick DA, Li Y, Zhang A, Smith AD, Geisler SC, Jacobo-Molina A, Lee W, Webster RG, Arnold E (1994) Design and construction of rhinovirus chimeras incorporating immunogens from polio, influenza, and human immunodeficiency viruses. Virology 198:703–708

Assaad F, Cockburn WC (1982) The relation between acute persisting spinal paralysis and poliomyelitis vaccine – results of a ten-year enquiry. Bull WHO 60:231–242

Bae YS, Yoon JW (1993) Determination of diabetogenicity attributable to a single amino acid, Ala776, on the polyprotein of encephalomyocarditis virus. Diabetes 42:435–443

Banatvala JE, Bryant J, Schernthaner G, Borkenstein M, Schober E, Brown D, De Silva LM, Menser MA, Silink M (1985) Coxsackie B, mumps, rubella, and cytomegalovirus specific IgM responses in patients with juvenile-onset insulin-dependent diabetes mellitus in Britain, Austria, and Australia. Lancet 1:1409–1412

Beck MA, Chapman NM, McManus BM, Mullican JC, Tracy S (1990) Secondary enterovirus infection in the murine model of myocarditis. Pathologic and immunologic aspects. Am J Pathol 136:669–681

Beck MA, Tracy S, Coller BA, Chapman NM, Hufnagel G, Johnson JE, Lomonossoff G (1992) Comoviruses and enteroviruses share a T cell epitope. Virology 186:238–246

Beck M, Kolbeck P, Rohr L, Shi Q, Morris V, Levander O (1994a) Vitamin E deficiency intensifies the myocardial injury of coxsackievirus B3 infection of mice. J Nutr 124:345–358

Beck M, Kolbeck P, Rohr L, Shi Q, Morris V, Levander O (1994b) Benign human virus becomes virulent in selenium-deficient mice. J Med Virol 43:166–170

Beck M, Kolbeck P, Shi Q, Rohr L, Morris V, Levander O (1994c) Increased virulence of a human enterovirus (coxsackievirus B3) in selenium-deficient mice. J Infect Dis 170:351–357

Beck M, Shi Q, Levander O (1995) Rapid genomic evolution of a non-virulent coxsackievirus B3 in selenium-deficient mice results in selection of identical viral isolates. Nat Med 1:433–436

Bienz K, Egger D, Troxler M, Pasamontes L (1990) Structural organization of poliovirus RNA replication is mediated by viral proteins of the P2 genomic region. J Virol 64:1156–1163

Blyn LB, Chen R, Semler BL, Ehrenfeld E (1995) Host cell proteins binding to domain IV of the 5' noncoding region of poliovirus RNA. J Virol 69:4381–4389

Bouchard MJ, Lam DH, Racaniello VR (1995) Determinants of attenuation and temperature sensitivity in the type 1 poliovirus Sabin vaccine. J Virol 69:4972–4978

Bowles NE, Richardson PJ, Olsen EG, Archard LC (1986) Detection of Coxsackie-B-virus-specific RNA sequences in myocardial biopsy samples from patients with myocarditis and dilated cardiomyopathy. Lancet 1:1120–1123

Burke KL, Dunn G, Ferguson M, Minor PD, Almond JW (1988) Antigen chimaeras of poliovirus as potential new vaccines. Nature 332:81–82

Caggana M, Chan P, Ramsingh A (1993) Identification of a single amino acid residue in the capsid protein VP1 of Coxsackievirus B4 that determines the virulent phenotype. J Virol 67:4797–4803

Calenoff MA, Faaberg KS, Lipton HL (1990) Genomic regions of neurovirulence and attenuation in Theiler murine encephalomyelitis virus. Proc Natl Acad Sci USA 87:978–982

Chapman NM, Tu Z, Tracy S, Gauntt CJ (1994) An infectious cDNA copy of the genome of a non-cardiovirulent coxsackievirus B3 strain: its complete sequence analysis and comparison to the genomes of cardiovirulent coxsackieviruses. Arch Virol 135:115–130

Chatterjee NK, Nejman C, Gerling I (1988) Purification and characterization of a strain of coxsackievirus B4 of human origin that induces diabetes in mice. J Med Virol 26:57–69

Chen HH, Kong WP, Zhang L, Ward PL, Roos RP (1995) A picornaviral protein synthesized out of frame with the polyprotein plays a key role in a virus-induced immune-mediated demyelinating disease. Nat Med 1:927–931

Chow LH, Beisel KW, McManus BM (1992) Enteroviral infection of mice with severe combined immunodeficiency. Evidence for direct viral pathogenesis of myocardial injury. Lab Invest 66:24–31

Clements GB, Galbraith DN, Taylor KW (1995) Coxsackie B virus infection and onset of childhood diabetes. Lancet 346:221–223

Couderc T, Christodoulou C, Kopecka H, Marsden S, Taffs LF, Crainic R, Horaud F (1989) Molecular pathogenesis of neural lesions induced by poliovirus type 1. J Gen Virol 70:2907–2918

Couderc T, Guedo N, Calvez V, Pelletier I, Hogle J, Colbere-Garapin F, Blondel B (1994). Substitutions in the capsids of poliovirus mutants selected in human neuroblastoma cells confer on the Mahoney type 1 strain a phenotype neurovirulent in mice. J Virol 68:8386–8391

Currey KM, Peterlin BM, Maizel JV Jr (1986) Secondary structure of poliovirus RNA: correlation of computer-predicted with electron microscopically observed structure. Virology 148:33–46
Dalldorf G (1955) The coxsackie viruses. Ann Rev Microbiol 9:277–296
Dalldorf G, Sickles, G.M. (1948) An unidentified, filtrable agent isolated from the feces of children with paralysis. Science 108:61–62
del Angel R, Papavassilou A, Fernandez-Tomas C, Silverstein S, Racaniello V (1989) Cell proteins bind to multiple sites within the 5' untranslated region of poliovirus RNA. Proc Natl Acad Sci USA 86:8299–8303
Duke GM, Osorio JE, Palmenberg AC (1990) Attenuation of Mengo virus through genetic engineering of the 5' noncoding poly(C) tract. Nature 343:474–476
Easton AJ, Eglin RP (1988) The detection of coxsackievirus RNA in cardiac tissue by in situ hybridization. J Gen Virol 69:285–291
Evans DM, Dunn G, Minor PD, Schild GC, Cann AJ, Stanway G, Almond JW, Currey K, Maizel JV Jr (1985) Increased neurovirulence associated with a single nucleotide change in a noncoding region of the Sabin type 3 poliovaccine genome. Nature 314:548–550
Evans DJ, McKeating J, Meredith JM, Burke KL, Katrak K, John A, Ferguson M, Minor PD, Weiss RA, Almond JW (1989) An engineered poliovirus chimaera elicits broadly reactive HIV-1 neutralizing antibodies. Nature 339:385–388
Filman DJ, Syed R, Chow M, Macadam AJ, Minor PD, Hogle JM (1989) Structural factors that control conformational transitions and serotype specificity in type 3 poliovirus. EMBO J 8:1567–1579
Finnegan A, Amburgey CF (1989) A single amino acid mutation in a protein antigen abrogates presentation of certain T cell determinants. J Exp Med 170:2171–2176
Fletcher E, Brennan CF (1957) Cardiac complications of coxsackie virus infection. Lancet i:913
Fohlman J, Pauksen K, Morein B, Bjare U, Ilback NG, Friman G (1993) High yield production of an inactivated coxsackie B3 adjuvant vaccine with protective effect against experimental myocarditis. Scand J Infect Dis Suppl 88:103–108
Fu J, Rodriguez M, Roos RP (1990) Strains from both Theiler's virus subgroups encode a determinant for demyelination. J Virol 64:6345–6348
Gauntt C, Pallansch M (1996) Coxsackievirus B3 clinical isolates and murine myocarditis. Virus Res 41:89–99
Gauntt CJ, Trousdale MD, LaBadie DR, Paque RE, Nealon T (1979) Properties of coxsackievirus B3 variants which are amyocarditic or myocarditic for mice. J Med Virol 3:207–220
Gauntt CJ, Paque RE, Trousdale MD, Gudvangen RJ, Barr DT, Lipotich GJ, Nealon TJ, Duffey PS (1983) Temperature-sensitive mutant of coxsackievirus B3 establishes resistance in neonatal mice that protects them during adolescence against coxsackievirus B3-induced myocarditis. Infect Immun 39:851–864
Gauntt CJ, Godeny EK, Lutton CW, Arizpe HM, Chapman NM, Tracy SM, Revtyak GE, Valente AJ, Rozek MM (1989) Mechanism(s) of coxsackievirus-induced acute myocarditis in the mouse. In: de la Maza LM, Peterson EM (eds) Medical Virology 8. Plenum, New York, pp 161–182
Gear J, Measroch V, Prinsloo FR (1956) The medical and public health importance of the coxsackie viruses. S Afr Med J 30:806–810
Gebhard J, Ehrenfeld E (1992) Specific interactions of HeLa cell proteins with proposed translation domains of the poliovirus 5' noncoding region. J Virol 66:3101–3109
Georgescu MM, Delpeyroux F, Tardy-Panit M, Balanant J, Combiescu M, Combiescu AA, Guillot S, Crainic R (1994) High diversity of poliovirus strains isolated from the central nervous system from patients with vaccine-associated paralytic poliomyelitis. J Virol 68:8089–8101
Godney EK, Arizpe HM, Gauntt CJ (1987) Characterization of the antibody response in vaccinated mice protected against Coxsackievirus B3-induced myocarditis. Viral Immunol 1:305–314
Grant RA, Filman DJ, Fujinami RS, Icenogle JP, Hogle JM (1992) Three-dimensional structure of Theiler's virus. Proc Natl Acad Sci USA 89:2061–2065
Grasso M, Arbustini E, Silini E, Diegoli M, Percivalle E, Ratti G, Bramerio M, Gavazzi A, Vigano M, Milanesi G (1992) Search for Coxsackievirus B3 RNA in idiopathic dilated cardiomyopathy using gene amplification by polymerase chain reaction. Am J Cardiol 69:658–664
Grist N, Bell E, Assaad F (1978) Enteroviruses and human disease. Prog Med Virol 24:114–157
Hahn H, Palmenberg AC (1995) Encephalomyocarditis viruses with short poly(C) tracts are more virulent than their mengovirus counterparts. J Virol 69:2697–2699
Haller AA, Stewart SR, Semler BL (1996) Attenuation stem-loop lesions in the 5' noncoding region of

poliovirus RNA: neuronal cell-specific translation defects. J Virol 70:1467–1474
Hellen C, Wimmer E (1992) Maturation of poliovirus capsid proteins. Virology 187:391–397
Hellen C, Wimmer E (1995) Enterovirus genetics. In: Rotbart H (ed) Human enterovirus infections, Amer Soc Microbiol, Washington, DC, pp 25–72
Hellen C, Pestova T, Litterst M, Wimmer E (1994) The cellular peptide p57 (pyrimidine tract binding protein) binds to multiple sites in the poliovirus 5' non-translated region. J Virol 68:941–950
Hogle JM, Chow M, Filman DJ (1985) Three-dimensional structure of poliovirus at 2.9 A resolution. Science 229:1358–1365
Horie H, Koike S Kurata T Sato-Yoshida Y, Ise I, Ota Y, Abe S, Hioki K, Kato H, Taya C, Nomura T, Hashizume S, Yonekawa H, Nomoto A (1994) Transgenic mice carrying the human poliovirus receptor: new animal models for study of poliovirus neurovirulence. J Virol 68:681–688
Huber SA, Job LP (1983) Differences in cytolytic T cell response of BALB/c mice infected with myocarditic and non-myocarditic strains of coxsackievirus group B, type 3. Infect Immun 39:1419–1427
Huber SA, Pfaeffle B (1994) Differential Th1 and Th2 cell responses in male and female BALB/c mice infected with coxsackievirus group B type 3. J Virol 68:5126–5132
Huber SA, Job LP, Woodruff JF (1980) Lysis of infected myofibers by coxsackievirus B-3-immune T lymphocytes. Am J Pathol 98:681–694
Huber SA, Polgar J, Schultheiss P, Schwimmbeck P (1994) Augmentation of pathogenesis of coxsackievirus B3 infections in mice by exogenous administration of interleukin-1 and interleukin-2. J Virol 68:195–206
Jarousse N, Grant RA, Hogle JM, Zhang L, Senkowski A, Roos RP, Michiels T, Brahic M, McAllister A (1994) A single amino acid change determines persistence of a chimeric Theiler's virus. J Virol 68:3364–3368
Javett SN, Heymann S, Mundel B, Pepler WJ, Lurie HI, Gear J, Measroch V, Kirsch Z (1956) Myocarditis in the newborn infant: a study of an outbreak associated with coxsackie group B virus infection in a maternity home in Johannesburg. J Pediatr 48:1–22
Jenkins O, Booth J, Minor P, Almond J (1987) The complete nucleotide sequence of coxsackievirus B4 and its comparison to other members of the Picornaviridae. J Gen Virol 68:1835–1848
Johnson KL, Sarnow P (1991) Three poliovirus 2B mutants exhibit noncomplementable defects in viral RNA amplification and display dosage-dependent dominance over wild-type poliovirus. J Virol 65:4341–4349
Johnson VH, Semler BL (1988) Defined recombinants of poliovirus and coxsackievirus: sequence-specific deletions and functional substitutions in the 5'-noncoding regions of viral RNAs. Virology 162:47–57
Kandolf R, Hofschneider PH (1989) Viral heart disease. Springer Semin Immunopathol 11:1–13
Kang Y, Chatterjee NK, Nodwell MJ, Yoon JW (1994) Complete nucleotide sequence of a strain of coxsackie B4 virus of human origin that induces diabetes in mice and its comparison with nondiabetogenic coxsackie B4 JBV strain. J Med Virol 44:353–361
Kaptur PE, Thomas DC, Giron DJ (1989) Differing attachment of diabetogenic and nondiabetogenic variants of encephalomyocarditis virus to beta-cells. Diabetes 38:1103–1108
Kawamura N, Kohara M, Abe S, Komatsu T, Tago K, Arita M, Nomoto A (1989) Determinants in the 5' noncoding region of poliovirus Sabin 1 RNA that influence the attenuation phenotype. J Virol 63:1302–1309
Kibrick S, Benirschke K (1958) Severe generalized disease (encephalohepatomyocarditis) occurring in the newborn period and due to infection with coxsackievirus, group B. Pediatrics 22:857–875
King ML, Shaikh A, Bidwell D, Boller A, Banatvala JE (1983) Coxsackie-B-virus-specific IgM responses in children with insulin-dependent (juvenile-onset; type I) diabetes mellitus. Lancet 1:1397–1399
Klump WM, Bergmann I, Muller BC, Ameis D, Kandolf R (1990) Complete nucleotide sequence of infectious Coxsackievirus B3 cDNA: two initial 5' uridine residues are regained during plus-strand RNA synthesis. J Virol 64:1573–1583
La Monica N, Racaniello VR (1989) Differences in replication of attenuated and neurovirulent polioviruses in human neuroblastoma cell line SH-SY5Y. J Virol 63:2357–2360
Lansdown ABG (1976) Pathological changes in the pancreas of mice following infection with coxsackie B viruses. Br J Exp Path 57:331–338
Le SY, Zuker M (1990) Common structures of the 5' non-coding RNA in enteroviruses and rhinoviruses. Thermodynamical stability and statistical significance. J Mol Biol 216:729–741
Le SY, Chen JH, Sonenberg N, Maizel JV (1992) Conserved tertiary structure elements in the 5' untrans-

lated region of human enteroviruses and rhinoviruses. Virology 191:858–866
Leslie K, Blay R, Haisch C, Lodge A, Weller A, Huber S (1989) Clinical and experimental aspects of viral myocarditis. Clin Microbiol Rev 2:191–203
Lipton HL (1975) Theiler's virus infection in mice: an unusual biphasic disease process leading to demyelination. Infect Immun 11:1147–1155
Lorch Y, Friedmann A, Lipton HL, Kotler M (1981) Theiler's murine encephalomyelitis virus group includes two distinct genetic subgroups that differ pathologically and biologically. J Virol 40:560–567
Loudon RP, Moraska AF, Huber SA, Schwimmbeck P, Schultheiss P (1991) An attenuated variant of Coxsackievirus B3 preferentially induces immunoregulatory T cells in vivo. J Virol 65:5813–5819
Luo M, Vriend G, Kamer G, Minor I, Arnold E, Rossmann MG, Boege U, Scraba DG, Duke GM, Palmenberg AC (1987) The atomic structure of Mengo virus at 3.0 A resolution. Science 235:182–191
Luo M, He C, Toth KS, Zhang CX, Lipton HL (1992) Three-dimensional structure of Theiler murine encephalomyelitis virus (BeAn strain). Proc Natl Acad Sci USA 89:2409–2413
Macadam AJ, Arnold C, Howlett J, John A, Marsden S Taffs F, Reeve P, Hamada N, Wareham K, Almond J, Cammick N, Minor PD (1989) Reversion of the attenuated and temperature-sensitive phenotypes of the Sabin type 3 strain of poliovirus in vaccinees. Virology 172:408–414
Macadam AJ, Pollard SR, Ferguson G, Skuce R, Wood D, Almond JW, Minor PD (1993) Genetic basis of attenuation of the Sabin type 2 vaccine strain of poliovirus in primates. Virology 192:18–26
Maisch B, Bauer E, Cirsi M, Kochsiek K (1993) Cytolytic cross-reactive antibodies directed against the cardiac membrane and viral proteins in coxsackievirus B3 and B4 myocarditis. Characterization and pathogenetic relevance. Circulation 87(5 suppl):IV49–IV65
Manolio TA, Baughman KL, Rodeheffer R, Pearson TA, Bristow JD, Michels VV, Abelmann WH, Harlan WR (1992) Prevalence and etiology of idiopathic dilated cardiomyopathy. Am J Cardiol 69:1458–1466
Martin A, Benichou D, Couderc T, Hogle JM, Wychowski C, Van der Werf S, Girard M (1991) Use of type 1/type 2 chimeric polioviruses to study determinants of poliovirus type 1 neurovirulence in a mouse model. Virology 180:648–658
Martin LR, Duke GM, Osorio JE, Hall DJ, Palmenberg AC (1996) Mutational analysis of the mengovirus poly(C) tract and surrounding heteropolymeric sequences. J Virol 70:2027–2031
Martino TA, Liu P, Petric M, Sole MJ (1995) Enteroviral myocarditis and dilated cardiomyopathy: a review of clinical and experimental studies. In: Rotbart HA (ed) Human enterovirus infections. Amer Soc Microbiol, Washington, DC, pp 291–351
Matsumori A (1992) Lessons from animal experiments in myocarditis. Herz 17:107–111
Mattion NM, Reilly PA, DiMichele SJ, Crowley JC, Weeks-Levy C (1994) Attenuated poliovirus strain as a live vector: expression of regions of rotavirus outer capsid protein VP7 by using recombinant Sabin 3 viruses. J Virol 68:3925–3933
McAllister A, Tangy F, Aubert C, Brahic M (1990) Genetic mapping of the ability of Theiler's virus to persist and demyelinate. J Virol 64:4252–4257
Meerovitch K, Nicholson R, Sonenberg N (1991) In vitro mutational analysis of cis-acting RNA translational elements within the poliovirus type 2 5' untranslated region. J Virol 65:5895–5901
Melnick JL (1990) Enteroviruses: polioviruses, coxsackieviruses, echoviruses, and newer enteroviruses. In: Fields BN, Knipe DM (eds) Fields virology. Raven, New York, pp549–605
Michiels T, Jarousse N, Brahic M (1995) Analysis of the leader and capsid coding regions of persistent and neurovirulent strains of Theiler's virus. Virology 214:550–558
Minor PD (1992) The molecular biology of poliovaccines. J Gen Virol 73:3065–3077
Minor PD, John A, Ferguson M, Icenogle JP (1986) Antigenic and molecular evolution of the vaccine strain of type 3 poliovirus during the period of excretion by a primary vaccinee. J Gen Virol 67:693–706
Minor PD, Macadam AJ, Stone DM, Almond JW (1993) Genetic basis of attenuation of the Sabin oral poliovirus vaccines. Biologicals 21:357–363
Montgomery J, Gear J, Prinsloo FR, Kahn M, Kirsch ZG (1955) Myocarditis of the newborn: an outbreak in a maternity home in Southern Rhodesia associated with coxsackie group B virus infections. S Afr Med J 29:608–612
Moscufo N, Yafal AG, Rogove A, Hogle J, Chow M (1993) A mutation in VP4 defines a new step in the late stages of cell entry by poliovirus. J. Virol 67:5075–5078
Mucklebauer J, Kremer M, Minor I, Diana G, Dutko F, Groarke J, Pevear D, Rossmann M (1995) The structure of coxsackievirus B3 at 3.5 angstrom resolution. Structure 3:653–667

Murray MG, Bradley J, Yang XF, Wimmer E, Moss EG, Racaniello VR (1988) Poliovirus host range is determined by a short amino acid sequence in neutralization antigenic site 1. Science 241:213–215

Nathanson N, Martin JR (1979) The epidemiology of poliomyelitis: enigmas surrounding its appearance, epidemicity, and disappearance. Am J Epidemiol 110:672–692

Nicholson R, Pelletier J, Le SY, Sonenberg N (1991) Structural and functional analysis of the ribosome landing pad of poliovirus type 2: in vivo translation studies. J Virol 65:5886–5894

Nkowane BM, Wassilak SG, Orenstein WA, Bart KJ, Schonberger LB, Hinman AR, Kew OM (1987) Vaccine-associated paralytic poliomyelitis. United States: 1973 through 1984. JAMA 257:1335–1340

Nomoto A, Omata T, Toyoda H, Kuge S, Horie H, Kataoka Y, Genba Y, Nakano Y, Imura N (1982) Complete nucleotide sequence of the attenuated poliovirus Sabin 1 strain genome. Proc Natl Acad Sci USA 79:5793–5797

Osorio J, Hubbard G, Soike K, Girard M, van der Werf S, Moulin J, Palmenberg A (1996) Protection of non-murine mammals against encephalomyocarditis virus using a genetically engineered Mengo virus. Vaccine 14:155–161

Pestova T, Hellen C, Wimmer E (1991) Translation of poliovirus RNA: role of an essential cis-acting oligopyrimidine element within the 5' nontranslated region and involvement of a cellular 57-kilodalton protein. J Virol 65:6194–6204

Pilipenko EV, Blinov VM, Romanova LI, Sinyakov AN, Maslova SV, Agol VI (1989) Conserved structural domains in the 5'-untranslated region of picornaviral genomes: an analysis of the segment controlling translation and neurovirulence. Virology 168:201–209

Pircher H, Moskophidis D, Rohrer U, Burki K, Hengartner H, Zinkernagel RM (1990) Viral escape by selection of cytotoxic T cell-resistant virus variants in vivo. Nature 346:629–633

Pollard SR, Dunn G, Cammack N; Minor PD; Almond JW (1989) Nucleotide sequence of a neurovirulent variant of the type 2 oral poliovirus vaccine. J Virol 63:4949–4951

Poyry T, Hyypia T, Horsnell C, Kinnunen L, Hovi T, Stanway G (1994) Molecular analysis of coxsackievirus A16 reveals a new genetic group of enteroviruses. Virology 202:982–987

Pritchard AE, Jensen K, Lipton HL (1993) Assembly of Theiler's virus recombinants used in mapping determinants of neurovirulence. J Virol 67:3901–3907

Ramsingh AI, Collins DN (1995) A point mutation in the VP4 coding sequence of coxsackievirus B4 influences virulence. J Virol 69:7278–7281

Ramsingh A, Slack J, Silkworth J, Hixson A (1989) Severity of disease induced by a pancreatropic Coxsackie B4 virus correlates with the H-2Kq locus of the major histocompatibility complex. Virus Res 14:347–358

Ramsingh A, Hixson A, Duceman B, Slack J (1990) Evidence suggesting that virulence maps to the P1 region of the Coxsackievirus B4 genome. J Virol 64:3078–3081.

Ramsingh A, Araki H, Bryant S, Hixson A (1992) Identification of candidate sequences that determine virulence in Coxsackievirus B4. Virus Res 23:281–292

Ramsingh AI, Caggana M, Ronstrom S (1995) Genetic mapping of the determinants of plaque morphology of coxsackievirus B4. Arch Virol 140:2215–2226

Reimann BY, Zell R, Kandolf R (1991) Mapping of a neutralizing antigenic site of Coxsackievirus B4 by construction of an antigen chimera. J Virol 65:3475–3480

Ren RB, Costantini F, Gorgacz EJ, Lee JJ, Racaniello VR (1990) Transgenic mice expressing a human poliovirus receptor: a new model for poliomyelitis. Cell 63:353–362

Ren RB, Moss EG, Racaniello VR (1991) Identification of two determinants that attenuate vaccine-related type 2 poliovirus. J Virol 65:1377–1382

Roesing TG, Landau BJ, Crowell RL (1979) Limited persistence of viral antigen in coxsackievirus B3 induced heart disease in mice. Proc Soc Exp Biol Med 160:382–386

Rohll JB, Percy N, Ley R, Evans DJ, Almond JW, Barclay WS (1994) The 5'-untranslated regions of picornavirus RNAs contain independent functional domains essential for RNA replication and translation. J Virol 68:4384–4391

Roos RP, Stein S, Routbort M, Senkowski A, Bodwell T, Wollmann R (1989) Theiler's murine encephalomyelitis virus neutralization escape mutants have a change in disease phenotype. J Virol 63:4469–4473

Ross ME, Hayashi K, Notkins AL (1974) Virus-induced pancreatic disease: alteration in concentration of glucose and amylase in blood. J Inf Dis 129:669–676

Sabin AB (1955) Characteristics and genetic potentialities of experimentally produced and naturally occurring variants of poliomyelitis virus. Ann NY Acad Sci 61:924–939

Sabin A, Boulger L (1973) History of Sabin attenuated poliovirus oral live vaccine strains. J Biol Stand 1:115–118

Satoh M, Tamura G, Segawa I, Hiramori K, Satodate R (1994) Enteroviral RNA in dilated cardiomyopathy. Eur Heart J 15:934–939

Schnurr DP, Cao Y, Schmidt NJ (1984) Coxsackievirus B3 persistence and myocarditis in N:NIH(S) II nu/nu and +/nu mice. J Gen Virol 65:1197–1201

Semler BL, Johnson VH, Tracy S (1986) A chimeric plasmid from cDNA clones of poliovirus and coxsackievirus produces a recombinant virus that is temperature-sensitive. Proc Natl Acad Sci USA 83:1777–1781

Simenhoff ML, Ulys CJ (1958) Coxsackie virus myocarditis and the newborn. A pathological survey of 4 cases. Med Proc 4:389–397

Skinner MA, Racaniello VR, Dunn G, Cooper J, Minor PD, Almond JW (1989) New model for the secondary structure of the 5' non-coding RNA of poliovirus is supported by biochemical and genetic data that also show that RNA secondary structure is important in neurovirulence. J Mol Biol 207:379–392

Street NE, Mosmann TR (1991) Functional diversity of T lymphocytes due to secretion of different cytokine patterns. FASEB J 5:171–177

Stroop WG, Baringer JR, Brahic M (1981) Detection of Theiler's virus RNA in mouse central nervous system by in situ hybridization. Lab Invest 45: 504–509

Svitkin YV, Maslova SV, Agol VI (1985) The genomes of attenuated and virulent poliovirus strains differ in their in vitro translation efficiencies. Virology 147:243–252

Svitkin YV, Pestova TV, Maslova SV, Agol VI (1988) Point mutations modify the response of poliovirus RNA to a translation initiation factor: a comparison of neurovirulent and attenuated strains. Virology 166:394–404

Svitkin Y, Meerovich K, Lee H, Dholakia J, Kenan D, Agol V, Sonenberg N (1994) Internal translation initiation on poliovirus RNA: further characterization of La function in poliovirus translation in vitro. J Virol 68:1544–1550

Tatem JM, Weeks-Levy C, Georgiu A, DiMichele SJ, Gorgacz EJ, Racaniello VR, Cano FR, Mento SJ (1992) A mutation present in the amino terminus of Sabin 3 poliovirus VP1 protein is attenuating. J Virol 66:3194–3197

Tilles J (1994) Efficacy of a polyvalent inactivated-virus vaccine in protecting mice from infection with clinical strains of group B coxsackieviruses. Scan J Infect Dis 26:739–74

Titchener PA, Jenkins O, Szopa TM, Taylor KW, Almond JW (1994) Complete nucleotide sequence of a beta-cell tropic variant of coxsackievirus B4. J Med Virol 42:369–373

Tracy S, Gauntt C (1987) Phenotypic and genotypic differences among naturally occurring coxsackie virus B3 variants. Europ Heart J 8(Suppl J):445–448

Tracy S, Liu HL, Chapman NM (1985) Coxsackievirus B3: primary structure of the 5' non-coding and capsid protein-coding regions of the genome. Virus Res 3:263–270

Tracy S, Chapman NM, McManus BM, Pallansch MA, Beck MA, Carstens J (1990) A molecular and serologic evaluation of enteroviral involvement in human myocarditis. J Mol Cell Cardiol 22:403–414

Tracy S, Chapman NM, Tu Z (1992) Coxsackievirus B3 from an infectious cDNA copy of the genome is cardiovirulent in mice. Arch Virol 122:399–409

Tracy S, Chapman NM, Rubocki RJ, Beck MA (1995) Host immune responses to enterovirus infections. In: Rotbart HA (ed), Human enterovirus infections. Amer Soc Microbiol, Washington, DC, pp175–191

Tracy S, Chapman NM, Romero J, Ramsingh AI (1996) Genetics of coxsackievirus B cardiovirulence and inflammatory heart muscle disease. Trends Microbiol 4:175–179

Trousdale MD, Paque RE, Gauntt CJ (1976) Isolation of Coxsackievirus B3 temperature-sensitive mutants and their assignment to complementation groups. Biochem Biophys Res Commun 76:368–375

Tu Z, Chapman NM, Hufnagel G, Tracy S, Romero JR, Barry WH, Zhao L, Currey K, Shapiro B (1995) The cardiovirulent phenotype of coxsackievirus B3 is determined at a single site in the genomic 5' nontranslated region. J Virol 69:4607–4618

Van Creveld S, De Jager H (1956) Myocarditis in newborns, caused by coxsackie virus: clinical and pathological data. Ann Pediat 187:100–112

Van Houten N, Bouchard PE, Moraska A, Huber SA (1991) Selection of an attenuated Coxsackievirus B3 variant, using a monoclonal antibody reactive to myocyte antigen. J Virol 65:1286–1290

van Kuppeveld FJ, Galama JM, Zoll J, Melchers WJ (1995) Genetic analysis of a hydrophobic domain of

coxsackie B3 virus protein 2B: a moderate degree of hydrophobicity is required for a cis-acting function in viral RNA synthesis. J Virol 69:7782–7790
van Kuppeveld FJM, Galama JMD, Zoll J, van den Hurk PJJC, Melchers WJG (1996) Coxsackie B3 virus protein 2B contains a cationic amphipathic helix that is required for viral RNA replication. J Virol 70:3876–3886
Vella C, Brown CL, McCarthy DA (1992) Coxsackievirus B4 infection of the mouse pancreas: Acute and persistent infection. J Gen Virol 73:1387–1394
Verlinde JD, van Tongeren HAE, Kret A (1956) Myocarditis in newborns due to group B coxsackievirus: virus studies. Ann Pediat 187:113–118
Wada Y; Pierce ML; Fujinami RS (1994) Importance of amino acid 101 within capsid protein VP1 for modulation of Theiler's virus-induced disease. J Virol 68:1219–1223
Weaver CT, Hawrylowicz CM, Unanue ER (1988) T helper cell subsets require the expression of distinct costimulatory signals by antigen-presenting cells. Proc Natl Acad Sci USA 85:8181–8185
Westrop GD, Wareham KA, Evans DM, Dunn G, Minor PD, Magrath DI, Taffs F, Marsden S, Skinner MA, Schild GC, Almond JW (1989) Genetic basis of attenuation of the Sabin type 3 oral poliovirus vaccine. J Virol 63:1338–1344
Wiegers K, Uhlig H, Dernick R (1989) NAg1B of poliovirus type 1: A discontinuous epitope formed by two loops of VP1 comprising residues 96–104 and 141–152. Virology 170:583–586
Williams D, Olsen E (1985) Prevalence of overt dilated cardiomyopathy in two regions of England. Br Heart J 54:153–155
Woodruff JF (1980) Viral myocarditis. A review. Am J Pathol 101:425–484
Woodruff JF, Kilbourne ED (1970) The influence of quantitated post-weaning undernutrition on coxsackievirus B3 infection of adult mice. I. Viral persistence and increased severity of lesions. J Infect Dis 121:137–163
Woodruff JF, Woodruff JJ (1974) Involvement of T lymphocytes in the pathogenesis of coxsackie virus B3 heart disease. J Immunol 113:1726–1734
Yeates TO, Jacobson DH, Martin A, Wychowski C, Girard M, Filman DJ, Hogle JM (1991) Three-dimensional structure of a mouse-adapted type 2/type 1 poliovirus chimera. EMBO J 10:2331–2341
Yim TJ, Tang S, Andino R (1996) Poliovirus recombinants expressing hepatitis B virus antigens elicited a humoral immune response in susceptible mice. Virology 218:61–70
Yoon JW, Onodera T, Notkins AL (1978) Virus-induced diabetes mellitus:Beta cell damage and insulin-dependent hyperglycemia in mice infected with Coxsackie virus B4. J Exp Med 148:1068–1080
Yoon JW, Austin M, Onodera T, Notkins AL (1979) Virus-induced diabetes mellitus:Isolation of a virus from the pancreas of a child with diabetic ketoacidosis. N Engl J Med 300:1173–1179
Yoon JW, McClintock PR, Onodera T, Notkins AL (1980) Virus-induced diabetes mellitus. XVIII. Inhibition by a nondiabetogenic variant of encephalomyocarditis virus. J Exp Med 152:878–892
Zhang H, Yousef G, Cunningham L, Blake N, Ouyang X, Bayston T, Kandolf R, Archard L (1993) Attenuation of a reactivated cardiovirulent coxsackievirus B3: the 5'-nontranslated region does not contain major attenuation determinants. J Med Virol 41:129–137
Zhang H, Blake N, Ouyang X, Pandolfino Y, Morgan-Capner P, Archard L (1995) A single amino acid substitution in the capsid protein VP1 of coxsackievirus B3 (CVB3) alters plaque phenotype in Vero cells but not cardiovirulence in a mouse model. Arch Virol 140:959–966
Zsofia O, Deak J, Pap A (1992) Possible role of coxsackie-B virus infection in pancreatitis. Int J Pancreatology 11:105–108
Zurbriggen A, Thomas C, Yamada M, Roos RP, Fujinami RS (1991) Direct evidence of a role for amino acid 101 of VP-1 in central nervous system disease in Theiler's murine encephalomyelitis virus infection. J Virol 65:1929–1937

Roles of the Humoral Response in Coxsackievirus B-Induced Disease

C.J. GAUNTT

1 Introduction

Antibodies to a coxsackie virus group B (CVB) may be regarded as a footprint record of trespass in an animal by one of these viruses that resulted in subsequent induction of a humoral immune response toward numerous immunogenic viral protein epitopes. These epitopes will be found in virion polypeptides, virion polypeptides in various stages of completion, nonstructural viral proteins required for replication and the latter proteins in stages of synthesis. Linear, conformational and combinatorial viral epitopes will stimulate production of anti-viral protein antibodies of several immunoglobulin classes and IgG subclasses. Natural challenges to humans by the CVB occur most frequently by ingestion from liquids or foods, contact with

Department of Microbiology, University of Texas Health Science Center at San Antonio, 7703 Floyd Curl Drive, San Antonio, Texas 78284-7758, USA

fingers or fomites, transplacental passage to the fetus and probably by aerosols during replication in the respiratory tract. The major outcomes of most CVB infections of humans are asymptomatic, but there are numerous transient infections that result in illnesses in the upper respiratory and gastrointestinal tracts and skin. Less frequently, CVB cause serious illnesses involving inflammatory responses in the heart, meninges, brain and pancreas that lead to chronic diseases, some of which terminate in death. To delineate parameters of infection involved in protection or in pathology, murine models of CVB-induced human diseases have been studied for nearly four decades. Reviews of studies on basic mechanisms involved in CVB-induced diabetes mellitus (BARRETT-CONNER 1985; TONIOLO et al. 1988; D'ALESSIO 1992) and in CVB3- and CVB2-induced myocarditis (LERNER and WILSON 1973; WOODRUFF 1980; LERNER 1985, GAUNTT 1987; GAUNTT et al. 1993a) show the value of these models in increasing our understanding of the contributions of innate resistance mechanisms and both humoral and cell-mediated immune systems to the resolution or continuation of disease. Factors which play major roles in deciding the outcome of these infections include: (1) genetic background of the host, (2) immune status of the host, which relates to age, functional capabilities of the immune systems and coincident infections by other microorganisms, (3) general health of the host, (4) host lifestyle-induced immunosuppressive factors and, (5) virulence capabilities of the challenge population of CVB. Within a host's capabilities, infections by a CVB will result in induction of synthesis of antibodies with numerous specificities. Some antibodies will target viral proteins and provide protection via participation in clearance of the virus (GAUNTT et al. 1988a, b, 1993a, b; LESLIE et al. 1989). Other CVB-induced antibodies will recognize epitopes shared by both viral and cellular epitopes (MAISCH et al. 1993; GAUNTT et al. 1993a, b, 1995). Viral infections will also result in lysis of cells and release of sequestered proteins normally not presented to the humoral immune system and some of these proteins can induce production of autoantibodies (WOLFGRAM et al. 1985, 1986; HUBER and LODGE 1986; WOLFGRAM and ROSE 1989). This brief review will focus on results from animal model and human patient studies relative to CVB-induced antibodies that recognize epitopes in viral polypeptides and/or in molecules within or on normal cells. The potential role(s) of antiviral and autoantibodies in contributing to chronic diseases in humans or mice will be evaluated for myocarditis, in particular, and insulin-dependent diabetes mellitus (IDDM) to a limited extent. The available data show that CVB infections can induce antibodies that: (1) presumably recognize only viral proteins, (2) definitively recognize epitopes on both viral and cellular macromolecules, and (3) presumably recognize only cellular macromolecules.

1.1 Humoral Immune Responses to Coxsackievirus B Infections

Natural infections of humans by a CVB serotype can induce production of both serotype- and CVB group-specific antibodies (PATTISON 1983; CHAN and HAM-

MOND 1985; TORFASON et al. 1987; FRISK et al. 1989). Titration of serotype-specific IgG or IgM (homotypic) antibodies requires use of a virus-neutralizing assay with a single serotype (KATZE and CROWELL 1980a; MELNICK 1985; KAPSENBERG 1988) or enzyme-linked immunosorbent assays (ELISA) (DÖRRIES and TER MEULEN 1983; TORFASON et al. 1988). Some caution must be used in selecting a cell line for CVB antibody neutralization assays. A Vero cell line was found to release a protein(s) factor that competed with CVB3 for antibodies (KONISHI et al. 1991). Heterotypic antibodies directed against CVB group-reactive epitopes have been detected by immunodiffusion, complement-fixation, immunofluorescence or ELISA (HALONEN at al. 1959; SCHMIDT et al. 1965; KATZE and CROWELL 1980a). The latter nonneutralizing antibodies generally recognize epitopes on capsid polypeptide VP1 (KATZE and CROWELL 1980b; DÖRRIES and TER MEULEN 1983; CELLO et al. 1993; HAARMANN et al. 1994). Infection of a human by a single CVB serotype presumably gives rise to homotypic IgM antibodies in the primary infection; homotypic IgG appear later and can persist for years (BENDINELLI et al. 1988). Some patients with chronic relapsing pericarditis or dilated cardiomyopathy have maintained IgM titers to a CVB for up to 1.5 and 10 years, respectively (MUIR et al. 1989). Heterotypic antibodies presumably arise by an anamnestic response due to a previous infection(s) by a different CVB serotype(s) (KATZE and CROWELL 1980b; PATTISON 1983). Secretory IgA_2, specific to several infecting CVB serotypes, has been detected in nasopharyngeal and intestinal washes (WELLIVER et al. 1982).

Challenge of infant or adolescent mice with a CVB routinely occurs by the intraperitoneal (i.p.) route, although peroral inoculation of virus was shown to induce a similar level of myocarditis (GRODUMS and DEMPSTER 1959; LORIA et al. 1976; LERNER and REYES 1985). Either virus inoculation route can establish infections in neonates, but adult mice exhibit resistance to infection because of a mucosal barrier and other virus clearance mechanisms that effectively reduce the peroral route challenge by at least 10^5-fold (LORIA et al. 1974). Within a few hours of i.p. challenge of mice with coxsackievirus B3 (CVB3), virus is dispersed within the blood and is detectable in heart tissues (HUBER et al. 1984; LYDEN et al. 1987; GAUNTT et al. 1988a;). Replication in an unidentified cell or tissue results in a substantial viremia due to nascent virus production within 24 h (GAUNTT et al. 1988a). Virus replication in heart tissues is also apparent within 24 h (HUBER et al. 1984; LYDEN et al. 1987; GODENY and GAUNTT 1987a) and in the pancreas, spleen and liver between 1 and 2 days postinoculation (p.i.) (WOODRUFF 1980; HUBER et al. 1984; LODGE et al. 1987; GAUNTT et al. 1988b). This rapid exposure to the immune system stimulates production of detectable anti-viral IgM antibodies by 2–4 days postinoculation and anti-viral IgG titers by 6 days postinoculation (WOODRUFF 1980; HUBER et al. 1984; LODGE et al. 1987; GODENY et al. 1988). As will be seen below, there are subtle but important murine strain differences in time of response of the humoral immune system to CVB particle antigens.

Persistent virus replication can occur in some murine strains to provide viral antigens continuously. The CVB can persist in mice for weeks to months in several murine strains, but with one exception, the hosts are either immunosuppressed or immunodeficient (KHATIB et al. 1988; SCHNURR and SCHMIDT 1988; CHOW et al. 1992; SATO et al. 1994). Infectious virus was isolated from heart tissues of these immune defective mice which produce considerable titers of anti-CVB3 antibodies, but cannot clear the virus. Virus was not recovered from euthymic mice. Several molecular analyses have used in situ hybridization reactions with cDNA probes or polymerase chain reaction (PCR) analyses to detect CVB3 genomes in heart tissues of virus-inoculated euthymic mice with chronic myocarditis (KANDOLF et al. 1987; TRACY et al. 1990b, 1991; KLINGEL et al. 1992; KLINGEL and KANDOLF 1993). In one study of persisting CVB3 genomes (KLINGEL et al. 1992), data was presented that suggested some viral genomes express capsid proteins. Human studies show that enteroviral RNA sequences have been detected in 20%–25% of patients with myocarditis or idiopathic dilated cardiomyopathy (IDC) (JIN et al. 1990; TRACY et al. 1990a, b; HILTON et al. 1993). These findings offer the possibility that some viral antigens may be produced continuously in hearts of the latter patients.

1.2 Protective Role of Humoral Immune Responses Against Coxsackievirus B-Induced Diseases

Studies of virus-induced heart disease in CVB3-infected inbred strains of mice suggest that those strains capable of producing virus-neutralizing antibodies by day 3 postinoculation will have a reduced level of myocardial pathology compared with strains of mice that delay production of these antibodies by one more day (WOLFGRAM et al. 1986; HERSKOWITZ et al. 1985, 1987; LODGE et al. 1987; WOLFGRAM and ROSE 1989). The action of the early antiviral antibodies that reduces severity of CVB3-induced myocarditis is likely a reaction with nascent virions which blocks establishment of new foci of virus-infected cells that would subsequently elicit a focal pro-inflammatory response in the myocardium and death of myocytes. Administration of neutralizing antibodies to neonatal mice 24 h prior to or 2 h postinoculation of CVB3 prevented death, but were not ameliorative if given 24 h after virus challenge (RAGER-ZISMAN and ALLISON 1973a, b; CHO et al. 1982; TAKADA et al. 1995). Another study showed that passive administration of an anti-CVB3 antibody preparation (to purified virus) within 24 h pre- or post-virus challenge reduced the number of CVB3-induced lesions formed in the heart, whereas injection of the same antibody preparation at 3 days postinoculation significantly increased lesion number (GODENY et al. 1988). This latter result was interpreted as indicating that the antibody population contained cardiopathologic antibodies. Studies in homozygous nu/nu C3H/HeN mice challenged with CVB3 show that antibody is insufficient to clear the virus, as was found in euthymic C3H/HeN (nu/+) mice, suggesting that virus-specific T cells are required (SATO et al. 1994). In addition to age, gender may play a role in affecting protection afforded mice by anti-CVB3 neutralizing antibodies. Female BALB/c mice challenged with CVB3 respond quicker and produce a higher titer anti-CVB3 neutralizing antibody response than males, and this response has been attributed to a faster clearance of virus from heart tissues (WONG et al. 1977). A recent study shows that the protective humoral immune response in virus-infected female BALB/c mice is greatly affected by hormones and is primarily associated with Th_2 cells and a predominant synthesis of anti-viral IgG_1, in contrast to virus-specific IgG_{2a} responses in male mice (HUBER and PFAEFFLE 1994). A role for virus-specific antibodies in clearance of virus in humans was suggested from studies of infants with congenital agammaglobulinemia who developed persistent CVB infections (GRIFFITH et al. 1977). Although a protective role for virus-specific anti-CVB antibodies in preventing or reducing subsequent serotype-specific induced diseases can be inferred from numerous epidemiological studies (PALLANSCH 1988), this protection is not absolute as contacts of CVB-infected individuals can excrete these viruses in the presence of existing serotype-specific antibodies (KOGON et al. 1969).

Data from vaccination studies performed in mice with temperature-sensitive (ts) mutants of CVB3 (TROUSDALE et al. 1979; GODENY et al. 1988), a ts revertant of CVB3 (LUTTON et al. 1985) and inactivated CVB3 viruses (FOHLMAN et al. 1993) show that protection against challenge with a myocarditic or fatal dose of a cardiovirulent CVB3 correlates with induction of anti-CVB3 antibodies. A formalin-inactivated CVB vaccine that contained all six CVB serotypes induced significant antibody titers only after three postinoculation doses; these antibodies reduced subsequent nascent virus titers in the blood and pancreas and did not bind to antigens on pancreas islet cells (SEE and TILLES 1994). Naturally occurring clinical isolates of CVB3 that can induce antibodies without inducing disease(s) (GAUNTT 1988; GAUNTT and PALLANSCH 1996) may be regarded as nature's vaccines at this time.

In addition to CVB3 vaccines altering (ameliorating) the outcome of a CVB3 challenge relative to severity of murine myocardial disease or rate of mortality, antecedent infections with heterologous CVB can affect the outcome of a CVB3 infection in mice. Challenge of mice having a pre-existing CVB4-induced myocardial disease with CVB3 inhibited the total (ELISA) CVB3 antibody produced and reduced CVB3-induced cardiopathology (KHATIB and GIRALDO 1993). Other work has confirmed that a prior CVB4 infection favors a reduction in the subsequent CVB3-induced cardiopathologic alterations (GAUNTT et al. 1993c). In contrast, mice challenged with a CVB2 that does not induce cardiomyopathy subsequently develop a more severe myocarditis and show a drop in anti-CVB2 antibody (ELISA) titer subsequent to infection with a myocarditic $CVB3_m$ strain (BECK et al. 1990). A different study found that whereas challenge of mice with one CVB2 strain can exacerbate $CVB3_m$-induced myocardial disease, prior challenge with another CVB2 strain has no effect on the outcome (Maull, Wood, Gauntt, unpublished data). It is not known whether the enteroviral group antigen(s) affects humoral (FRISK et al. 1989) or cell-mediated (BECK and TRACY 1989, 1990) immune responses and is a factor in the disease process. Because humans are constantly exposed to sequential CVB challenges (MELNICK 1985; PALLANSCH 1988), any protective or harmful effects induced by a previous CVB serotype infection on a subsequent heterologous CVB serotype infection must be understood before serious development of a CVB vaccine can proceed.

1.3 Autoantibodies Induced by Coxsackievirus B Infections

Maisch and colleagues (MAISCH 1984a, b, 1986, 1990; MAISCH et al. 1979, 1980, 1982, 1983) were the first to call attention to the presence of autoantibodies in the sera of patients with myocarditis associated with the CVB or with IDC. The latter condition is a likely sequela in some cases of virus-induced myocarditis (ABELMANN 1973, 1984; MCMANUS et al. 1987; REYES and LERNER 1988; LESLIE et al. 1989; O'CONNELL 1987; SEE and TILLES 1991; OLSEN 1992). About half of the clinical cases of myocarditis are associated with the CVB and 40%–100% of these patients have autoantibodies to heart tissue antigens (MAISCH et al. 1982, 1984b; ROSE et al. 1988a, 1992). Overall the clinical data suggest that 50% of myocarditis cases and 10%–20% of IDC cases may have a CVB as the etiologic agent (MAISCH 1984a, 1989; NEUMANN et al. 1990; MAISCH et al. 1993). Among the many self-antigens recognized by autoantibodies from patients with either disease, a few have been identified, including myosin (NEUMANN et al. 1990), laminin (WOLFF et al. 1989), branched chain keto acid dehydrogenase (BCKD; ANSARI et al. 1991), adenine nucleotide translocator (ANT; SCHULTHEISS and BOLTE 1985; SCHULTHEISS et al. 1987, 1989; SCHULTHEISS 1989; ANSARI et al. 1991), calcium channel (SCHULTHEISS 1989), β-adrenergic receptor (LIMAS et al. 1989; WALLUKAT et al. 1991), actin (DE SCHEERDER et al. 1991; MAISCH et al. 1993) and fibrillary proteins (MAISCH et al. 1993). Other studies detected serum antibodies directed against a multitude of unidentified antigens isolated from myocardial tissues and separated by electrophoresis in polyacrylamide gels (CAFORIO et al. 1990; NEUMANN et al. 1990; MAISCH et al. 1993). In the acute phase of CVB-associated myocarditis in patients, IgM autoantibodies to myolemmal antigens were detected, whereas in the chronic phase, IgG or IgA autoantibodies to myolemmal antigens were detected in 90% of patients (MAISCH et al. 1993). Autoantibodies were bound to tissues in endomyocardial biopsy specimens in many of the aforementioned studies and some were deposited in conjunction with complement components C1q, C3 and C5b9 complex (MAISCH et al. 1993). Autoantibodies to sarcolemmal and myolemma membrane proteins were of diagnostic relevance in patients with CVB3- or CVB4-associated myocarditis, i.e., higher titers were associated with more severe disease and a poor prognosis for recovery (MAISCH et al. 1993). Many of the early studies with human materials provided a focus for subsequent studies in murine models of CVB3-induced myocarditis. Studies in infected mice confirmed the results of a portion of the data generated in the human studies.

Studies of autoantibodies to heart tissue antigens induced in mice during infection with a CVB3 were initiated by Rose and colleagues (WOLFGRAM et al. 1985, 1986; WOLFGRAM and ROSE 1989). They found that a significant portion of these autoantibodies were directed against epitopes in the H chain of cardiac myosin

(ALVAREZ et al. 1987). These autoantibodies did not recognize epitopes on virus particles of the pediatric CVB3 strain used in their studies (NEU et al. 1987c). Moreover, induction of these autoantibodies was dependent on the murine strain under study (ROSE et al. 1988b). Only those strains that subsequently developed CVB3-induced chronic myocarditis, in contrast to strains that resolved the acute disease, produced significant titers of IgG autoantibodies (NEU et al. 1987b). Initially during the acute phase of disease, autoantibodies to heart tissues were of the IgM class in strains that both resolved the disease or were programmed to develop chronic disease. However, only those murine strains which developed chronic disease switched to producing IgG autoantibodies (NEU et al. 1987b). Deposited autoantibodies that were recovered from infected mouse heart tissues were found to recognize epitopes on molecules with a broad range of molecular sizes (NEUMANN et al. 1991). Similar to the human studies, CVB3-infected mice also developed autoantibodies which recognized epitopes in ANT and BCKD (NEUMANN et al. 1994). In the latter models, genetic control of CVB3-induced production of autoantibodies to heart tissue antigens was found to be multigenic (ROSE et al. 1988b; WOLFGRAM and ROSE 1989). Subsequent inheritance pattern mapping studies of H-2 congenic murine strains and their F_1 hybrid offspring showed the anti-sarcolemma autoantibody response was controlled by a gene linked to the Np-2 and Tcrα loci on chromosome 14, whereas linkage for anti-cardiac myosin antibodies could not be determined (BEISEL and TRAYSTMAN 1988; TRAYSTMAN et al. 1991). Using a different CVB3 strain, Huber and colleagues also pioneered research on detection of virus-induced anti-heart autoantibodies in murine strains of CVB3-induced myocarditis and showed a correlation in appearance of these autoantibodies in sera with development of chronic disease and severity of disease induced by CVB3 (HUBER and LODGE 1986; LODGE et al. 1987; HUBER et al. 1988; LESLIE et al. 1989). Challenge of mice with purified CVB3 particles was also found to induce anti-heart tissue antibodies (GAUNTT et al. 1991b). A very interesting study showed that cardiac sarcoplasma reticulum calcium ATPase is an excellent autoantigen for inducing myocarditis in mice, but data on induction of autoantibodies were not presented (SHARAF et al. 1994). The above studies conclusively show that a CVB3 infection of certain murine strains can induce autoantibodies directed at numerous epitopes on cells in heart tissues. In the next section, in vitro mechanism studies will be presented to show how some of the autoantibodies could participate in immunopathologic reactions in vivo.

IDDM has long been associated with CVB infections as the triggering event that leads to a long subclinical incubation period and eventually manifestations of the disease (JENISTA et al. 1984; YOON 1990; D'ALESSIO 1992; FOHLMAN and FRIMAN 1993). Genetic predisposition to development of IDDM is also a major factor in this disease (FOHLMAN and FRIMAN 1993). Both the seasonal incidence of CVB infections and the onset of IDDM peak in the late summer and autumn (YOON 1990). CVB have been isolated from patients with acute signs and symptoms of IDDM and viruses isolated from patients can induce diabetes in susceptible mice (YOON 1990; HYÖTY 1995). Anti-CVB IgM antibody titers show a significant association between a CVB infection in utero or infancy (SZOPA et al. 1993) and onset of IDDM in patients of <3 years of age (HYÖTY et al. 1995) or ≥10 years of age in HLA-DR3 positive persons (D'ALESSIO 1992).

Detection of autoantibodies in IDDM has a voluminous publications history. About 70%–80% of newly diagnosed IDDM patients have autoantibodies again the cell enzyme glutamic acid decarboxylase (GAD), which synthesizes GABA in the pancreas. While this finding is clearly associated with an increased risk of IDDM, T cell responses are considered the main effector mechanism (TODD 1990; ATKINSON et al. 1994; TIAN et al. 1994). The principal autoantibodies found in patients with IDDM are directed against GAD65 (BAEKKESKOV et al. 1990; CHATTERJEE 1994) and islet cell cytoplasmic (ICA) autoantigens (BOTTAZZO et al. 1974; RABIN et al. 1994; HYÖTY et al. 1995; MYERS et al. 1995). Anti-GAD65 autoantibodies are persistent and appear frequently in both childhood (Verge et al. 1994) and adult-onset cases of IDDM (ROLL et al. 1994), in contrast to a transient appearance of ICA autoantibodies in childhood-onset cases of IDDM (BRUINING et al. 1989). Anti-GAD65 autoantibodies are likely a substantial subset contribution to the total ICA autoantibody titer (MYERS et al. 1995), but a combination of data on these two serologic markers may provide better diagnostic screening for prediction of IDDM development in genetically susceptible populations. The antigens that induce ICA autoantibodies in addition to GAD are unidentified at this time.

1.4 Potential Contributions of Coxsackievirus B3-Induced Autoantibodies to Disease

Detection and/or isolation of autoantibodies from sera (WOLFGRAM et al. 1985; BEISEL and ROSE 1988) and histologic analyses of endomyocardial biopsies of human heart tissues or the entire murine myocardium show that CVB infections can stimulate B cells to produce antibodies that recognize epitopes on self-antigens and structures. It must be noted that most normal mice of strains that can develop CVB3-induced chronic myocarditis (mice of C3H and A genetic backgrounds), but not strains that resolve the acute disease (e.g., CD-1, C57Bl/6J) have autoantibodies in low but detectable titers to human cardiac myosin (GAUNTT et al. 1993b) or murine cardiac myosin (NEU et al. 1987a) prior to challenge. These normal mice rarely have evidence of any cardiopathology. These "natural autoantibodies" are not the result of subclinical enterovirus infections in mice obtained from commercial breeding facilities in the U.S., because the CVB have never been isolated from normal mice, as they have from rats, dogs and pigs (MAHY 1988). The murine picornaviruses, Theiler's murine encephalomyelitis and encephalomyocarditis viruses are unlikely passengers in these mice; moreover, the latter two cardioviruses do not share epitopes with any viral proteins of the human enteroviruses (RUECKERT 1996). In summary, the capacity for production of autoantibodies against epitopes on heart tissues during a CVB3 infection is murine strain-specific and induction of specific autoantibodies appears to co-vary with capacity for development of chronic myocarditis in select murine strains (BEISEL and ROSE 1988; TRAYSTMAN and BEISEL 1991).

Given that some CVB3-infected murine strains will produce autoantibodies against epitopes in molecules within or on heart tissues, the question becomes: "How might these autoantibodies contribute to pathology in the myocardium?" Studies of human sera from patients with myocarditis or a presumed sequelae, IDC, show a variety of reactions in vitro that could have potential pathologic consequences in vivo. Sera containing antibodies against myolemmal antigens on isolated adult human myocytes were shown to participate in complement-mediated lysis of normal rat myocytes (MAISCH et al. 1993). Antibodies in the sera of some patients with myocarditis can affect functions of isolated human cardiac myocytes, including: a disturbance of energy metabolism (SCHULZE et al. 1989), inhibition of nucleoside transport (SCHULTHEISS et al. 1987), impairment of contractility (ULRICH et al. 1988), increase Ca^{2+} influx (MORAD et al. 1988; SCHULTHEISS et al. 1988) or by direct cytotoxicity on cardiomyocytes (MAISCH et al. 1983, 1991; ULRICH et al. 1988; SCHULZE et al. 1989). A high proportion of patients with myocarditis have complement and immunoglobulins deposited in their heart tissues (HAMMOND et al. 1988; NEUMANN et al. 1990; MAISCH et al. 1993).

Pathophysiologic studies of autoantibodies from CVB3-murine models have not been conducted in the aforementioned manner. Although available in 1990 (NEU et al. 1990), only recently has a reliable method for obtaining isolated adult murine cardiac myocytes been exploited to study murine cardiocytes (HERZUM et al. 1994). Indirect conclusions have been drawn from analyses of heart tissue-reactive autoantibodies in sera from CVB3-infected mice with myocarditis. The presence of cardiac-specific anti-sarcolemma antibodies in sera or deposited on heart tissues correlates with severity of chronic myocarditis induced by CVB3, whereas the presence of serum anti-cardiac myosin autoantibodies does not correlate with virus-induced chronic disease (TRAYSTMAN and BEISEL 1991; TRAYSTMAN et al. 1991). In another study (NEUMANN et al. 1994), the extent of cardiopathologic severity induced by CVB3 in a murine model correlated with titers of both anti-cardiac myosin and anti-BCKD autoantibodies in sera and with concentrations of anti-ANT and anti-BCKD autoantibodies deposited in heart tissues. A role for anti-cardiac myosin autoantibodies appears to be lacking in murine models of acute myocarditis in which the inducing agent is purified mouse cardiac myosin (NEU et al. 1987a). Passive transfer of high titer anti-cardiac myosin autoantibodies into normal mice failed to induce myocarditis (NEU et al. 1990). It seems that potential mechanisms by which CVB3-induced autoantibodies could contribute to chronic murine myocarditis must await in vitro studies with isolated adult murine myocytes, cardiac endothelial cells (HUBER et al. 1990) or cardiac fibroblasts (GAUNTT et al. 1987). Data from studies of identities and quantities of soluble mediators (cytokines, chemokines) induced by autoantibodies from the latter cells and unelicited cardiac macrophages could be most instructive concerning pathophysiologic mechanisms.

2 Molecular Mimicry Among Coxsackievirus B and Host Cell Proteins

Studies by Maisch and colleagues (MAISCH et al. 1982, 1993) of human sera from patients with CVB-associated myocarditis showed that cross-adsorption of sera containing autoantibodies to myolemmal proteins and sarcolemmal structures with purified CVB3 or CVB4 abolished both binding to rat and human cardiocytes and direct lysis of either isolated cell type. Immunoblot analysis provided indirect evidence that virus particles removed autoantibodies capable of recognizing lower molecular weight polypeptides (35–28 kDa) that are of the size of CVB particle capsid proteins (MAISCH et al. 1993). These data on cross-reactive antibodies suggest that the myolemma, sarcolemma and CVB particles share epitopes in proteins of the latter size.

Neutralizing monoclonal antibodies (mAbs) directed against CVB3 or CVB4 have been used in a variety of studies that suggest molecular mimicry as an alternative explanation for the autoantibody induction mechanism. One neutralizing mAb generated against CVB4 strain JVB (PRABHAKAR et al. 1982) was found to recognize an epitope on capsid polypeptide VP1 (PRABHAKAR 1988) and on molecules in sectioned heart tissues of several animals (SAEGUSA et al. 1986; BEISEL et al. 1990). The vast majority of myocytes in these sectioned tissues would have been sliced open to expose internal structures and fibrils. Further work identified an epitope in the murine α-cardiac myosin heavy chain (BEISEL et al. 1990) and specifically within residues 1632–1647 (BEISEL et al. 1991). Unfortunately, it is not known if this virus or mAb is associated with myocarditis in the mouse. One neutralizing mAb generated against CVB3 was also found to bind an epitope on normal mouse myocytes, HeLa cells and Hep2 cells (HUBER et al. 1988; WELLER et al. 1989). In a reverse of the latter situation, mAbs were generated against murine cardiac myocytes (WELLER et al. 1989). One of the latter mAbs (mAb10A1) could significantly inhibit binding of CVB3 to these myocytes and recognized epitopes on CVB3 particles without neutralizing the virus. In addition, this mAb could also partially protect BALB/c mice from myocarditis or death induced by a highly myocarditic strain of CVB3 in these mice (LOUDON et al. 1991). The data from studies of this latter mAb suggest that it may be recognizing an epitope near or on the viral receptor. The latter multitarget mAb also recognizes an epitope on ANT (LOUDON et al. 1991; SCHWIMMBECK et al. 1993), the inner mitochondrial membrane protein that shares an epitope(s) with the surface calcium channel (ULRICH et al. 1988). Thus this mAb links shared epitopes among CVB3 particles, a highly immunogenic mitochondrial protein and a surface protein that can control calcium levels in cardiac cells, a most interesting collection of proteins with potential pathologic significance. Our studies of a panel of neutralizing mAbs directed against a highly myocarditic CVB3 ($CVB3_m$) provide further support for molecular mimicry. All of these mAbs recognize discontinuous epitopes on $CVB3_m$ virions (GAUNTT et al. 1995). Several of these mAbs recognized epitopes on murine and human cardiac myosins by ELISA and/or immunoblot assays (GAUNTT et al. 1991a, 1993a, b, 1995). Analyses of these mAbs showed that $CVB3_m$ also shares at least two neutralizing epitopes with eight clinical isolates of CVB3 and at least one well-studied cardiovirulent strain of CVB3 (GAUNTT and PALLANSCH 1996) whose entire sequence has been determined (TRACY et al. 1992). These studies also showed that several mAbs can participate in complement-mediated lysis of normal adolescent mouse cardiac fibroblasts and stimulate increased synthesis of a chemoattractant for unelicited macrophages from the latter fibroblasts. Furthermore, about one third of these mAbs can, subsequent to inoculation of normal mice of two strains, induce cardiopathologic alterations (GAUNTT et al. 1995). The mAbs with cardiopathologic potential could best be identified as those that recognized epitopes on α-cardiac myosin and elicited increased production of the chemoattractant for macrophages. One of the cardiopathologic mAbs binds to an epitope on the surface of normal cardiac fibroblasts and relative binding is murine strain-specific: approximately 50% of fibroblasts from CD-1 mice that resolve CVB3-induced acute myocarditis bind this mAb at reduced levels compared to extensive binding of the mAb to 90% of fibroblasts cultured from heart tissues of C3H/HeJ mice that develop CVB3-induced chronic myocarditis (GAUNTT et al. 1995). These data provide one possible genetic reason for development of chronic disease via binding of autoantibodies, i.e., concentration of an autoantigen(s) on the surface of cardiac fibroblasts. The latter mAb also immunostained sectioned normal mouse heart tissues in a manner indistinguishable from a photomicrograph of the CVB4-neutralizing mAb that was shown previously to recognize cardiac myosin in A bands (GAUNTT et al. 1993a). Different anti-$CVB3_m$ mAbs were also shown to recognize epitopes on laminin, insulin, actin, elastin, phosphorylase B, vimentin and thyroglobulin in ELISA (GAUNTT et al. 1993a, b).

mAbs and polyvalent antisera against heterologous normal cell molecules or synthetic peptides of that molecule, prepared in labs not working with CVB3, were used to demonstrate the presence of shared epitopes with $CVB3_m$ by several types of assays. Antibodies against ANT (SCHWIMMBECK et al. 1989, 1993; LOUDON et al. 1991;), the M protein of group A streptococci (CUNNINGHAM et al. 1992) and collagen type I and rabbit, bovine or human myosins (GAUNTT et al. 1993a) recognized an epitope(s) on $CVB3_m$ in ELISA or by virus neutralization assays. A novel study (HUBER et al. 1994) showed how epitopes shared between a CVB3 and M protein of group A streptococci could alter the course of murine myocarditis. mAbs to the latter M protein (that also shared epitopes with human cardiac myosin) were used to generate escape mutants. Two such mutants were found to be altered in their host range, i.e., they induced more severe disease in congenic murine strains in which the parent CVB3 induced only mild myocarditis.

Molecular mimicry is likely to be a mechanism operative in inducing autoantibodies in humans with IDDM and in murine models of the disease that show cross-reactivity between a CVB and GAD65. Strong sequence similarities between GAD65 and the nonstructural viral protein PC2 have led to identification of a shared epitope(s) (CHATTERJEE 1994). Hyperimmune antisera to synthetic peptides representing GAD65 or P2C and IDDM patients' sera show extensive cross-reactivity between these antigens (CHATTERJEE 1994). Epitope sharing between these two antigens was also demonstrated by indirect immunofluorescence using IDDM patients' sera and competitive blocking experiments with purified antigens (MYERS et al. 1995). IDDM patient's sera also contain autoantibodies to insulin and carboxypeptidase H (CASTAÑO et al. 1991; ATKINSON and MACLAREN 1993). Human carboxypeptidase H, VP1 of CVB4 and the HLA-DQ 3.2 -chain share significant homology in a nine amino acid sequence, an example of triple molecular mimicry that offers one possible explanation for why IDDM may be restricted to individuals with particular haplotypes (BAUM et al. 1995). ELISA studies with anti-CVB3m neutralizing mAbs found that several mAbs recognize epitopes on insulin, whereas other mAbs recognize epitopes on GABA (Gauntt and Wood, unpublished data). Thus several kinds of evidence show that the CVB and pancreatic autoantigens share epitopes.

3 Hypotheses on Mechanisms by Which Coxsackievirus B3-Induced Antibodies Contribute to Myocardial Disease

Mechanistically, how can one account for a nonenveloped virus inducing acute or chronic focal inflammatory lesions in heart tissues? Several major hypotheses have evolved from the considerable data generated over the past 20 years. In regard to the initial acute phase of disease, infection results in lysis of myocytes prior to infiltration of leukocytes (WOODRUFF 1980). Natural killer (NKs) cells and virus-specific cytotoxic T lymphocytes (CTLs) can lyse virus-infected cells (GODENY and GAUNTT 1987a; VAN HOUTEN and HUBER 1989). Both NKs and CTLs can be found in focal myocarditic lesions near myocytes undergoing necrosis in the acute stage of disease (GODENY and GAUNTT 1987b; CHOW et al. 1991). The chronic phase of CVB3-induced myocarditis, i.e., beyond day 21 postinoculation, occurs only in certain murine strains of select genetic backgrounds (ROSE et al. 1988a, b). Heart tissues of these mice with CVB3-induced chronic disease show a more diffuse interstitial mononuclear cell infiltrate, evidence of myocyte injury with minimal myocyte necrosis and significant deposition of autoantibodies (HERSKOWITZ et al. 1987; WOLFGRAM and ROSE 1989; CHOW et al. 1991; NEUMANN et al. 1991). Induction of these autoantibodies derives from sensitization of IgG-producing B cells to epitopes in or on two types of antigens. The first type is host cell cryptic antigens released from myocytes and other heart tissue cells as a result of virus-, NK cell- or immunologically mediated cell lysis (WOLFGRAM and ROSE 1989). The second type of antigens is found in or on virus particle capsid proteins that mimic molecules found in or on normal heart tissue cells. These antigens are presented early in infection but titers of pathologic antibodies are not attained until the acute phrase wanes (GAUNTT et al. 1991b, 1993a, b, 1995). The hearts of patients with chronic myocarditis often show global impairment in function, and severity of disease generally correlates with levels of autoantibodies bound to these tissues (NEUMANN et al. 1990). A working hypothesis of several labs is that these bound autoantibodies stimulate the production of pro-inflammatory soluble mediators and/or induce metabolic disruptions of myocytes and other resident heart tissue cells. Both in vitro and in vivo studies show that a CVB3 infection can stimulate production of several pro-inflammatory cytokines such as interferon-γ(IFN-γ) interleukin^{-1} (IL-1) and tumor necrosis factor-α(TNFα) (GODENY and GAUNTT 1986; GAUNTT et al. 1988a, b; LANE et al. 1992; ROSE et al. 1992). These molecules could then signal the cell-mediated immune system which responds to establish the interstitial infiltrate and produce additional cytokines (NEUMANN et al. 1993). Cytokines can induce up-regulation of numerous adhesion molecules (CALLARD and GEARING 1994) and it seems plausible that the role of autoantibody-induced cytokines could be to make the myoctes "sticky" for adhering leukocytes. Either IFN-γ or TNFα treatment of mouse fetal myoctes induced expression of intercellular adhesion molecule 1 (ICAM-1; SEKO et al. 1993). Induction of leukocyte chemotaxis by chemokines (BAGGIOLINI et al. 1994), low molecular weight proteins that induce chemotaxis of leukocytes to focal areas (IMHOF and DUNON 1995), may play a role(s) in chronic myocarditis. Macrophage inflammatory protein-1α(MIP-1α) is a chemokine with pro-inflammatory properties in vitro (Cook 1996). A comparison of MIP-1α-knockout mice with their normal counterparts for induction of myocarditis by CVB3 showed that the mutant mice were resistant to virus-induced disease (COOK et al. 1995). These results imply that MIP-1α is an important participant in the molecular evolution of CVB3-induced myocarditis. Therefore, in CVB3-infected murine strains that develop chronic disease, myocardium-bound autoantibodies induce cytokines and provide the stimulus for induction of chemokines that indirectly contribute to the continued inflammation found in chronic myocarditis.

4 Data Needed for Further Confirmation of Coxsackievirus B-Induced Autoantibody Contributions to Myocarditis

In numerous studies cited above, researchers have attempted to identify proteins containing autoantibody-targeted epitopes *on* myocytes and *on* other cells in heart tissues. Indirect studies of human heart tissues thus far have identified only the calcium channel, which shares an epitope(s) with the mitochondrial ANT protein and the β-adrenergic receptor protein, as a human surface protein that induces autoantibodies during myocarditis and IDC. Suggestive studies that the CVB3 receptor protein on murine heart tissues may be a target of autoantibodies have been published but need confirmation. Surface counterpart proteins/structures sharing epitopes with cardiac myosin, actin, tropomyosin and BCKD remain to be identified. Abundant binding of autoantibodies to sarcolemmal/myolemmal membranes suggests a tissue source with many proteins/structures available for cross-linking studies that could resolve the antigenic identities and ascertain any immunopathologic role(s) in myocarditis.

Identification of surface target proteins/structures and the pathologic epitope should have high priority. Of obvious critical importance will be to ascertain how binding of autoantibodies contributes to any type of cellular dysfunction and perhaps necrosis in isolated viable myocytes. Does binding of a specific autoantibody molecule(s) activate a kinase(s), a transcription factor(s), an apoptosis sequence of events, Ca^{2+} dysregulation, or lead to events that alter the ATP/ADP ratio? It is also possible that most autoantibodies may have only minor, if any, roles in chronic myocarditis or IDC and only a minor fraction of the autoantibodies have specificities that will lead to serious consequences for normal myocytes in either humans or mice infected with the CVB. Dissecting such a hypothesis will require analyses of polyclonal populations of autoantibodies, affinity purified against the isolated putative autoantigen epitope(s). However, a major pitfall in such studies is that denaturation of targeted epitopes on proteins/structures isolated from myocytes or other heart tissue cells may result in loss of binding of important autoantibodies, and mimicking of the actual reactants may require use of intact, gently fixed, surface membranes as the affinity material. Synthetic peptides as mimics of potential immunopathologic autoantibody epitope sequence(s) could become very important as competitive inhibitors in sorting out which epitope(s) on a surface protein or structure are important in binding an autoantibody that alters a myocyte function(s). Perhaps a limited number of specific synthetic peptides could be of considerable therapeutic benefit in myocarditis and/or IDC.

5 Potential Diagnostic Value of Coxsackievirus B-Induced Autoantibodies as Surrogate Indicators of Myocarditis

At this time, there are no laboratory tests for confirmation of a clinical diagnosis of myocarditis. It follows that identification of synthetic peptides that mimic sequences on myocyte surface proteins or structures, or other cardiac cells, that are targeted by pathologic autoantibodies would be extremely useful as antigens in ELISA. Detection of chronic myocarditis, severity of disease and recovery could be monitored by such assays in conjunction with heart function studies. SCHWIMMBECK et al. (1993) found that 60% of patients with myocarditis have autoantibodies against whole isolated ANT. Synthetic peptides representing predicted antigenic sequences in ANT identified one highly immunogenic sequence that deserves further analysis using human sera from patients (SCHWIMMBECK et al. 1993). Studies with synthetic peptides spanning the light meromyosin region (LMM) of human cardiac -myosin heavy chain found that some LMM region synthetic peptides could cause cardiopathologic alterations in normal mice of three strains, others could induce antibodies which could be titrated against purified CVB3 particles (ELISA) and one LMM synthetic peptide could titrate anti-IgG_{2a} antibodies in mice with CVB3_m-induced chronic myocarditis and also mice challenged with a noncardiovirulent CVB3 (GAUNTT et al. 1996). These data suggest that differentiation of mice with chronic myocarditis from mice with a subclinical infection must await additional studies with subpeptides of the latter LMM synthetic peptide which is 18 amino acids in length.

6 Future Directions for Studies of Coxsackievirus B-Induced Autoantibodies

Are B lymphocytes present in myocardial tissues or in focal myocarditic lesions of patients with chronic myocarditis or IDC? There are no data on whether B cells are present in murine heart tissues of mice with CVB3-induced chronic myocarditis. If these cells are indeed present, it would be quite useful to determine whether they are producing autoantibodies and to identify the targeted epitopes. We need to know if sensitized B cells are located in heart tissues and are producing anti-heart tissue autoantibodies that could quickly bind to cells in the proximity and induce cardiopathology. In such a scenario, autoantibodies produced locally could bind within minutes to affect myocytes and other heart tissue cells. If heart tissue-localized production of cardiac tissue-specific autoantibodies occurs in patients with CVB-induced myocarditis or in CVB3-infected murine models, current assays for autoantibodies in sera would likely detect only high titer subpopulations of antibodies directed against total heart tissue antigens (internal and surface) and overlook minor subpopulations which become bound to myocyte epitopes soon after synthesis and contribute to cardiopathology. One way of detecting these minor subpopulations of potentially important autoantibodies would be to clone the B cells producing them (or establish hybridomas) and then identify their antigenic targets and the specific epitopes being recognized.

Studies of antigenic targets for the CVB-induced autoantibodies have focused on protein antigens. Ample data exists to show that this approach is incomplete. One mAb (49.8.9) generated against a streptococcal M protein was found to recognize a linear epitope in the VP1 capsid protein of CVB3 and recently shown to cross-react with N-acetyl-β-D-glucosamine (GlcNAc; SHIKHMAN et al. 1993). Other work has shown that the latter carbohydrate moiety also shares epitopes with cytokeratin peptides (SHIKHMAN and CUNNINGHAM 1994). These results may explain earlier data on binding of a lectin only to CVB3 producing cells in cultures of virus-challenged murine fibroblasts (LUTTON and GAUNTT 1986): wheat germ agglutinin, which recognizes GlcNAc, bound to CVB3-infected fibroblasts and probably to a peptide sequence (RRKLEFF) found on VP1 as part of emerging nascent virus particles (SHIKHMAN et al. 1993). Thus some autoantibodies may be recognizing carbohydrate moieties on targeted cardiac cells, but the original sensitizing antigen was protein. Phospholipids as targets on cardiac cells could also bear examination, as antibodies against phospholipids are found in many infectious diseases (McNeil et al. 1991). Aminoacyl-tRNA synthetases serve as antigens to induce antibodies in patients with idiopathic myositis (Naparstek and Plotz 1993) and there are no reports on the presence of these antibodies in CVB-induced myocarditis. Finally, specific sequences in RNA molecules that mimic proteins, e.g., an RNA sequence that mimics the human insulin receptor (Doudna et al. 1995), open a new possibility for explaining the problem of understanding how the 5' nonstructural (noncoding) region (5'-NTR) of the CVB3 genome contributes to cardiovirulence (Tracy et al. 1991; Tu et al. 1995; Gauntt and Pallansch et al. 1996). Is it possible that a 5'-NTR CVB3 sequence induces autoantibodies that react with myocytes to cause dysfunction?

Polyclonal anti-idiotypic (anti-Id) antibodies and an anti-Id mAb to CVB3 antibodies have been generated and shown in pre-CVB3 challenge experiments to significantly reduce the severity of CVB3-induced myocarditis in mice (Paque and Miller 1988, 1989a). An additional study involving a mAb (10A1) against the putative CVB3 receptor on murine cardiac tissue cells suggested that reduction of myocarditis by mAb10A1 may have been mediated by an anti-Id (Weller et al. 1989). Anti-Ids that bind the virus receptor may induce several cytokines, an avenue not explored with the counterpart, inactivated intact virus. Additionally, anti-Ids against anti-CVB3 antibodies to viral capsid polypeptide epitopes that mimic epitopes on host cell proteins could serve as antigens during the chronic phase and induce production of anti-anti-Ids that could also bind to the original shared epitope(s) on normal cells and further induce synthesis of cytokines or chemokines. Murine strains that resolve CVB3-induced acute myocarditis may not produce anti-Ids to antiviral antibodies, whereas murine strains that transit from CVB3-induced to chronic myocarditis may produce high titer anti-Ids. There are no published data on this point. The anti-Ids may be of particular importance if they are found to induce cytokines or chemokines, because administration of IL-1 and TNFα to a strain of mouse that is resistant to CVB3-induced myocarditis caused induction of myocarditis and synthesis of anti-heart antibodies that deposited in the myocarditic heart tissues (Neumann et al. 1993). Polyclonal anti-Ids (to polyclonal anti-CVB3 antibodies) were found to enhance chemotaxis of murine peritoneal macrophages (Paque and Miller 1989b), suggesting that anti-Ids produced by sensitized B cells in heart tissues could have a role in maintaining the presence of macrophages in heart tissues. Thus murine strains that resolve CVB3-induced acute myocarditis may have Id-anti-Id control mechanisms in place whereas murine strains that develop CVB3-induced chronic myocarditis may have control flaws in their idiotype network.

Acknowledgements. This research was supported by a grant from the ERACE Foundation, Los Angeles, CA. The excellent typing skills and patience of Annette Mandujano are sincerely appreciated.

References

Abelmann WH (1973) Viral myocarditis and its sequelae. Ann Rev Med 24:145–152

Abelmann WH (1984) Classification and natural history of primary myocardial disease. Prog Cardiovasc Dis 27:73–94

Alvarez FL, Neu N, Rose NR, Craig SW, Beisel KW (1987) Heart-specific autoantibodies induced by Coxsackievirus B3: Identification of heart autoantigens. Clin Immunol Immunopathol 43:129–139

Ansari AA, Yang YC, Danner DJ, Gravanis MB, Neckelmann N, Sell KW, Herskowitz A (1991) Abnormal expression of histocompatibility and mitochondrial antigens by cardiac tissue from patients with myocarditis and dilated cardiomyopathy. Am J Pathol 139:1–12

Atkinson MA, Maclaren NK (1993) Islet cell autoantigens in insulin-dependent diabetes. J Clin Invest 92:1608–1616

Atkinson MA, Bowman MA, Campbell L, Darrow BL, Kaufman DL, Maclaren NK (1994) Autoimmunity to two forms of glutamate decarboxylase in insulin-dependent diabetes mellitus. J Clin Invest 94:2125–219

Baekkeskov S, Aanstoot H, Christgau S, Reetz A, Solimena M, Cascalho M, Folli F, Richter-Olesen H, De Camilli P (1990) Identification of the 64 K autoantigen in insulin-dependent diabetes as the GABA-synthesizing enzyme glutamic acid decarboxylase. Nature 347:151–156

Baggiolini M, Dewald B, Moser B (1994) Interleukin-8 and related chemotactic cytokines-CXC and CC chemokines. Adv Immunol 55:97–179

Barrett-Conner E (1985) Is insulin-dependent diabetes mellitus caused by coxsackievirus B infection? A review of the epidemiologic evidence. Rev Infect Dis 7:207–215.

Baum H, Brusic V, Choudhuri K, Cunningham P, Vergani D, Peakman M (1995) MHC molecular mimicry in diabetes. Nature Med (Letters) 1:388

Beck MA, Tracy SM (1989) Murine cell-mediated immune response recognizes an enterovirus group-specific antigen(s). J Virol 63:4148–4156

Beck MA, Tracy SM (1990) Evidence for a group-specific enteroviral antigen(s) recognized by human T cells. J Clin Microbiol 28:1822–1827

Beck MA, Chapman NM, McManus BM, Mullican JC, Tracy S (1990) Secondary enterovirus infection in the murine model of myocarditis. Am J Pathol 136:669–681

Beisel KW, Rose NR (1988) Relationship of coxsackievirus to cardiac autoimmunity. In: Bendinelli M, Friedman H (eds), Coxsackieviruses. A general update. Plenum Press, New York pp 271–292

Beisel KW, Traystmann MD (1988) Viral myocarditis: immunogenetic and autoimmune aspects. In: Schultheiss H-P (ed) New Concepts in Viral Heart Disease, Springer-Verlag, Berlin pp 148–159

Beisel KW, Srinivasappa J, Olsen MR, Stiff AC, Essani K, Prabhakar (1990) A neutralizing monoclonal antibody against coxsackievirus B4 cross-reacts with contractile muscle proteins. Microbial Pathogenesis 8:151–156

Beisel KW, Srinivasappa J, Prabhakar BS (1991) Identification of a putative shared epitope between coxsackievirus B4 and α cardiac myosin heavy chain. Clin Exp Immunol 86:49–55

Bendinelli M, Conaldi PG, Matteucci D (1988) Interactions with the immune system. In: Bendinelli M, Friedman H (eds) Coxsackieviruses. A general update. Plenum Press, New York pp 81–102

Bottazzo GJ, Florin-Christensen A, Doniach (1974) Islet-cell antibodies in diabetes mellitus with autoimmune polyendocrine deficiencies. Lancet ii:1279–1283

Bruining GJ, Molenaar JL, Grobbee DE, Hofman A, Scheffer GJ, Bruining HA, de Bruyn AM, Valkenburg HA (1989) Ten year follow up study of islet cell antibodies and childhood diabetes mellitus. Lancet i:1100–1103

Caforio ALP, Bonfacio E, Stewart JT, Neglia D, Parodi O, Bottazzo GF, McKenna WJ (1990) Novel organ-specific circulating autoantibodies in dilated cardiomyopathy. J. Am Coll Cardiol 15:1527–1534

Callard R, Gearing A (1994) The cytokine facts book. Academic Press, Inc, San Diego, CA, p 1–246

Castaño L, Russo E, Zhou L, Lipes MA, Eisenbarth GS (1991) Identification and cloning of a granule autoantigen (carboxypeptidase H) associated with type 1 diabetes. J Clin Endocrinol Metab 73:1197–1201

Cello J, Sammuelson A, Stalhandske, Svennerholm B, Jeansson S, Forgren M (1993) Identification of group-common linear epitopes in structural and nonstructural proteins of enteroviruses by using synthetic peptides. J Clin Micro 31:911–916

Chan D, Hammond GW (1985) Comparison of serodiagnosis of group B coxsackievirus infections by an immunoglobulin M capture enzyme immunoassay virus microneutralization. J Clin Micro 21:830–834

Chatterjee NK (1994) Antibodies to glutamic acid decarboxylase and P2-C peptides in sera from coxsackievirus B4-infected mice and IDDM patients. Diabetes 43:1260–1266

Cho CT, Feng KK, McCarthy VP, Lenahan MF (1982) Role of antiviral antibodies in resistance against coxsackievirus B3 infection: interaction between preexisting antibodies and an interferon inducer. Infect Immun 37:720–727

Chow LH, Gauntt CJ, McManus BM (1991) Differential effects of myocarditic variants of coxsackievirus B3 in inbred mice: A pathologic characterization of heart tissue damage. Lab Invest 64:55–64

Chow LH, Beisel KW, McManus BM (1992) Enteroviral infection of mice with severe combined immunodeficiency. Lab Invest 66:24–31

Cook DN, Beck MA, Coffman TM, Kirby SL, Sheridan JF, Pragnell IB, Smithies O (1995) Requirement of MIP-1α for an inflammatory response to viral infection. Science 269:1583–1585

Cook DN (1996) The role of MIP-1α in inflammation and hematopoiesis. J Leukoc Biol 59:61–66

Cunningham MW, Antone SM, Gulizia JM, McManus BM, Fischetti VA, Gauntt CJ (1992) Cytotoxic and viral neutralizing antibodies crossreact with streptococcal M protein, enteroviruses and human cardiac myosin. Proc Natl Acad Sci USA 89:1320–1324

D'Alessio DJ (1992) A case-control study of group B coxsackievirus immunoglobulin M antibody prevalence and HLA-DR antigens in newly diagnosed cases of insulin-dependent diabetes mellitus. Am J Epidemiol 135:1331–1338

De Scheerder IK, De Buyzere M, Delanghe J, Maas A, Clement DL, Wieme R (1991) Humoral immune response against contractile proteins (actin and myosin) during cardiovascular disease. European Heart J 12(Suppl D):88–94

Dörries R, Ter Meulen V (1983) Specificity of IgM antibodies in acute human coxsackievirus B infections analyzed by indirect solid phase enzyme immunoassay and immunoblot technique. J Gen Virol 64:159–167

Doudna JA, Cech TR, Sullenger BA (1995) Selection of an RNA molecule that mimics a major autoantigenic epitope of human insulin receptor. Proc Natl Acad Sci USA 92:2355–2359

Fohlman J, Friman G (1993) Is juvenile diabetes a viral disease? [Review] Ann Med 25:569–574

Fohlman J, Paulssen K, Morein B, Bjare U, Ilback N-G Friman G (1993) High yield production of an inactivated coxsackie B3 adjuvant vaccine with protective effect against experimental myocarditis. Scand J Infect Dis 88[Suppl]:103–108

Frisk G, Nilsson E, Ehrnst A, Diderholm H (1989) Enterovirus IgM detection: Specificity of μ-antibody-capture radioimmunoassays using virions and procapsids of coxsackievirus B virus. J Vir Meth 24:191–202μ

Gauntt CJ (1987) Cellular and humoral immune responses in coxsackievirus myocarditis. In: Kawai C, Abelmann WH, Matsumori A (eds), Pathogenesis of Myocarditis and Cardiomyopathy; Recent Experimental and Clinical Studies. Cardiomyopathy 1987. Univ of Tokyo Press, Tokyo, pp 49–61

Gauntt CJ (1988) The possible role of viral variants in pathogenesis. In: Bendinelli M, Friedman H (eds), Coxsackieviruses. A general update. Plenum, New York pp 159–179

Gauntt CJ, Pallansch MA (1996) Coxsackievirus B3 clinical isolates and murine myocarditis. Virus Res 41:89–99

Gauntt CJ, Lutton CW, Godeny EK, Witherspoon SM, Arizpe HM, Lanford RE (1987) Murine coxsackievirus B3 myocarditis. European Heart J 8 [Suppl] J: 393–397

Gauntt C J, Godeny EK, Lutton CW (1988a) Host factors regulating viral clearance. Pathol Immunopathol Res 7:251–265

Gauntt CJ, Godeny EK, Lutton, CW, Arizpe HM, Chapman NM, Tracy SM, Revtyak GE, Valente AJ, Rozek MM (1988b) Mechanism(s) of coxsackievirus-induced acute myocarditis in the mouse. In: de la Maza LM, Peterson E (eds), Medical Virology VIII, Lawrence Erlbaum Associates, Inc, Hillsdale, NJ p 161–182

Gauntt CJ, Arizpe HM, Higdon AL, Rozek MM, Crawley R, Cunningham MW (1991a) Anti-coxsackievirus B3 neutralizing antibodies with pathologic potential. European Heart J 12:[Suppl D]:124–129

Gauntt CJ, Lutton CW, Arizpe HM, Higdon AH, Tracy SM (1991b) Autoimmune reactions in coxsackievirus B3-induced murine myocarditis. Life Sci Adv 10:23–31

Gauntt C, Higdon A, Bowers D, Maull E, Wood J, Crawley R (1993a) What lessons can be learned from animal model studies in viral heart diseases? Scand J Infect Dis 88(Suppl):49–65

Gauntt CJ, Higdon AL, Arizpe HM, Tamayo MR, Crawley R, Henkel RD, Perera MEA, Tracy SM, Cunningham MW (1993b) Epitopes shared between coxsackievirus B3 (CVB3) and normal heart tissue contribute to CVB3-induced murine myocarditis. Clin Immunol Immunopathol 68:129–134

Gauntt CJ, Higdon, AL, Arizpe HM, Maull EA, Lutton CW, Beck MA, Chapman NM, McManus B, Mullican JC, Tracy SM (1993c) Specific and nonspecific heart defenses in enteroviral infections. In: Figulla HR, Kandolf R, McManus B (eds), Idiopathic Dilated Cardiomyopathy. Springer-Verlag, Berlin pp 310–321

Gauntt CJ, Arizpe HM, Higdon AL, Wood HJ, Bowers DF, Rozek MM, Crawley R (1995) Molecular mimicry, anti-coxsackievirus B3 neutralizing monoclonal antibodies and myocarditis. J Immunol 154:2983–2995

Gauntt CJ, Winfrey CL, Wood HJ, Karaganis AG, Lee CH, Cunningham MW (1996) Anti-coxsackievirus group B antibodies and inflammatory heard disease. In: Pandaloi G (ed), Recent Research Developments in Antimicrobial Agents and Chemotherapy. Research Signpost, Trivandrum 1:257–270

Godeny EK, Gauntt CJ (1986) Involvement of natural killer cells in coxsackie B3 virus-induced murine myocarditits. J Immunol 137:1695–1702

Godeny EK, Gauntt CJ (1987a) *In situ* autoradiographic identification of cells in heart tissues of mice with coxsackievirus B3-induced myocarditis. Am J Pathol 129:267–276

Godeny EK, Gauntt CJ (1987b) Murine natural killer cells limit coxsackievirus B3 replication. J Immunol 139:913–918

Godeny EK, Arizpe HM, Gauntt CJ (1988) Characterization of the antibody response in vaccinated mice protected against coxsackievirus B3-induced myocarditis. Viral Immunol 1:305–314

Griffith JS, Katz SL, Moore M (1977) Persistent enterovirus infections in agammaglobulinemia. In: Schlessinger D (ed), Microbiology-1977. Am Soc Microbiol, Washington, DC pp 488–493

Grodums EI, Dempster G (1959) Myocarditis in experimental coxsackie B3 infection. Can J Micro 5:605–615

Haarmann CM, Schwimmbeck PL, Mertens T, Schultheiss HP (1994) Identification of serotype-specific and nonserotype-specific B cell epitopes of coxsackie B virus using synthetic peptides. Virol 200:381–389

Halonen P, Rosen L, Hubener RL (1959) Homologous and heterologous complement fixing antibody in persons infected with echo, coxsackie and poliomyelitis viruses. Proc Soc Exp Biol Med 101:236–241

Hammond EH, Menlove RL, Anderson JL (1988) Immunofluorescence microscopy in the diagnosis and follow-up of patients suspected of having inflammatory heart disease. In: Schultheiss H-P (ed),. New Concepts in Viral Heart Disease. Springer-Verlag, Berlin pp 303–311

Herskowitz A, Beisel KW, Wolfgram LJ, Rose NR (1985) Coxsackievirus B3 murine myocarditis: wide pathologic spectrum in genetically defined inbred strains. Human Pathol 16:671–673

Herskowitz A, Wolfgram LJ, Rose NR, Beisel KW (1987) Coxsackievirus B3 murine myocarditis-marked strain differences in histopathologic features of early and late phase myocarditis J Am Coll Cardiol 9:1311–1319

Herzum M, Ruppert V, Kuytz B, Jomaa H, Nakamura I, Maisch B (1994) Coxsackievirus B3 infection leads to cell death of cardiac myocytes. J Mol Cardiol 26:907–913

Hilton DA, Variend S, Pringle JH (1993) Demonstration of coxsackievirus RNA in formalin-fixed tissue sections from childhood myocarditis cases by in situ hybridization and the polymerase chain reaction. J Pathol 170:45–55

Huber SA, Lodge PA (1986) Coxsackievirus B3 myocarditis. Identification of different pathogenic mechanisms in DBA/2 and BALB/c Mice. Am J Pathol 122:284–291

Huber SA, Pfaeffle B (1994) Differential Th_1 and Th_2 cell responses in male and female BALB/c mice infected with coxsackievirus group B type 3. J Virol 68:5126–5132

Huber SA, Lodge PA, Job LP (1984) The role of virus and immune mediated cardiocyte injury in coxsackievirus B3-induced myocarditis. *In*: Bolte H-D (ed), Viral Heart Disease. Springer-Verlag, Berlin pp 64–73

Huber SA, Weller A, Herzum M, Lodge PA, Estrin M, Simpson K, Guthrie M (1988) Immunopathogenic mechanisms in experimental picornavirus-induced autoimmunity. Pathol Immunopathol Res 7:279–291

Huber SA, Harsch C, Lodge PA (1990) Functional diversity in vascular endothelial cells: role in coxsackievirus tropism. J Virol 64:4516–4522

Huber S, Moraska A, Cunningham MW (1994) Alterations in major histocompatability complex association of myocarditis induced by coxsackievirus B3 mutants selected with monoclonal antibodies to group A streptococci. Proc Natl Acad Sci USA 91:5543–5547

Hyöty H, Hiltunen M, Knip M, Laakkonen M, Vähäsalo P, karjalainen J, Koskela P, Roivainen M, Leinikki P, Hovi T, Åkerblom HK, The Childhood Diabetes in Finland (DiMe) Study Group (1995). A prospective study of the role of coxsackievirus B and other enterovirus infections in the pathogenesis of IDDM. Diabetes 44:652–657

Imhof BA, Dunon D (1995) Leukocyte migration and adhesión. Adv Immunol 58:345–416

Jenista JA, Powell KR, Menegus MA (1984) Epidemiology of neonatal enterovirus infection. J. Pediatr 104:685–690

Jin O, Sole MJ, Butany JW, Chia WK, McLaughlin PR, Fli P, Liew CC (1990) Detection of enterovirus RNA in myocardial biopsies from patients with myocarditis and cardiomyopathy using gene amplification by polymerase chain reaction. Circ 82:8–16

Kandolf R, Ameis D, Kirschner P, Canu A, Hofschnieder PH (1987) *In situ* detection of enteroviral genomes in myocardial cells by nucleic acid hybridization: An approach to the diagnosis of viral heart disease. Proc Natl Acad Sci USA 84:6272–6276

Kaspenberg J (1988) Picornaviridae: the enteroviruses (poliviruses, coxsackieviruses, echoviruses). In: Lennette EH, Halonen P, Murphy FA (eds). Laboratory Diagnosis of Infectious Diseases, Principles and Practice, Vol II. Viral, rickettsial and chlamydial diseases. Springer-Verlag, New York, pp 692–722

Katze MG, Crowell RL (1980a) Indirect enzyme-linked immunosorbent assay (ELISA) for the detection of coxsackievirus group B antibodies. J Gen Virol 48:225–229

Katze MG, Crowell RL (1980b) Immunological studies of the group B coxsackieviruses by the sandwich enzyme-linked immunosorbent assay (ELISA) and immunoprecipitation. J Gen Virol 50:357–367

Khatib R, Probert A, Reyes MP, Khatib G, Chason JL (1988) Mouse strain-related variation as a factor in the pathogenesis of coxsackievirus B3 murine myocarditis. J Gen Virol 68:2981–2988

Khatib G, Giraldo A (1993) The effects of pre-existing coxsackievirus B4 myocardial disease on the expression of coxsackievirus B3 myocarditis. Can J Cardiol 9:444–447

Klingel K, Hohenadl C, Canu A, Albrecht M, Seemann M, Mall G, Kandolf R (1992) Ongoing enterovirus-induced myocarditis is associated with persistent heart muscle infection: Quantitative analysis of virus replication, tissue damage and inflammation. Proc Natl Acad Sci USA 89:314–318

Klingel K, Kandolf R (1993) The role of enterovirus replication in the development of acute and chronic heart muscle disease in different immunocompetent mouse strains. Scan J Infect Dis 88(Suppl):79–85

Kogon A, Spigland I, Frothingham TE, Elveback, Williams C, Hall CE, Fox JP (1969) The virus watch program: a continuing surveillance of viral infections in metropolitan New York families. VII Observations on viral excretion, seroimmunity, intrafamilial spread and illness association in coxsackie and echovirus infections. Am J Epidemiol 89:51–61

Konishi K, Kawabata M, Ohtaki K (1991) Neutralization activity of some coxsackie B3 antibodies is compromised by protein factor(s) secreted by Vero cells.. Arch Virol 116:197–207.

Lane JR, Neumann DA, Lafond-Walker A, Herskowitz A, Rose NR (1992) IL-1 or TNF can promote CB3-induced myocarditis in resistant BIO.A mice J Exp Med 175:1123–1131

Lerner AM (1985) Myocarditis and pericarditis. In: Mandell GL, Douglas RG, Bennett (eds), Principles and Practice of Infectious Diseases. John Wiley and Sons, New York pp 544–551

Lerner AM, Wilson FM (1973) Virus myocardiopathy. Prog Med Virol 15:63–91AM, Reyes MP (1985) Coxsackievirus myocarditis – with special reference to acute and chronic effects. Prog Cardiovascular Dis 27:373–394

Leslie K, Blay R, Haisch C, Lodge P, Weller A, Huber S (1989) Clinical and experimental aspects of viral myocarditis. Clin Microbiol Rev 2:191–203

Limas CJ, Goldenberg IF, Limas C (1989) Autoantibodies against -adrenoreceptors in human idiopathic dilated cardiomyopathy. Cir Res 64:97–103

Lodge PA, Herzum M, Olszewski J, Huber SA (1987) Coxsackievirus B3 myocarditis. Acute and chronic forms of the disease caused by different immunopathologic mechanisms. Am J Path 128:455–463

Loria RM, Kibrick S, Broitman SA (1974) Peroral infection with group B coxsackieviruses in the newborn mouse: a model for human neonatal infection. J Infect Dis 130:225–230

Loria RM, Shadoff N, Kibrick S, Broitman S (1976) Maturation of intestinal defenses against perioral infection with group B coxsackieviruses in mice. Infect Immun 13:1397–1401

Loudon RP, Morasha AF, Huber SA, Schwimmbeck P, Schultheiss H-P (1991) An attenuated variant of coxsackievirus B3 preferentially induces immunoregulatory T cells in vivo J Virol 65:5813–5819

Lutton CW, Gudvangen RJ, Nealon TJ Paque RE, Gauntt CJ (1985) Cellular Immune responses in mice challenged with an amyocarditic variant of coxsackievirus B3. J Med Virol 17:345–357

Lutton CJ, Gauntt CJ (1986) Coxsackievirus B3 infection alters plasma membrane of neonatal skin fibroblasts. J Virol 60:294–296
Lyden DC, Feran M, Olszewski J, Job LP, Huber, SA (1987) Coxsackievirus B-3-induced myocarditis. Effect of sex steriods on viremia and infectivity of cardiocytes. Am J Pathol 126:432–438
Mahy BWJ (1988) Classification and general properties. In: Bendinelli M, Friedman H (eds) Coxsackieviruses. A general update. Plenum Press, New York pp 1–18
Maisch B (1984a) Cardiocytolysis by sera of patients suffering from acute perimyocarditis. In: Bolte HD (ed), Viral Heart Disease. Springer-Verlag, Basel, Switzerland pp 121–130
Maisch B (1984b) Diagnostic relevance of humoral and cell mediated immune reactions in patients with acute myocarditis and congestive cardiomyopathy. In: Chazov EI, Smirnov VN, Oganov RG (eds), Cardiology. Plenum ,New York pp 1327–1338
Maisch B (1986) Immunologic regulator and effector functions in pericarditis, post-myocarditic heart muscle disease and dilated cardiomyopathy. Basic Res Cardiol 81(Suppl 1):217–242
Maisch B (1989) Autoreactivity to the cardiac myocyte, connective tissue and the extracellular matrix in heart disease and postcardiac injury. Springer Semin Immunopathol 11:369–396
Maisch B (1990) Myocarditis. Curr Opin Cardiol 5:320–327
Maisch B, Berg PA, Kochsiek K (1979) Clinical significance of immunopathological findings in patients with post-pericardiotomy syndrome. I Relevance of antibody pattern. Clin Exp Immunol 38:189–197
Maisch B, Berg PA, Kochsiek K (1980) Autoantibodies and serum inhibition factors (sif) in patients with myocarditis. Klin Wochenschr 58:219–225
Maisch B, Trostel-Soeder R, Stechemesser E, Berg PA, Kochsiek K (1982) Diagnostic relevance of humoral and cell-mediated immune reactions in patients with acute viral myocarditis. Clin Exp Immunol 48:533–545
Maisch B, Deeg P, Liebau G, Kochsiek K (1983) Diagnostic relevance of humoral and cytotoxic immune reactions in primary and secondary dilated cardiomyopathy. Am J Cardiol 52:1072–1078
Maisch B, Outzen H, Roth D, Hiby A, Herzum M, Hengstenberg C, Hufnagel G, Schöman U, Kochsiek K (1991) Prognostic determinants in conventionally treated myocarditis and perimyocarditis – focus on antimyolemmal antibodies. European Heart J 12(Suppl D):81–87
Maisch B, Bauer E, Cirsi M, Kocksiek K (1993) Cytolytic cross-reactive antibodies directed against the cardiac membrane and viral proteins in coxsackievirus B3 and B4 myocarditis. Circulation 87 (Suppl IV):49–65
McManus BM, Gauntt CJ, Cassling RS (1987) Immunopathologic basis of myocardial injury. Cardiovasc Clinics 18:163–184
McNeil HP, Chesterman CN, Krilis SA (1991) Immunology and clinical importance of antiphospholipid antibodies. Adv Immunol 49:193–280
Melnick JL (1985) Enteroviruses: Polioviruses, coxsackieviruses, echoviruses, and newer enteroviruses. In: Fields B (ed). Raven Press, New York pp 739–794
Morad M, Näbauer M, Schultheiss H-P (1988) Antibodies and autoantibodies against ADP/ATP carrier enhance calcium current in isolated ventricular myocytes. In: Schultheiss H-P (ed), New Concepts in Viral Heart Disease. Springer-Verlag, Berlin pp 236–242
Muir P, Nicholson F, Tilzey AJ, Signy M, English TAH, Banatvala JE (1989) Chronic relapsing pericarditis and dilated cardiomyopathy: serological evidence of persistent enterovirus infection. Lancet i:804–807
Myers MA, Rabin DV, Rowley MJ (1995) Pancreatic islet cell cytoplasmic antibody in diabetes is represented by antibodies to islet cell antigen 512 and glutamic acid decarboxylase. Diabetes 44:1290–1295
Naparstek Y, Plotz PH (1993) The role of autoantibodies in autoimmune disease. Annu Rev Immunol 11:79–104
Neu N, Rose NR, Beisel KW, Herskowitz A, Gurri-Glass G, Craig SW (1987a) Cardiac myosin induces myocarditis in genetically predisposed mice. J Immunol 139:3630–3636
Neu NK, Beisel W, Traystman MD, Rose NR, Craig SW (1987b) Autoantibodies specific for cardiac myosin isoform are found in mice susceptible to coxsackievirus B3-induced myocarditis. J. Immunol. 138:2488–2492
Neu N, Craig SW, Beisel KW, Rose NR (1987c) Coxsackievirus induced autoimmune myocarditis in mice: Cardiac myosin autoantibodies do not cross react with virus. Clin Exp Immunol 69:566–574
Neu N, Ploier B, Öfner C (1990) Cardiac myosin-induced myocarditis. Heart autoantibodies are not involved in the induction of the disease. J Immunol 145:4094–4100.

Neumann DA, Burek CL, Baughman KL, Rose NR, Herskowitz A (1990) Circulating heart-reactive antibodies in patients with myocarditis or cardiomyopathy. J. Am. Coll. Cardiol. 16:839–846

Neumann DA, Lane JR, Lafond-Walker A, Allen GS, Wulff SM, Herskowitz A, Rose N R (1991) Heart-specific autoantibodies can be eluted from the hearts of coxsackievirus B3-infected mice. Clin Exp Immunol 86:405–413

Neumann DA, Lane JR, Allen GS, Herskowitz A, Rose NR (1993) Viral myocarditis leading to cardiomyopathy: Do cytokines contribute to pathogenesis. Clin Immunol Immunopath 68:181–190

Neumann DA, Rose NR, Ansari AA, Herskowitz A (1994) Induction of multiple heart autoantibodies in mice with coxsackievirus B3- and cardiac myosin-induced autoimmune myocarditis. J Immunol 152:43–350

O'Connell JB (1987) The role of myocarditis in end-stage dilated cardiomyopathy. Texas Heart Inst J 14:268–275

Olsen EG (1992) The pathogenesis of dilated cardiomyopathy. Postgrad Med J 68 [Suppl1]:S7-10

Pallansch MA (1988) Epidemiology of group B coxsackievirus. In: Bendinelli M, Friedman H (eds), Coxsackievirus-A General Update. Plenum Press, New York pp 399–417

Paque RP, Miller R (1988) Modulation of murine coxsackievirus-induced myocarditis utilizing anti-idiotypes. Viral Immunol 1:207–224

Paque RP, Miller R (1989a) Monoclonal anti-idiotypic antibodies regulate the expression of virus-induced murine myocarditis. Infect Immun 57:2864–2871.

Paque RP, Miller R (1989b) Polyclonal anti-idiotypes influence macrophage chemotaxis in coxsackievirus-induced murine myocarditis. J Leuko Biol 45:79–86

Pattison JR (1983) Tests for coxsackie B virus-specific IgM. (Editorial) J Hyg 90:327–332

Prabhakar BS (1988) Application of monoclonal antibodies to the study of coxsackieviruses. In: Bendinelli M, Friedman H (eds) Coxsackieviruses. A general update. Plenum, New York pp 117–134

Prabhakar BS, Haspel MV, McClintock PR, Notkins AL (1982) High frequency of antigenic variants among naturally occurring human coxsackie B4 virus isolates identified by monoclonal antibodies. Nature (Lond) 300:374–376

Rabin DU, Pleasic SM, Shapiro JA, Yoo-Warren H, Oles J, Hicks JM, Goldstein DE, Rae PMM (1994) Islet cell antigen 512 is a diabetes-specific autoantigen related to protein tyrosine phosphatases. J Immunol 152:3183–3188

Rager-Zisman B, Allison AC (1973a) The role of antibody and host cells in the resistance of mice against infection by coxsackie B3 virus. J Gen Virol 19:329–338

Rager-Zisman B, Allison AC (1973b) Effects of immunosuppression on coxsackie B3 virus infection in mice and passive protection by circulating antibody. J Gen Virol 19:339–351

Reyes MP, Lerner AM (1988) Myocarditis. Clinical and experimental correlates. In: Bendinelli M, Friedman H (eds), Coxsackieviruses. A general update. Plenum, New York pp 253–270

Roll V, Christie MR, Standl E, Ziegler A (1994) Associations of anti-GAD antibodies with islet cell antibodies and insulin autoantibodies in first-degree relatives of Type I diabetic patients. Diabetes 43:154–160

Rose NR, Herskowitz A, Neumann DA, Neu N (1988a) Autoimmune myocarditis: a paradigm of post-infection autoimmune disease. Immunol Today 9:117–120

Rose NR, Newmann DA, Herskowitz A (1988b) Genetics of susceptibility to viral myocarditis in mice. Pathol Immunopathol Res 7:266–278

Rose NR, Neumann DA, Herskowitz A (1992) Coxsackievirus myocarditis. Adv Internal Med 37:411–429

Rueckert RR (1996) Picornaviruses. In: Fields BN, Knipe DM, Howley PM (eds), Fundamental Virology, 3rd Ed. Lippincott-Raven, Philadelphia pp 477–522

Saegusa J, Prabhakar BS, Essani K, McClintock PR, Fukuda Y, Ferrans VJ, Notkins AL (1986) Monoclonal antibody to coxsackievirus B4 reacts with myocardium. J Inf Dis 153:372–373

Sato S, Tsutsumi R, Burke A, Carlson G, Porro V, Seko Y, Oksumura K, Kawana R, Virmani R (1994) Persistence of replicating coxsackievirus B3 in the athymic murine heart is associated with development of myocarditic lesions. J Gen Virol 75:2911–2924

Schmidt NJ, Dennis J, Lennette EJ, Ho HH, Shinomoto TT (1965) Antibody response of rhesus (Macaca mulatta) monkeys experimentally infected with coxsackieviruses of group B and group A, type 9. Antibody responses within the coxsackievirus group. J Immunol 95:54–69

Schnurr DP, Schmidt NJ (1988) Persistent infections. In: Bendinelli M, Friedman H (eds), Coxsackieviruses. A general update. Plenum Press, New York pp 181–201

Schultheiss H-P (1989) The significance of auto-antibodies against the ADP/ATP carrier for the pathogenesis of myocarditis and dilated cardiomyopathy: clinical and experimental data. Springer Sem Immunopathol 11:15–30

Schultheiss HP, Bolte HD (1985) Immunological analysis of autoantibodies against the adenine nucleotide translocator in dilated cardiomyopathy. J Mol Cell Cardiol 17:603–617

Schultheiss H-P, Schulz K., Kühl V, Ulrich G, Klingenberg M (1987) The ADP/ATP carrier as a mitochondrial autoantigen-facts and perspectives. Ann NY Acad Sci 488:44–68

Schultheiss H-P, Kühl U, Janda I, Melzner B, Ulrich G, Morad M (1988) Antibody-mediated enhancement of calcium permeability in cardiac myocytes. J Exp Med 168:2105–2119

Schulze K, Becker B, Schauer R, Schultheiss H-P (1989) Antibodies to the ADP/ATP carrier–an autoantigen in myocarditis and dilated cardiomyopathy, penetrate into myocardial cells and disturb energy metabolism *in vivo*. Cir Res 64:179–186

Schwimmbeck PL, Schultheiss H-P, Strauer BE (1989) Identification of a main auto-immunogenic epitope of the adenine nucleotide translocator which crossreacts with coxsackievirus B3: use in the diagnosis of myocarditis and dilated cardiomyopathy. Circulation 80(Suppl II):665–669

Schwimmbeck PL, Schwimmbeck NK, Schultheiss H-P, Strauer BE (1993) Mapping of antigenic determinants of the adenine-nucleotide translocator and coxsackie B3 virus with synthetic peptides: Use for the diagnosis of viral heart disease. Clin Immunol Immunopathol 68:135–140

See DM, Tilles JG (1991) Viral myocarditis. Rev Infect Dis 13:951–956

See DM, Tilles JG (1994) Efficacy of a polyvalent inactivated-virus vaccine in protecting mice from infection with clinical strains of group B. Scand J Infect Dis 26:739–747

Seko Y, Matsuda H, Kato K, Hashimoto Y, Yagita H, Okumura K, Yazaki Y (1993) Expression of intercellular adhesion molecule-1 in murine hearts with acute myocarditis caused by coxsackievirus B3. J Clin Invest 91:1327–1336

Sharaf A-e R, Narula J, Nicol PD, Southern JF, Khaw B-A (1994) Cardiac sarcoplasmic reticulum calcium ATPase, an autoimmune antigen in experimental cardiomyopathy. Circulation 89:1217–1222

Shikhman AR, Greenspan NS, Cunningham MW (1993) A subset of mouse monoclonal antibodies cross-reactive with cytoskeletal proteins and group A streptococcal M proteins recognizes N-acetyl-D-glucosamine. J Immunol 151:3902–3913

Shikhman AR, Cunningham MW (1994) Immunological mimicry between N-acetyl-D-glucosamine and cytokeratin peptides.: Evidence for a microbially driven anti-keratin antibody response. J Immunol 152:4375–4387

Szopa TM, Titchener PA, Portwood ND, Taylor KW (1993) Diabetes mellitus due to viruses: Some recent developments. Diabetologia 36:687–695

Takada H, Kishimoto C, Hiraoka Y (1995) Therapy with immunoglobulin suppresses myocarditis in a murine coxsackievirus B3 model. Antiviral and anti-inflammatory effects. Circ 92:1604–1611

Tian J, Lehmann PV, Kaufman DL (1994) T cell cross-reactivity between coxsackievirus and glutamate decarboxylase is associated with a murine diabetes susceptibility. J Exp Med 180:1979–1984

Todd JA (1990) Genetic control of autoimmunity in type 1 diabetes. Immun Today 11:122–128

Toniolo A, Federico G, Basolo F, Onodera T (1988) Diabetes mellitus. In: Bendinelli M, Friedman H (eds), Coxsackieviruses. A general update. Plenum Press, New York pp 351–382

Torfason EG, Reimer CB, Keyserling HL (1987) Subclass restriction of human enterovirus antibodies J Clin Micro 25:1376–1379

Torfason EG, Galindo R, Keyserling HL (1988) Comparison of five ELISA assays for IgG antibody against coxsackievirus B1. J Med Virol 25:53–60

Tracy S, Chapman NM, McManus BM, Pallansch MA, Beck, MA, Carstens J (1990a) A molecular and serologic evaluation of enteroviral involvement in human myocarditis. J. Mol Cell Cardiol 22:403–414

Tracy S, Wiegand V, McManus B, Gauntt C, Pallansch M, Beck M, Chapman N (1990b) Molecular approaches to enteroviral diagnosis in idiopathic cardiomyopathy and myocarditis. J Am Coll Cardiol 15:1688–1694

Tracy S, Chapman NM, Beck MA (1991) Molecular biology and pathogenesis of coxsackie B viruses. Rev Med Virol 1:145–154

Tracy S, Chapman NM, Tu Z (1992) Coxsackievirus B3 from an infectious cDNA copy of the genome is cardiovirulent in mice. Arch Virol 122:398–409

Traystman MD, Beisel KW (1991) Genetic control of coxsackievirus B3-induced heart-specific autoantibodies associated with chronic myocarditis. Clin Exp Immunol 86:291–299

Traystman MD, Chow LH, McManus BM, Herskowitz A, Nesbitt MN, Beisel KW (1991) Susceptibility to coxsackievirus B3-induced chronic myocarditis maps near the murine Tcr α and Myhc α loci on chromosome 14. Am J Pathol 138:721–732

Trousdale MD, Paque RE, Nealon T, Gauntt CJ (1979) Assessment of coxsackievirus B3 *ts* mutants for induction of myocarditis in a murine model. Infect Immun 23:486–495

Tu Z, Chapman NM, Hufnagel G, Tracy S, Romero JR, Barry WH, Zhao L, Currey K, Shapiro B (1995) The cardiovirulent phenotype of coxsackievirus B3 is determined at a single site in the genomic 5' non-translated region. J Virol 69:4607–4618

Ulrich G, Kühl U, Melzner B, Janda I, Schäfer B, Schultheiss H-P (1988) Antibodies against the adenosine di-/triphosphate carrier cross-react with the Ca channel-functional and biochemical data. In: Schultheiss H-P (ed). New Concepts in Viral Heard Disease. Springer-Verlag, Berlin pp 226–235

van Houten N, Huber S (1989) Role of cytotoxic T cells in experimental myocarditis. Springer Sem Immunopathol II:61–68

Verge CF, Howard NJ, Rowley MJ, Mackay IR, Zimmet PZ, Egan M, Hulinska H, Hulinsky I, Silvestrini RA, Kamath S, Sharp A, Arundel T, Silink M (1994) Anti-GAD and other autoantibodies at the onset of childhood IDDM: a large population based study. Diabetologia 37:1113–1120

Wallukat G, Morwinski M, Kowal K, Förster A, Boewer V, Wollenberger A (1991) Autoantibodies against the β-adrenergic receptor in human myocarditis and dilated cardiomyopathy: β-adrenergic agnoism without desensitization. Eur Heart J 12(Suppl D):178–181

Weller AH, Simpson K, Herzum M, van Houten N, Huber SA (1989). Coxsackievirus-B3-induced myocarditis: virus receptor antibodies modulate myocarditis. J Immunol 143:1843–1850

Welliver RC, Drucker MM, Olgra PL (1982) Immunology of enteroviruses. In: Nahmias AJ, O'Reilly RJ (eds), Comprehensive Immunology – Immunology of Human Infection. Part II. Viruses and Parasites; Immunodiagnosis and Prevention of Infectious Disease. Plenum Medical, New York pp 185–203

Wolff PG, Kühl U, Schultheiss H-P (1989) Laminin distribution and autoantibodies to laminin in dilated cardiomyopathy and myocarditis. Am Heart J 117:303–311

Wolfgram LJ, Beisel KW, Rose NR (1985) Heart-specific autoantibodies following murine coxsackievirus B3 myocarditis. J Exp Med 161:1112–1121

Wolfgram LJ, Beisel KW, Herskowitz A, Rose NR (1986) Variations in the susceptibility to coxsackievirus B3-induced myocarditis among different strains of mice. J Immunol 136:1846–1852

Wolfgram LJ, Rose NR (1989) Coxsackievirus infection as a trigger of cardiac autoimmunity. Immunol Res 8:61–80

Wong CY, Woodruff JJ, Woodruff JF (1977) Generation of cytotoxic T lymphocytes during coxsackievirus B3 infection. III. Role of sex. J Immunol 119:591–597

Woodruff JF (1980) Viral myocarditis. A review. Am J Pathol 101:425–483

Yoon JW (1990) The role of viruses and environmental factors in the induction of diabetes. Curr Top Microbiol Immunol 164:95–123

Roles of T Cells in Coxsackievirus B-Induced Disease

P. L. SCHWIMMBECK[1], S. A. HUBER[2] AND H.-P. SCHULTHEISS[1]

1 Introduction

Enteroviral infection, for example by coxsackieviruses, is thought to be the initiating event of human myocarditis. This is supported by epidemiological studies showing an increased incidence of viral myocarditis following endemics of coxsackie B virus (CVB) infections (HELIN et al. 1968). However, the assumed pathogenetic role of enteroviral infection in the pathogenesis of viral myocarditis was established only in the last decade, by the demonstration of enteroviral RNA in the myocardium of patients with myocarditis and dilated cardiomyopathy using molecular biological methods (BOWLES et al. 1986). The percentage of patients with myocarditis having enteroviral sequences in the myocardium differs widely between different studies, possibly due to different probes used for hybridization, differences in

[1]Department of Internal Medicine/Cardiology, Benjamin Franklin Hospital, Free University, Hindenburgdamm 30, 12200 Berlin, Germany

[2]Department of Pathology, University of Vermont, College of Medicine, Burlington, VT, USA

the primers selected for amplification or unspecific contamination with viral RNA especially in the polymerase chain reaction, thus giving false positive results. In general and in studies testing more than 100 patients for the presence of enteroviral RNA in the myocardium, about 30%–40% of the patients were positive.

The viral etiology of myocarditis can also be simulated in animal models. Using CBVs, it was possible to induce inflammatory heart disease very much resembling human myocarditis in several strains of susceptible mice (HUBER and LODGE 1984). In these experiments, an acute viremic state of the disease could be differentiated from a more chronic stage that was characterized by clearance of the virus and an autoimmune response against uninfected host tissue by sensitized T cells. These lymphocytes reacted both with coxsackievirus B3 (CVB3) and also with host proteins like myosin. However, the location of the antigen within the viral protein has not been identified so far.

In human myocarditis (MC) and dilated cardiomyopathy (DCM), an autoimmune response seems to play a role besides in the initial viral infection. This is suggested by the demonstration of autoantibodies against different autoantigens such as cardiac myosin (DE SCHEERDER et al. 1991), the β-adrenergic receptor (LIMAS and LIMAS 1991) and the adenine nucleotide translocator (ANT) (SCHULTHEISS et al. 1983) in sera of patients with MC and DCM. Especially autoantibodies against the ANT have been shown to cause an impairment of the energy metabolism, both in vitro and in vivo in the animal model (SCHULZE et al. 1989, 1990). It was possible in these experiments to induce autoantibodies in guinea pigs after immunization with the isolated ANT, leading to a shortage of energy-rich nucleotides in the cytosol of myocytes and a decrease of left ventricular function in a working heart model.

A characteristic finding in human MC are infiltrating lymphocytes in the myocardium. These can be found both in the acute and in the chronic stages of disease that is clinically diagnosed as DCM. Infiltrating lymphocytes can also be demonstrated in the myocardium of many patients in whom the routine histological examination was not able to detect a relevant cellular infiltration by using immunhistological methods with monoclonal antibodies against lymphocyte surface markers (KÜHL et al. 1992, 1996). In addition, using monoclonal antibodies against HLA class I and II antigens and against adhesion molecules, an up-regulation of these proteins as an indicator of a chronic immune response could be demonstrated in such hearts. This immune reaction is independent of the presence or persistence of the initial enteroviral infection and may be triggered by a cross-reactive autoimmune response (FUJINAMI et al. 1983).

However, many questions regarding the function and significance of sensitized T cells in MC and DCM remain unanswered. Among these are the role of T lymphocytes in the animal model of CVB-induced MC, the identification of epitopes recognized by lymphocytes in MC, the specificity and phenotype of infiltrating cells in the myocardium of patients with MC and DCM, and the pathogenetic significance of the T cell response in human MC and DCM.

2 T Cells in Murine Myocarditis

2.1 Beneficial or Pathogenic?

The one fact made abundantly clear by experimental CVB-induced MC models is that myocarditis is a conglomeration of distinct diseases sharing the common characteristic of inflammatory cells infiltrating the heart but having dramatically different pathogenic mechanisms of myocyte injury. A number of studies report that the inflammation in experimental viral MC is beneficial in clearing the virus and induces minimal direct or autoimmune damage to the myocardium (KANDOLF et al. 1993; CHOW et al. 1992; SATO et al. 1994). In these studies, infection of mice deficient in T lymphocytes or in both T and B cells (SCID mice) results in dramatically increased cardiac necrosis and virus titers in the heart (up to 10^4-fold increase in virus compared to infected immunocompetent mice). Additionally, inflammatory cells cluster around foci of infected cardiocytes and are not observed independently of infected heart cells, suggesting that lymphoid cells enter the heart only to control the infection (KLINGEL and KANDOLF 1993). While such studies demonstrate the potential for direct viral injury to the myocardium independent of the immune system, the relevance of these results to the understanding of the pathogenesis of MC under normal conditions is uncertain. The extraordinary increase of the total viral load and the prolonged persistence of replicating virus in immunocompromised animals might make direct virus-induced injury much more pronounced, while such damage might be a more modest factor in the cardiac injury of infected, immunocompetent mice. Thus, the act of immunosuppressing the animal might have changed the dominant mechanism for the induction of tissue injury.

Other studies report that cardiac injury during viral MC results from mixed virus and immune-mediated mechanisms or immune mechanisms only (WOODRUFF and WOODRUFF 1974; HUBER and LODGE 1984; KISHIMOTO and ABELMAN 1990; ROSE et al. 1992; GAUNTT et al. 1993; KISHIMOTO et al. 1994; COOK et al. 1995; HENKE et al. 1995). WOODRUFF and WOODRUFF (1974) first demonstrated that CVB infection of mice selectively depleted of T lymphocytes induced substantially less cardiac inflammation and myocyte necrosis than infection of T cell-sufficient animals. Unlike those studies discussed above, immunocompromised mice infected with CVB3 in these experiments show no significant increase in cardiac virus replication. Using the same virus isolate, other groups have confirmed dominant T cell-dependent mechanisms of myocyte lysis (HUBER and LODGE 1984; HENKE et al. 1995).

2.2 T Lymphocytes in Immunopathogenic Myocarditis

T lymphocytes can be divided into various subgroups based on expression of CD4 and CD8 cell surface molecules (KUPFER and SINGER 1989), the type of T cell receptor (TcR) (ALLISON and HAVRAN 1991; DOHERTY et al. 1992), or the types of cytokines they produce (FITCH et al. 1993). Although originally CD4+ cells were designated as "helper" cells and CD8+ cells were designated as "cytolytic/suppressor" cells; it is now known that all helper, cytolytic and suppressor activities can be mediated by either lymphocyte population (KUPFER and SINGER 1989; LASARTE et al. 1995). The major difference between CD4+ and CD8+ cells is that primarily CD4+ cells recognize antigens presented on class II major histocompatibility complex (MHC) antigens while CD8+ cells recognize peptides presented by class I MHC antigens (KUPFER and SINGER 1989). Even this dichotomy is not absolute, as class II restricted CD8+ and class I restricted CD4+ T cells have been reported (MARUSIC-GALESIC and WALDEN 1995). Both CD4+ and CD8+ can provide T helper (Th) cell activity. Th cells can be subdivided according to the types of cytokines they produce. Naive or early activated (Th0) cells ,which produce various cytokines including interferon-γ (IFN-g>, interleukin-2 (IL-2) and IL-4, differentiate during antigenic stimulation into either Th1 or Th2 cells. Th1 cells, which primarily produce IL-2 and IFN-γ, are associated with delayed hypersensitivity reactions (cellular immunity) while Th2 cells, which primarily produce IL-4, IL-5, IL-6 and IL-10 are associated with providing helper activity for most T cell-dependent B cell responses (humoral immunity). Various factors present during antigenic stimulation will influence differentiation of Th0 cells into either Th1 or Th2 pathways. The presence of IFN-γ, IL-12 or activated CD8+ T cells promote Th1 cell differentiation (FITCH et al. 1993). In various diseases, e.g., leishmaniasis (SCHARTON and SCOTT 1993); allergy (MCMENAMIN et al. 1994); or tuberculosis (ZHANG et al. 1995), the type of Th cell response that occurs will determine disease susceptibility or resistance. Finally, T cells can be categorized according to their TcR. TcR are comprised of either α and β (αβ+) or γ and δ (γδ+) polypeptide chains. Generally, αβ+ T cells constitute the classical antigen-specific T cell populations reacting to specific epitopes presented in the grooves of MHC antigens, γδ+ T cells show broader antigen reactivities and may not be MHC restricted. A high percentage of γδ+ T cells in inflammatory sites react to heat-shock or stress proteins (HSPs) produced by injured/stressed eukaryotic cells and microbial organisms (DOHERTY et al. 1992; HAAS et al. 1993).

One of the major observations using the Woodruff CVB3 strain is that distinct T cell-dependent immunopathogenic mechanisms cause myocyte damage in genetically different strains of mice (HUBER and LODGE 1984). Thus, BALB/c and DBA/2 mice, which are MHC identical (H-2^{d}), both develop severe myocarditis but cardiac injury depends upon CD8+ T cells in the former and CD4+ T cells in the latter strain. Furthermore, DBA/2 mice show substantial deposits of heart reactive IgG autoantibodies in the myocardium and complement depletion of infected mice prevents MC. No immunoglobulin deposits were found in BALB/c myocardium and complement depletion had no effect on heart disease in this strain. Dr. Betty Diamond and her colleagues, using a myosin-induced myocarditis model, have confirmed that BALB/c and DBA/2 mice develop heart disease through cellular and humoral immunopathogenic mechanisms, respectively (LIAO et al. 1995).

The reason why these two mouse strains develop different immunopathogenic diseases is primarily that CVB infection of BALB/c mice preferentially activates CD4+ Th1 cell differentiation while the same virus activates CD4+ Th2 cell differentiation in DBA/2 mice. Th1 cells provide helper activity for CD8+ cytolytic T lymphocyte (CTL) stimulation, while Th2 cells provide the cytokines needed for B cell proliferation and differentiation.

CVB infection of BALB/c myocytes up-regulates 65 kDa HSP expression and stimulates γδ+ T cell activation (HUBER 1992). Infiltration of these γδ+ T cells into the myocardium correlates with a decrease in CD4+ T cells and a shift from Th2 to Th1 phenotype suggesting that γδ+ cells modulate CD4+ Th subset responses (Huber at al. 1996a). This observation corresponds well with published reports that γδ+ T cells selectively suppress Th2 cells (MCMENAMIN et al. 1994). CVB infection of DBA/2 myocytes does not up-regulate HSP expression which most likely explains why γδ+ T cells are absent from their inflammatory infiltrate and why Th2 cell differentiation preferentially occurs. Why CVB infections of BALB/c and DBA/2 myocytes diverge in HSP up-regulation is not known.

γδ+ cells likely modulate Th subset responses in one of two ways. First, γδ+ cells often produce substantial quantities of IFN-γ and IL-2, cytokines which favor Th1 cell differentiation (YAMAMOTO et al. 1993). Secondly, γδ+ cells express high levels of FasL mRNA and should be potent mediators of Fas-dependent apoptosis (SUDA et al. 1995). Although apoptosis is rarely cited as a regulatory mechanism for Th cell responses, progressive differentiation of human T cells toward high IL-4 and low IL-2 production correlates with an increased susceptibility of these T cells to Fas-dependent apoptosis (SALMON et al. 1994). Furthermore, studies have shown that infection of BALB/c mice with a specific CVB strain (designated H310A1), which infects the heart but induces minimal myocarditis (amyocarditic CVB), preferentially activates Th2 cells in vivo (HUBER et al. 1996a). These Th2 cells act as "suppressor" T cells which inhibit the induction of Th1 cell responses needed to activate CD8+ CTLs, the pathogenic effector cell in BALB/c mice (LOUDON et al. 1991). Adoptive transfer of activated γδ+ T cells from the hearts of myocarditic BALB/c mice causes apoptosis of the CD4+ cell population, a shift from the Th2 to Th1 cytokine phenotype and induction of severe myocarditis (HUBER et al. 1996a).

Both the genetics of the host and the CVB strain can determine MC susceptibility. As stated above, specific amyocarditic CVB strains have been isolated which induce minimal cardiac disease despite replicating in the myocardium. Similarly, certain inbred mouse strains infected with myocarditic CVB strains fail to develop MC (WOLFGRAM et al. 1986). Host genetic susceptibility is complex and may differ with the specific CVB variant used. For example, studies done with the Woodruff CVB strain demonstrate a strong linkage between myocarditis susceptibility and the MHC haplotype (HUBER et al. 1994a). In contrast, in studies done by WOLFGRAM et al. (1986), TRAYSTMAN and BEISEL (1991) and TRAYSTMAN et al. (1991) using a different CVB strain, the MHC is much less important in controlling disease susceptibility. Rather, genetic control maps to chromosome 14 of the mouse near the TcRα and myosin heavy chain α loci. In most cases, "myocarditis resistance" may be a relative rather than absolute condition in that immunomodulation of "resistant" mice can completely restore disease susceptibility. HENKE et al. (1995) showed that normally resistant C57B1 mice become highly susceptible to MC when genetically altered to lack CD4+ cells (CD4ko mice). As in BALB/c mice, MC in C57B1 CD4ko animals is mediated by CD8+ T cells. Thus, suppressor CD4+ T cells, which are likely to be Th2 cells, must inhibit CD8+ CTL responses in the intact animal. MC resistance can also be reversed by administering specific cytokines during CBV infection. Tumor necrosis factor-α (TNFα)/IL-1 (LANE et al. 1992; HUBER et al. 1994b), IL-2 (HUBER et al. 1994b; KISHIMOTO et al. 1994), IFNα/β (LUTTON and GAUNTT 1985), and IFNγ (HUBER et al. 1994b) have all been reported to augment MC susceptibility depending upon dose or time of administration relative to infection. In each case, these cytokines could affect the disease by altering Th1/Th2 balances in the animal.

Identifying the type of immunopathogenic mechanism in MC is important for two reasons. First, specific immunopathogenic mechanisms might further the development of DCM. CVB infected BALB/c mice show a substantial reduction in the velocity of sarcomere shortening using isolated papillary muscles (HUBER et al. 1996b). T lymphocytes are necessary for this effect since infected T cell-deficient mice show normal contractility. The decreased shortening velocity is accompanied by a shift in myosin isoform expression from the fast to the slow type, and a substantial increase in left ventricular atrial natriuretic factor (ANF) mRNA expression. Although DBA/2 mice develop equivalent amounts of cardiac inflammation, none of these signs of physiological stress are present. Stating that humoral immunopathogenic reactions fail to induce cardiomyopathy-like disease might be premature. Clinically, a variety of autoantibodies to heart antigens, including the calcium channel (KÜHL et al. 1991), adenine nucleotide translocator (ANT) (SCHULTHEISS et al. 1983, SCHULTHEISS 1989), myosin (DE SCHEERDER et al. 1991), and creatine kinase (PANKUWEIT et al. 1995), have been identified in various MC patients. Some of these autoantibodies have been shown to alter the physiological function of cultured myocytes even if they do not lyse the cardiocytes. Thus, certain types of humoral autoimmunity might predispose toward DCM while others do not. Since the autoantigen targeted in DBA/2 mice appears to be myosin, this particular humoral response may not cause a cardiomypathy-like condition.

The type of immunopathogenic response in MC correlated with the effectiveness of immunosuppression. Cyclosporin A, FK506 and prednisone/azathioprine can have both diffferential time- and dose-dependent effects on Th1 and Th2 cell responses (WANG et al. 1993; QASIM et al. 1995). Thus, immunosuppressive drugs given before or soon after antigen exposure, as would occur with organ transplantation, may be highly effective in suppressing all T cell responses while the same treatments given long after antigen stimulation may or may not be effective depending upon several factors including the predominant immunopathogenic (Th cell) mechanism involved. Similarly, immunomodulating therapies such as giving specific cytokines or thymic hormones (MIRIC et al. 1994) might alter the balance between T cell subtypes resulting in immunosuppression of a pathogenic immune response while preserving beneficial (anti-viral) immunity. In the experimental CVB-induced MC model, treatment with cyclosporin A alone usually is ineffective in preventing cardiac damage (O'CONNELL et al. 1986; HERZUM et al. 1991). Yet when BALB/c mice were simultaneously treated with both cyclosporin A and anti-T cell antibody, MC was prevented indicating that the pathogenic T cell in this mouse strain was largely resistant to immunosuppression. In contrast, cyclosporin A or prednisone treatment of DBA/2 mice inhibited CVB-induced MC, indicating that the pathogenic T cell response was suceptible to immunosuppression.

2.3 Autoimmunity in Experimental Coxsackievirus B -Induced Myocarditis

Antigenic mimicry is usually cited as a mechanism by which infections induce autoimmunity (FUJINAMI et al. 1983; BARNETT and FUJINAMI 1992; CUNNINGHAM 1993). The concept is that the infectious organism expresses cross-reactive T or B cell epitopes to self molecules. Although these self epitopes are not able to induce an autoimmune response themselves due to tolerance, when present during active immunization to microbial antigens, tolerance may be broken. Thus, the immunity generated to the viral or bacterial antigens may potentiate antigen presentation and T/B cell proliferation responses to the mimicking epitopes as well. Antigenic mimicry is well-documented at the level of the humoral immune response, and adoptive transfer of some cross-reactive antibodies between CVB and heart antigens can cause MC in recipient animals (GAUNTT et al. 1995; Chap. 12). Indeed, in some cases, the cross-reactive epitopes may be more broadly distributed. Monoclonal antibodies produced by immunizing against the M5 protein of group A streptococcus were found to bind to heart proteins and neutralize CVB (CUNNINGHAM et al. 1992). Thus, bacteria that induce autoimmune (rheumatic) heart disease and a virus that induces autoimmune (MC) heart disease share common cross-reactive epitopes to the heart. Furthermore, one anti-streptococcal antibody was able to differentiate between myocarditic and amyocarditic CVB strains, binding to the former but not the latter. This last observation implies that the antibody must recognize an antigen determining the myocarditic potential of the virus.

Antigenic mimicry at the T lymphocyte level has only recently been shown. The evidence for cardiac myosin as an important autoantigen in viral myocarditis was initially demonstrated by Rose and colleagues using genetically susceptible mouse strains (NEU et al. 1987a, b). These investigators detected anti-myosin antibodies in CVB-infected mice and correlated the presence of the immunological mediators to disease susceptibility. Subsequently, severe MC which histologically resembled the virus-induced disease was induced by immunizing with cardiac but not skeletal myosin. Recently, specific peptide epitopes of the myosin heavy chain have been found which can induce MC (PUMMERER et al. 1996). Several T cell hybridomas have been derived from the hearts of CVB infected mice which respond by IL-2 production to CVB, cardiac myosin heavy chain α and a streptococcal M5 peptide (NT4) (S. Huber, unpublished results). NT4 contains both T and B cell epitopes cross-reactive to cardiac myosin (M. Cunningham, personal communication). Tolerization of mice to NT4 prior to infection with CVB successfully inhibited MC (HUBER and CUNNINGHAM 1996). Thus, a limited number of antigen-specific T cell clones must be immunopathogenic in virus-infected mice since tolerization to a single peptide prevented most cardiac injury, and the pathogenic T cells probably react to cardiac myosin due to the cross-reactivity noted between the virus, myosin and NT4 peptide. Demonstration of a limited antigen-specific clonality to the pathogenic T cell response correlates well with studies by SEKO et al. (1994), showing restricted TcR Vβ gene usage in experimental CBV-induced MC.

3 Identification of T and B Cell Epitopes of Coxsackievirus B

3.1 T Cell Epitopes of Coxsackievirus B

Previous experiments using western blot methodology had identified the viral coat protein VP1 of CVB3 as harboring main B cell epitopes of the virus (MERTENS et al. 1983). Therefore overlapping peptides of 15 amino acids each, representing the entire sequence of VP1 of CVB3, were synthesized. Altogether 31 peptides were prepared using the simultaneous multiple peptides synthesis method.

The infection of mice with CVB3 strain Nancy is a well-studied animal model of MC. After the infection of susceptible mice such as BALB/c, an inflammation in the hearts of the animals appears that is characterized by myocyte necrosis and massive cellular infiltrates. Previous experiments have shown that these infiltrating lymphocytes, at least in the initial state of the disease, are directed against the virus used for the infection (WOLFGRAM et al. 1986). Therefore, we used the sera of such infected BALB/c-mice to identify epitopes recognized by the B lymphocytes in these mice. When the peptides derived from VP1 of CVB3 were tested in an ELISA system, the amino acids 1–15, 21–35, 229–243 and 239–253 were identified as antigenic determinants, recognized by the anti-CVB3 antibodies (HAARMANN et al. 1994) (see Fig. 1). In additional studies with type-specific antisera raised against the serotypes 1–6 of CVBs, it was possible to differentiate between serotype-specific (amino acids 1–15), semi-specific (amino acids 21–35) and non-serotype-specific epitopes (amino acids 229–243) (HAARMANN et al. 1994). This is of major interest for the development of a serological test system for the CVBs. Using the different peptides, it seems to be possible to design both non-serotype-specific antibody screening tests and serotype-specific assays for the further identification of the infecting virus.

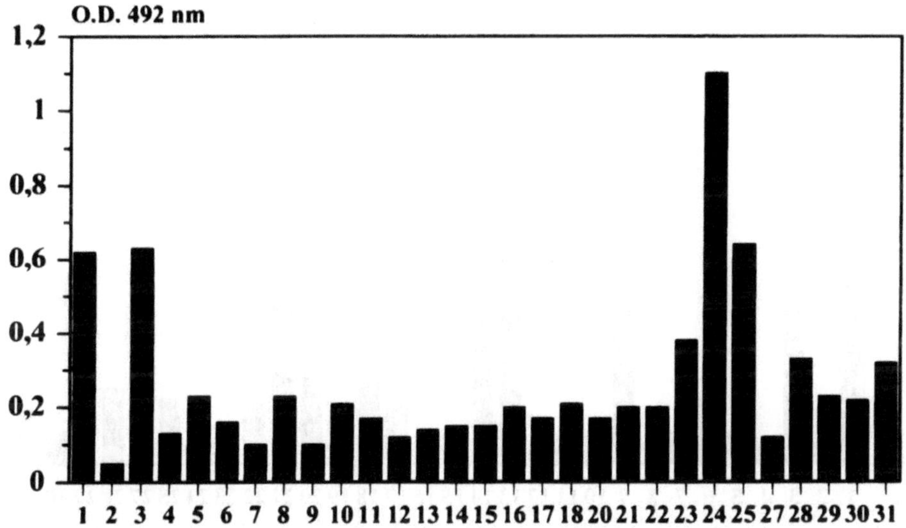

Fig. 1. Binding of sera from CVB-infected mice to synthetic peptides derived from VP1 of CB3. The x-axis gives the number of the peptide whereas the y-axis shows the optical density in an ELISA system. An O.D. of more than 0.5 was considered to be significant

The antigenic epitopes represented by the peptides are accessible on the native virus, since antibodies raised against the peptides can be used for the immunohistological demonstration of the virus in the myocardium of infected mice. In addition, the anti-peptide antibodies bind the isolated virus in a western blot system.

When the overlapping synthetic peptides derived from VP1 were subsequently tested against sera from patients with clinically suspected and immunohistologically proven MC, about 50% of sera bound to amino acids 21–35 when tested for IgG antibodies. In additional experiments, IgM antibodies could be demonstrated in 11% of patients after the removal of IgG antibodies. In contrast, IgM antibodies against the peptides were absent in the sera of healthy controls or in the sera of infants having had no previous contact with CVBs.

3.2 Identification of T Cell Epitopes

For the identification of T cell epitopes on VP1 of CVB3, we used the overlapping peptides and tested them in a proliferation assay with [3H]thymidine incorporation for reactivity with lymphoid cells from CVB3 infected mice. At least six different peptides induced significant proliferation of the lymphocytes, identifying them as T cell epitopes of CVB3 virus (Fig. 2). These contained the amino acids 1–15, 21–35, 79–93, 119–133, 129–143 and 199–213. The proliferating cells were shown to be T lymphocytes, since treatment of the animals with monoclonal antibodies against pan-T lymphocytes resulted in complete loss of reactivity.

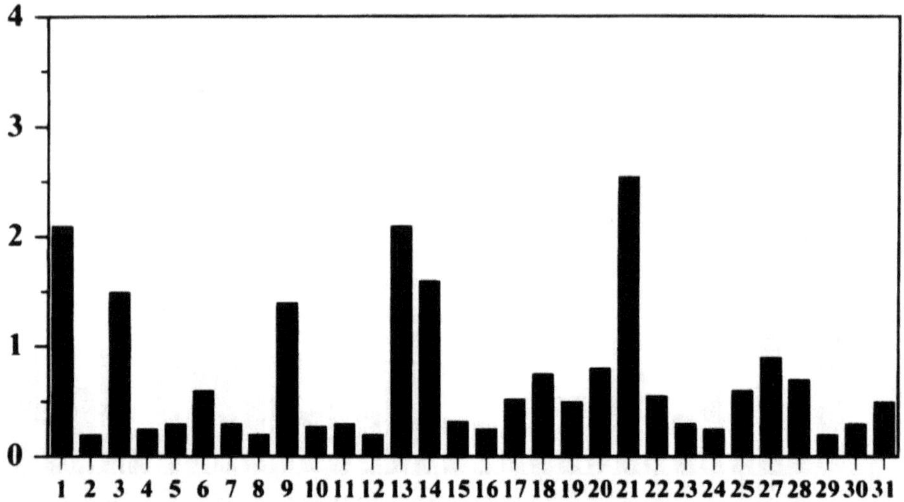

Fig. 2. Proliferation of lymph node cells from CVB3-infected mice after stimulation with synthetic peptides derived from the viral coat protein VP1 of CVB3. The x-axis gives the number of the peptide. The y-axis gives the incorporated [3H]thymidine (×1000 cpm) in a proliferation assay

In subsequent studies, the pathogenetic significance of the T cell epitopes was investigated (Huber et al. 1993). Mice were immunized with the peptides and subsequently infected with CVB 2 weeks later. After pre-immunization with the peptide representing amino acids 1–15 of CVB3, a significant reduction of virus titer in the heart and decreased inflammation was observed. In contrast, after immunization with amino acids 21–35, the viral titer in the myocardium after infection was unchanged when compared to animals without pre-immunization. However, both the degree of inflammation of the heart and mortality of the animals (normally rather low in this model of infection) were significantly increased. Thus, these experiments showed that an immune response against different T cell epitopes of CVB3 virus can either have a protective or a detrimental effect in the pathogenesis of MC. Therefore, experiments testing possible vaccines based on peptides or viral proteins or parts thereof should be evaluated carefully in order to avoid the induction of an immune response causing additional damage in the pathogenesis of MC.

In order to identify the antigens recognized by the T cells in patients with MC and especially in order to see if they are identical to the epitopes in murine MC, we tested the peripheral blood leukocytes (PBLs) of patients with immunhistologically proven MC for reactivity with the overlapping peptides representing the viral coat protein VP1 of CVB3 in a proliferation assay with thymidine incorporation. When we tested more than 30 patients, we found different reactivity patterns. About half of the patients showed reactivity with the peptides representing amino acids 41–55, 139–153, 239–253 and 269–283. The epitopes recognized by the T cells from patients with MC also differed between the individual patient and were different from the epitopes recognized by the T cells in murine MC after the infection of BALB/c mice with CVB3. This is most likely due to the differences in the antigen presenting cells (MHC) between mice and humans, in which only restricted peptides are presented to the immune system and can thus be recognized by the effector cells. This may also be one of the reasons for different reactivity patterns between the various patients due to different HLA types. However, our population of patients studied so far is not yet large enough to allocate a certain reactivity pattern to an individual MHC type. It is conceivable that the reported HLA class II correlations in DCM are due to the restricted presentation of pathogenetically important antigens (BJORKMANN et al. 1987). Another reason for the differences in the reactivity pattern to CVB3 viruses between mice and humans is certainly that, in humans, it is mostly impossible to determine what serotype of virus caused the infection, since there is a high rate of immunity against entroviruses in the adult population and there is a considerable cross-reactivity between the single types of coxsackieviruses (MERTENS et al. 1983).

4 Culture of Antigen-Specific T Cells from Endomyocardial Biopsies

A characteristic finding of human MC is the presence of infiltrating leukocytes in the myocardium of patients. However, so far neither the phenotype nor the specificity of the infiltrating cells is known. We therefore examined whether it is possible to culture the infiltrating mononuclear cells present in endomyocardial, right ventricular biopsies of patients with clinically suspected MC and DCM. Using co-cultivation with irradiated autologous peripheral blood leukocytes in the presence of recombinant IL-2 and restimulation with myocardial tissue, we were able to establish primary lymphocyte cultures in about 50% of the biopsies from MC and DCM patients (see Table 1) within 30 days. There was no major difference in the percentage of positive biopsies between both diseases. As controls, we cultured the biopsies of patients with arterial hypertension or hypertrophic cardiomyopathy. In none of these cases were we able to obtain positive cultures.

Table 1. Lymphocytes cultured from myocardial biopsies of patients with myocarditis and dilated cardiomyopathy

Patients	Positive culture		Negative culture	
	n	%	n	%
Myocarditis (n =71)	38	54	33	46
DCM (n=47)	25	53	22	47
Controls (n=8)	0		8	100

Cultures were considered to be positive if more than 1×10^6 cells could be grown within 30 days. Biopsies from patients with arterial hypertension or hypertrophic cardiomyopathy were used as controls.

For phenotyping of the cells, we used FACS analysis with double labeling of the cells. The majority of the cells were CD3-positive, identifying them as T lymphocytes. Most of these cells were either CD4 (helper/inducer) or CD8 (suppressor/cytotoxic) positive (see Table 2). The CD8-positive cultures had a tendency to be present more frequently in patients with healing MC; however, this did not reach statistical significance. Only in a very low percentage of patients were macrophages or natural killer cells seen in the cultures. Despite the fact that our culture conditions may have influenced the phenotype and number of growing cells, our experiments demonstrated that there are infiltrating, activated leukocytes in the myocardium of patients with MC as well as patients with DCM.

In order to identify the specificity of the cultured cells, we tested them for proliferation after stimulation with CVB3 and with peptides derived from the VP1 protein. Using a proliferation assay with thymidine incorporation, we found proliferation with a stimulation index of >3 in nine cultures. Similar to the results obtained with PBLs, we found proliferation against different epitopes of CVB3. However, most of the cultures reacted with amino acids 51–56 and 109–153 of VP1. Of course, there is the question of why did not more or even all of the cultures react with CB3 virus. One explanation is certainly that different entroviruses are thought to be involved in the pathogenesis of MC and DCM and that those may not share T cell epitopes with CVB3. In addition, microbes other than enteroviruses may also cause MC, since in only 20%–50% of biopsies from patients with MC and an even lower percentage of biopsies from patients with DCM enteroviral sequences can be found. In addition, the cells cultured from the endomyocardial biopsies may have an impaired capacity of proliferation due to the long culture times, thus resulting in low levels of proliferation or even none. Another problem in these proliferation assays is the quality of the antigen-presenting cells. A high proliferation rate could perhaps be achieved by using irradiated EBV-transformed immortal autologous leukocytes as antigen-presenting cells, since the quality of antigen-presenting cells is a crucial factor for proliferation. However, in our experiments we used only irradiated PBLs as antigen-presenting cells, as we desired to avoid a possible unspecific proliferation due to possible active EBV.

Table 2. Immune phenotype of cells cultured from myocardial biopsies

Marker	Percentage	Range (%)
CD3-positive (pan T) cells	95	84–99
CD4-posititve (helper/inducer) cells	42	18–89
CD8-positive (suppressor/cytotoxic) cells	51	13%–95
MA100-positive cells (macrophages)	0	0
CD57-positive (NK) cells	3%	(0%-10%)

The average number of positive cells for the different markers and the range of positive cells are shown. Cells were evaluated by FACS analysis after staining with monoclonal antibodies.

5 Significance of T Cell Responses: Transfer of Human Myocarditis into SCID Mice

Despite the demonstration of autoimmune phenomena in patients with MC and DCM, there is no proof of the autoimmune pathogenesis of these diseases. This is due to the lack of a transfer model for the human diseases. In the animal experiment, it is possible to transfer MC from CVB3-infected mice after clearance of the virus to uninfected mice by lymphocyte transfer (HUBER et al. 1984). Similarly, autoimmune MC can be transfered from animals immunized with cardiac myosin to untreated animals by sensitized T lymphocytes (SMITH et al. 1991).

SCID mice lack functional T and B lymphocytes in the peripheral blood (BOSMA et al. 1983). Therefore, these animals can be used in transfer experiments to receive allogeneic or even xenogeneic transplants (PHILLIPS et al. 1991). In previous reports, it was possible to transfer both normal human leukocytes and PBLs from patients with, for example, lupus erythmatosis (DUCHUSAL et al. 1990), primary biliary cirrhosis (KRAMS et al. 1989), rheumatoid arthritis (TIGHE et al. 1990), and Grave´s disease into SCID mice (MACHT et al. 1991). In order to test the pathogenetic potential of the PBLs present in patients with chronic MC, we investigated the possibility of transfering the PBLs into SCID mice, to examine the animals for the presence of autoimmune phenomena normally seen in the human disease and to measure the left ventricular function in the animals after transfer. As donors, we isolated the PBLs from 13 patients with immunhistologically proven MC that persisted from more than 6 months. Viral persistance in the myocardial biopsies of these patients was excluded by in situ hybridization and PCR using enterovirus-specific primers. Only patients having a negative serology for EBV were included since EBV-positive donors can induce lymphomas in the animals after transfer (ROWE et al. 1991). All the patients included had positive titers of autoantibodies against the ANT and the left ventricular function of the heart was impaired. As controls, we isolated PBLs from seven healthy blood donors also having negative EBV serology. For the transfer, 50×10^6 PBLs were injected intraperitonially into 8–10 week old male CB17 SCID mice. After 60 days, the animals were evaluated for the presence of immunoglobulins and autoantibodies in the peripheral blood, human mononuclear cells in the peripheral blood and different tissues, and the left-ventricular function was determined by measuring the slope of the left ventricular pressure pulse after direct puncture of the left ventricle of the heart under ether anesthesia.

At 60 days after transfer, human immunoglobulin levels of up to 4 mg/ml of human IgG and IgM were present in the sera of the SCID mice. There was no difference in the human immunoglobulin levels in the peripheral blood of the animals receiving PPLs from either controls or patients with MC. These immunoglobulins were due to a polyclonal immune response as shown by the distribution of the human IgG subclasses in the SCID mouse sera. Quite similar to the IgG subclass distribution in the donor sera with the usual predominance of IgG1, the SCID mouse sera in the reconstituted animals showed all four human IgG subclasses. Due to the exclusion of EBV-positive donors no monoclonal immunoglobulins in the transfused SCID mice were observed. When we subsequently tested for the presence of autoantibodies against the ANT in the SCID mice sera, no relevant autoantibody titers were observed in the animals receiving PBLs from controls. In contrast, more than 90% of the animals receiving PBLs from MC patients had significant levels of autoantibodies against the isolated ANT, both IgG and IgM (SCHWIMMBECK et al. 1994). These results suggest an ongoing human immune response against the ANT in the animals after transfer of PBLs.

When we tested for the presence of human PBLs in the peripheral blood of the SCID mice by FACS analysis, human mononuclear cells, representing up to 10% of the total leukocytes in the peripheral blood of the SCID mice, could be demonstrated. These cells consisted in about equal parts of human B and T lymphocytes. The T lymphocytes could be further differentiated into CD4-positive helper/inducer cells and CD8-positive suppressor/cytotoxic cells. There was no difference in the level of human leukocytes between the animals receiving cells from patients with MC or healthy controls. The human leukocytes in the peripheral blood of the SCID mice were activated, since the sera of animals receiving PBLs from patients with MC contained elevated levels of human soluble IL-2 receptor, which is released during the activation of T cells. In comparison, mice receiving cells from controls had normal levels of IL-2 receptor (SCHWIMMBECK et al. 1994).

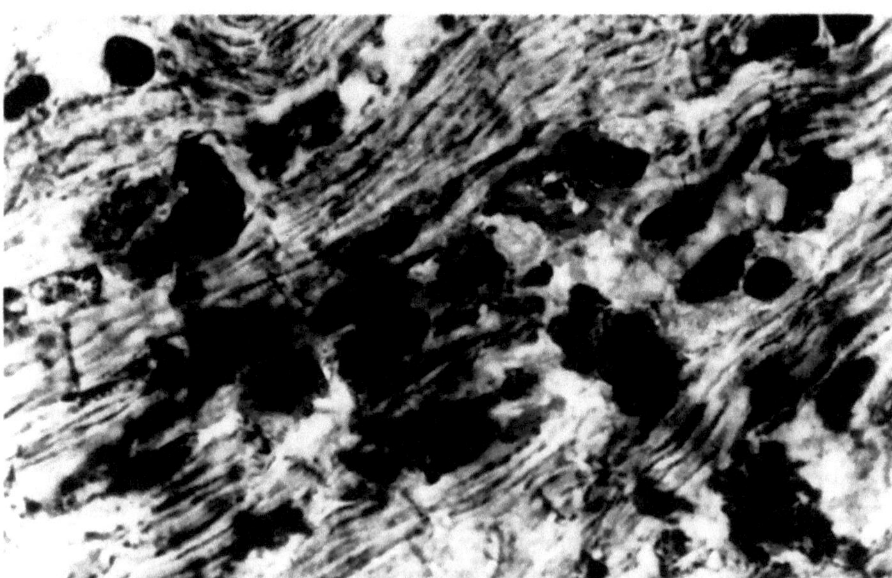

Fig. 3. Immunhistological staining of a section of a SCID mouse receiving PBLs from a patient with myocarditis. The section is stained with a monoclonal antibody against human CD3, identifying the infiltrating cells as human T lymphocytes. Magnification ×400

Staining of the myocardium of the SCID mice showed no relevant infiltrating human cells in the animals receiving PBLs from controls. However, animals receiving leukocytes from patients with MC had infiltrating CD3-positive human T cells in the myocardium of up to 6.1 infiltrating cells per 10 high power light microscopic fields. These cells formed focal infiltrations in the heart quite similar to the findings observed in endomyocardial biopsies from patients with MC (see Fig. 3). The peripheral skeletal muscle of the animal transfused with the PBLs from MC patients or from controls was free of infiltrating human leukocytes. Additional sections were obtained from the liver of the reconstituted animals, but also here no relevant levels of infiltrating human leukocytes were observed, thus excluding clinically apparent graft-versus-host-disease. Another indicator for the absence of a relevant graft-versus-host-disease was the survival of all the animals after transfer and the obviously healthy look of the animals without runting, necrosis of the tips of the toes or diarrhea (MURPHY at al. 1992).

The antigen(s) recognized by the infiltrating human leukocytes in the SCID mouse hearts is not known. However, since obviously an ongoing B cell response against the ANT is present in the SCID mice, it can be speculated that the T cell response is directed, at least in part, against the ANT (SCHWIMMBECK et al. 1991). In order to investigate the antigen specificity of the human infiltrating cells in the myocardium of the SCID mice, we are now attempting to culture the T cells in the SCID mouse hearts in order to determine their antigen specificity.

In additional experiments, SCID mice were transfused with either CD4-positive cells or with CD4-depleted PBLs from patients with MC. In these mice, we observed only low levels of human immunoglobulins and no infiltrating cells in the myocardium. These results are consistent with the necessity of an ongoing human immune response after the transfer of sensitized T cells for the full transfer of MC.

The mechanism by which sensitized human T cells are directed to the heart and recognize their antigens are not known so far. However, adhesion molecules could play an important role for the T cell trafficking as also shown for other diseases (BANKERT at al. 1989).

When we tested the myocardial function of the SCID mouse-hearts after transfer of PBLs, a normal left ventricular function was observed in the animals receiving PBLs from healthy controls. The slope of the left ventricular pressure pulse (dp/dt) was on the average 2450 mm Hg/s and thus not significantly different from the values measured in untreated animals (2570 mm Hg/s). In contrast, animals receiving PBLs from patients with MC showed a decreased left ventricular pressure pulse of about 1750 mmHg/s (see Fig. 4). The exact mechanism leading to the impairment of the left ventricular function is not known. Regarding the relatively low number of infiltrating human leukocytes, it is obviously not due to direct damage of the myocytes by the infiltrating cells but either to the release of cytokines from the infiltrating cells or to another humoral factor. In order to address this question, it will be necessary to determine the levels of human cytokines in the myocardium of SCID mice. Although questions remain, the transfer experiments support the hypothesis of an autoimmune pathogenesis for MC and DCM.

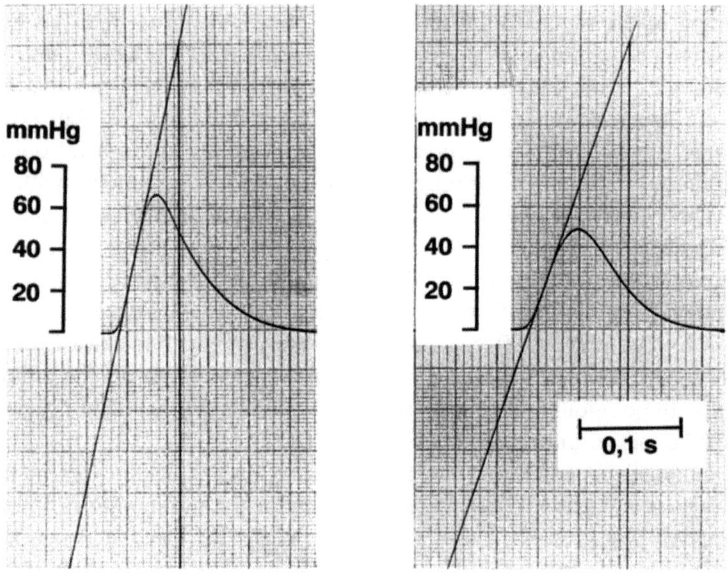

Fig. 4. Measurement of the left ventricular pressure pulse in SCID mice after direct puncture of the left ventricle. A tangent was added to the curve in order to determine the slope of the left ventricular pressure pulse. The left panel gives a representative example of the measurement of a SCID mouse after transfer of PBLs from a healthy control, showing a dp/dt of 2360 mm Hg/s. The right panel gives an example of the SCID mouse-heart after transfer of PBLs from a patient with myocarditis with a dp/dt of 1480 mm Hg/s. The heart rates of the animals were comparable

6 Clinical Considerations

The pathogenesis of human MC is still poorly understood. However, epidemiological studies and recent demonstration of enteroviral sequences in endomyocardial tissue of patients with MC and DCM suggest coxsackieviruses as one of the possible initiating agents in the human disease. Besides the viral infection, the immune reaction is also important for the pathogenesis of human myocarditis. This is supported by the demonstration of autoantibodies against different cardiac autoantigens in the sera of patients with myocarditis and by the presence of infiltrating human leukocytes in the myocardial biopsies (KÜHL et al. 1996). Since both viral infection and the immune system may play important roles in the pathogenesis of the human disease, a careful assessment of viral persistence and the immunological reaction in the hearts of patients is necessary in order to suggest specific therapy. Using immunohistochemistry and PCR on endomyocardial biopsies of patients with MC, different entities of the disease can be differentiated (see Fig. 5). The acute phase of human MC is characterized by a cellular infiltrate and cell necrosis. In the chronic stage of MC, which is clinically often diagnosed as DCM, no myocytolysis is observed. Using immune histology, cellular infiltrates of the

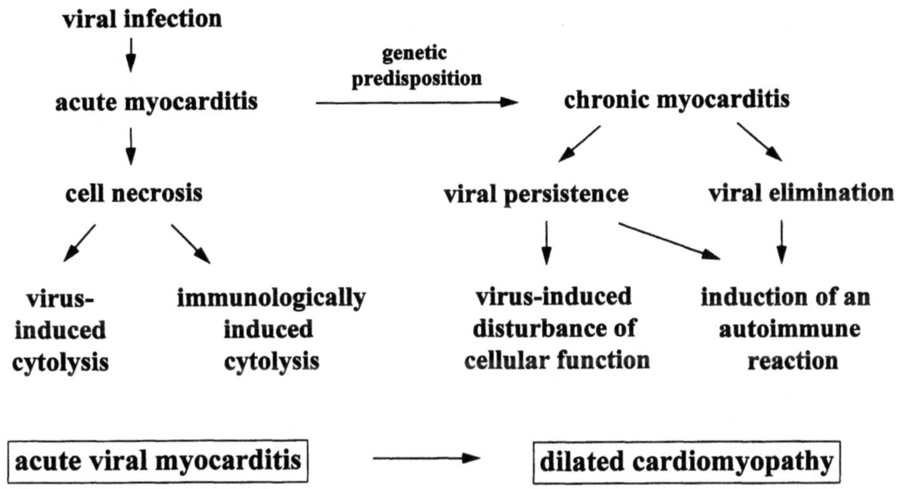

Fig. 5. Clinical stages of human myocarditis

myocardium associated with an increased expression of HLA class I and II antigens and/or adhesion molecules can be observed in about 50% of these cases (KUHL et al. 1996). A further differentiation is possible by the demonstration or absence of enteroviral RNA in the myocardial tissue. Conceptually, patients with a persistent cellular immune response in the myocardium in the absence of enteroviral RNA can be classified as having chronic autoimmunological MC. These patients would be candidates for immune suppression in clinical trials. In contrast, patients with persisting enteroviral genome could be categorized as having chronic persistent viral MC. These patients could perhaps instead be treated with an antiviral therapy, for example with interferon (WHY et al. 1994). These different disease entities with and without viral persistence may account for the general lack of effect of immunosuppressive therapy in acute human MC, quite similar to the results of an immune suppression in acute murine MC (MAISCH et al. 1995; MASON et al. 1995). Therefore, a precise immunological and virological diagnosis in patients with human MC is necessary in order to design a specific and effective therapy.

7 Conclusions

T cells play an important role in CVB-induced diseases. This can be shown in different animal models after the infection of susceptible mice with CVBs. Also in the human disease, infiltrating lymphocytes can be demonstrated using histological and immunhistological methods. These cells can be cultivated from endomyocardial biopsies of patients with MC and their specificity for CVB can be demonstrated.

As suggested from transfer experiments, T lymphocytes seem to have a pathogenetic significance in human disease since the autoimmune phenomena of human MC can be transfered into SCID mice, leading to impairment of left ventricular function in the animals. These results are of major interest both for the understanding of the pathogenesis of human myocarditis and the medical treatment of patients.

References

Allison JP, Havran WL (1991) The immunobiology of T cells with invariant γδ antigen receptors. Ann Rev Immunol 9:679–705

Bankert RB, Umemoto T, Sugiyama Y, Chen FA, Repasky E, Yokota S (1989) Human lung tumors, patients'peripheral blood lymphocytes and tumor infiltrating lymphocytes propagated in SCID mice. Curr Top Micobiol Immunol 152:201–210

Barnett LA, Fujinami RS (1992) Molecular mimicry: a mechanism for autoimmune injury. FASEB J 6:840–844

Bjorkmann PJ, Saper MA, Samraouni B, Bennett WS, Stominger JL, Wiley DC (1987). Structure of the human class I histocompatibility antigen HLA-A2. Nature 329:506–512

Bosna GC, Custer RP, Bosma MJ (1983) A severe combined immunodeficiency mutation in the mouse. Nature 301:527–530

Bowles NE, Richardson PJ, Olsen EGJ, Archard LC (1986) Detection of coxsackie-B-virus-specific RNA sequences in myocardial biopsy samples from patients with myocarditis and dilated cardiomyopathy. Lancet 1:1120–1123

Chow LH, Beisel KW, McManus BM (1992) Enterviral infection of mice with severe combined immunodeficiency: evidence for direct viral pathogenesis of myocardial injury. Lab Invest 66:24–31

Cook DN, Beck MA, Coffman TM, Kirby SL, Sheridan JF, Pragnell IB, Smithies O (1995) Requirement of MIP-1 alpha for an inflammatory response to viral infection. Science 269:1583–1585

Cunningham MW (1993) Bacterial antigen mimicry. In: Bona CA, Siminovitch K, Theophiloupoulos AN (eds), The pathology of autoimmunity. Academic, New York, pp 245–263

Cunningham MW, Antone SM, Gulizia JM, McManus BM, Fischetti VA, Gauntt CJ (1992) Cytotoxic and viral neutralizing antibodies crosscact with streptococcal M protein, enteroviruses, and human cardiac myosin. Proc Natl Acad Sci USA 8:1320–1324

De Scheerder IK, De Buyzere M, Delanghe J, Maas A, Clement DL, Wieme R (1991) Humoral immune response against contractile proteins (actin and myosin) during cardiovascular disease. Eur Heart J 12 (Suppl D):88–94

Doherty PC, Allan W, Eichelberger M, Carding SR (1992) Roles of αβ and γδ T cell subsets in viral immunity. Ann Rev Immunol 10:123–152

Duchusal MA, McConahey PJ, Robinson CA, Dixon FJ (1990) Transfer of human lupus erythmatodes in severe combined immunodeficient (SCID) mice. J Exp Med 172:985–988

Estrin M, Herzum M, Buie C, Huber SA (1987) Immunosuppressives in murine myocarditis. Eur Heart J 8:259–264

Fitch FW, McKisic MD, Lancki DW, Gajewski TF (1993) Differential regulation of murine T lymphocyte subsets. Ann Rev Immunol 11:29–48

Fujinami RS, Oldstone MBA, Wroblewska Z, Frankel ME, Koprowski H (1983) Molecular mimicry in virus infection: crossreaction of measles virus phosphoprotein or of herpes simplex virus protein with human intermediate filaments. Proc Natl Acad Sci USA 80:2346–2350

Gauntt C, Higdon A, Bowers D, Maull E, Wood J, Crawley R (1993) What lessons can be learned from animal model studies in viral heart disease? Scand J Infect Dis 88:49–65

Gauntt CJ, Arizpe HM, Higdon AL, Wood HJ, Bowers DF, Rozek MM, Crawley R (1995) Molecular mimicry, anti-coxsackievirus B3 neutralizing monoclonal antibodies and myocarditis. J Immunol 154:2983–2995

Haarmann CM, Schwimmbeck PL, Mertens T, Schultheiss HP, Strauer B (1994) Identification of serotype-specific and nonserotype-specific B-cell epitopes of coxsackie B viruses using synthetic peptides. Virology 200:381–389

Haas W, Pereira P, Tonegawa S (1993) Gamma/delta cells. Ann Rev Immunol 11:637–686

Helin M, Savola J, Lapinleimu K (1968) Cardiac manifestations during a coxsackie B5 epidemic. British Med J 3:97–99

Henke A, Huber S, Stelzner A, Whitton JL (1995) The role of CD8+ T lymphocytes in coxsackievirus B3-induced myocarditis. J Virol 69:6720–6728

Herzum M, Huber SA, Weller R, Grebe R, Maisch B (1991) Treatment of experimental murine coxsackie B3 myocarditis. Eur Heart J 12(Suppl D):200–202

Huber SA (1992) Heat-shock protein induction in adriamycin and picornavirus-infected cardiocytes. Lab Invest 67:218–224

Huber SA, Cunningham MA (1996) Streptococcal M protein peptide with similarity to myosin induces CD4+ T cell dependent myocarditis in MRL/++ mice and induces partial tolerance against coxsackieviral myocarditis. J Immunol 156:3528–3534

Huber SA, Lodge PA (1984) Coxsackievirus B3 myocarditis: identification of different pathogenic mechanissms in DBA/2 and Balb/c mice. Am J Pathol 122:284–291

Huber SA, Mortensen A, Moulton G (1996a) Modulation of cytokine expression by CD4+ T cells during coxsackievirus B3 infections of Balb/c mice initiated by cells expressing the $\gamma\delta$ + T cell receptor. J Virol 70:3039–3044

Huber SA, Hamrell BB, Knowlton KU (1996) Lessons from animal models of viral myocarditis. In: Schultheiss HP, Schwimmbeck P(eds), The role of immune mechanisms in cardiovascular disease. Springer, Berlin, Heidelberg, New York:35–43

Huber SA, Moraska A, Cunningham M (1994a) Alterations in major histocompatibiity complex association of myocarditis induced by coxsackievirus B3 (CVB3) mutants selected with monoclonal antibodies to group A steptococci. Proc Natl Acad Sci USA 91:5543–5547

Huber SA, Polgar J, Schultheiss P, Schwimmbeck P (1994b) Augmentation of pathogenesis of coxsackievirus B3 infections in mice by exogenous administration of interleukin-1 and interleukin-2. J Virol 68:195–206

Huber SA, Polgar J, Moraska A, Cunningham M, Schwimmbeck P, Schultheiss HP (1993) T lymphocyte responses in CVB3-induced murine myocarditis. Scand J Infect Diseases 88:67–78

Kandolf R, Klingel K, Zell R, Selinka HC, Raab U, Schneider-Brachert W, Bultmann B (1993) Molecular pathogenesis of enterovirus-induced myocarditis: virus persistence and chronic inflammation. Intervirology 35:140–151

Kishimoto C, Abelmann WH (1990) In vivo significance of T cells in the development of coxsackievirus B3 myocarditis in mice. Immature but antigen specific T cells aggravate cardiac injury. Circ Res 67:589–589

Kishimoto C, Kuroki Y, Hiraoka Y, Ochiai H, Kurokawa M, Sasayama S (1994) Cytokine and murine coxsackievirus B3 myocarditis. Interleukin-2 suppressed myocarditis in the acute stage but enhanced the condition in the subsequent stage. Circulation 89:2836–2842

Klingel K, Kandolf R (1993) The role of enterovirus replication in the development of acute and chronic heart muscle disease in different immunocompetent mouse strains. Scand J Infect Dis 88(suppl):79–85

Krams SM, Dorshkind K, Gershwin ME (1989) Generation of biliary lesions after transfer of human lymphocytes into severe combined immunodeficient (SCID) mice. J Exp Med 170:1919–1930

Kühl U, Noutsias M, Seeberg B, Schultheiss HP (1996) Immunohistological evidence for a chronic intramyocardial inflammatory process in dilated cardiomyopathy. Heart 75:295–300

Kühl U, Daun B, Seeberg B, Schultheiss HP, Strauer BE (1992) Dilated cardiomyopathy -a chronic myocarditis? Herz 17(2):97–106

Kühl U, Melzner B, Schafer B, Schultheiss HP, Strauer BE (1991) The Ca-channel as cardiac autoantigen. Eur Heart J 12(Suppl D):99–104

Kupfer A, Singer SJ (1989) Cell biology of cytotoxic and helper T cell functions. Ann Rev Immunol 7:309–337

Lane JR, Neumann DA, LaFond-Walker A, Herskowitz A, Rose NR (1992) Interleukin l or tumor necrosis factor can promote coxsackie B3-induced myocarditis in resistant B10A mice. J Exp Med 175:1123–1129

Lasarte JJ, Sarobe P, Prieto J, Borras-Cuesta F (1995) In vivo cytotoxic T-Lymphocyte induction may take place via CD8 T helper lymphocytes. Res Immunol 146:35–44

Liao O, Sindhwani R, Rojkind M, Factor S, Leinwand L, Diamond B (1995) Antibody-mediated autoimmune myocarditis depends on genetically determined target organ sesitivity. J Exp Med 181:1123–1131

Limas CJ, Limas C (1991) Beta-adrenoreceptor antibodies and genetics in dilated cardiomyopathy: An overview and review. Eur Heart J 12(suppl D):175–177

Loudon RP, Moraska AF, Huber SA, Schwimmbeck P, Schultheiss P (1991) An attenuated variant of coxsackievirus B3 preferentially induces immunoregulatory T cells in vivo. J Virol 65:5813–5819

Lutton CW, Gauntt CJ (1985) Ameliorating effect of IFN-β and anti-IFN-β on coxsackievirus B3-induced myocarditis in mice. J Interferon Res 5:137–146

Macht L, Kukuma N, Leader K, Sarsero D, Pegg CAS, Phillips DIW, Yates P, McLachlan SM, Elson C, Smith BR (1991) Severe combined immunodeficient (SCID) mice: A model for investigating human thyroid autoantibody synthesis. Clin Exp Immunol 84:34–42

Marusic-Galesic S, Walden P (1995) Increased number of CD4-CD8+ HMC class II-specific T cells in MHC class II-deficient mice. Immunology 85:442–446

Maisch B, Camerini F, Schultheiss HP (1995) Immunosuppressive therapy for myocarditis. N Engl J Med 333:1713

Mason JW, O´Connell JB, Herskowitz A, Rose NR, McManus BM, Billingham ME, Moon TE, The Myocarditis Treatment Trial Investigators (1995) A clinical trial of immunosuppressive therapy for myocarditis. N Engl J Med 333:269–275

McMenamin C, Pimm C, McKersey M, Holt PG (1994) Regulation of IgE responses to inhaled antigen in mice by antigen-specific δ T cells. Science 265:1869–1871

Mertens T, Pika U, Eggers HJ (1983) Cross antigenicity among enteroviruses as revealed by immunoblot technique. Virology 129:431–442

Miric M, Miskovic A, Brkic S, Vasiljevic J, Keserovic N, Pesic M (1994) Long-term follow-up of patients with myocarditis and idiopathic dilated cardiomyopathy after immunomodulating therapy. FEMS Immunol Med Microbiol 10:65–74

Murphy WJ; Bennett M, Anver MR, Baseler M, Longo DL (1992) Human-mouse lymphoid chimeras: Host-vs.-graft and graft-vs.-host reactions. Eur J Immunol 22:1421–1427

Neu N, Rose NR, Beisel KW, Herskowitz A, Gurri-Glass G, Craig SW (1987a) Cardiac myosin induces myocarditis in genetically predisposed mice. J Immunol 139:3630–3636

Neu N, Beisel KW, Traystman MD, Rose NR, Craig SW (1987) Antibodies specific for the cardiac myosin isoform are found in mice susceptible to coxsackievirus B3-induced myocartitis. J Immunol 138:2488–2492

O´Connel JB, Reap EA, Robinson JA (1986) The effects of cyclosporin on acute murine Coxsackie B3 myocarditis. Circulation 73:353–359

Pankuweit S, Lottspeich F, Maisch B (1995) Characterization of relevant membrane antigens by two-dimensional immunoblot and n-terminal sequence analysis in patients with myocarditis. Eur Heart J 16(Suppl O):81–84

Phillips RA, Spaner DE (1991) The SCID mouse: Mutation in a DNA repair gene creates recipients useful for studies on stem cells, lymphocyte development and graft-versus-host disease. Immunol Rev 124:63–73.

Pummerer CL, Luze K, Grasl G, Bachmair K, Offner F, Burell SK, Lenz DM, Zamborella TJ, Penninger JM, and Neu N (1996) Identification of cardiac myosin peptides capable of inducing autoimmune myocarditis in Balb/c mice. J Clin Invest 97:2057–2062

Qasim FJ, Mathieson PW, Thiru S, Oliveira DB (1995) Cyclosporin A exacerbates mercuric chloride-induced vasculitis in the brown Norway rat. Lab Invest 72:183–190

Rowe M, Young LS, Crocker J, Stokes H, Henderson S, Rickinson AB (1991) Epstein-Barr virus (EBV) -associated lymphoproliferative disease in the SCID mouse model: Implications for the pathogenesis of EBV-positive lymphomas in man. J Exp Med 173:147–158

Rose NR, Neumann DA, Herskowitz A (1992) Coxsackievirus myocarditis. Adv Intern Med 37:411–429

Salmon M, Pilling D, Borthwick NJ, Viner N, Janossy G, Bacon PA Akbur AN (1994) The progressive differentiation of primed T cells is associated with an increasing susceptibility to apoptosis. Eur J Immunol 24:892–899

Sato S, Tsutsumi R, Burke A, Carson G, Porro V, Seko Y, Okumura K, Kawana R, Virmani $ (1994) Persistence of replicating coxsackievirus B3 in the athymic murine heart is associated with development of myocarditic lesions. J Gen Virol 75:2911–2924

Scharton TM, Scott P (1993) Natural killer cells are a source of interferon gamma that drives differentiation of CD4+ T cell subsets and induces early resistance to Leishmania major in mice. J Exp Med 178:567–577

Schultheiss HP (1989) The significance of autoantibodies against the ADP/ATP carrier for the pathogenesis of myocarditis and dilated cardiomyopathy – clinical and experimental data. Springer Semin Immunopathol 11:15–30

Schultheiss HP, Bolte HD, Schwimmbeck P, Klingenberg M (1983) The antigenic characteristics and the significance of the adenine nucleotide translocator (ANT) as a major autoantigen for antimitochondrial antibodies in myocarditis and congestive cardiomyopathy. J Mol Cell Cardiol 15:85–93

Schulze K, Becker BF, Schultheiss HP (1989) Antibodies to the ADP-ATP carrier – an autoantigen in myocarditis and dilated cardiomyopathy – penetrate into myocardial cells and disturb energy metabolism in vivo. Circ Res 64:179–192

Schulze K, Becker BF, Schauer R, Schultheiss HP (1990) Antibodies to ADP-ATP carrier -an autoantigen in myocarditis and dilated cardiomyopathy – impair cardiac function. Circ 81:959–969

Schwimmbeck PL, Badorff C, Schultheiss HP, Strauer BE (1994) Transfer of human myocarditis into severe combined immundeficiency mice. Circ Res 75:156–164

Schwimmbeck PL, Schultheiss HP, Strauer BE (1991) Demonstration of antigen-specific lymphocytes cultured from myocardial tissue of patients with viral heart disease. Circ 84(Suppl II):II-440 Abstract

Seko Y, Yagita H, Okumura K, Yazaki Y (1994) T cell receptor V beta gene expression in infiltrating cells in murine hearts with acute myocarditis caused by coxsackievirus B3. Circulation 89:170–175

Smith SC, Allen PM (1991) Myosin-induced acute myocarditis is a T cell-mediated disease. J Immunol 147:2141–2147

Suda T, Okazaki T, Naito Y, Yokota T, Arai N, Ozaki S, Nakao K, Nagata S (1995) Expression of the Fas ligand in cells of the T cell linage. J Immunol 154:3806–3813

Tighe H, Silverman GJ, Kozin F, Tucker R, Gulizia R, Peebles C, Lotz M, Rhodes G, Machold K, Mosier DE, Carson DA (1990) Autoantibody production by severe combined immunodeficient mice reconstituted with synovial cells from rheumatoid arthritis patients. Eur J Immunol 20:1843–1848

Traystman MD, Beisel KW (1991) Genetic control of coxsackievirus B3-induced heart-specific autoantibodies associated with chronic myokarditis. Clin Exp Immunol 86:291–298

Traystman, MD, Chow LH, McManus BM, Herskowitz A, Nesbitt MN, Beisel KW (1991) Susceptibility to coxsackievirus B3-induced chronic myocarditis maps near the murine Tar alpha and Myhc alpha loci on chromonsome 14. Am J Pathol 138:721–726

Wang SC, Mrel PA, Wang Q, Jordan ML, Simmons RL, Tweardy DJ (1993) A dual mechanism of immunosuppression by FK-506. Differential suppression of IL-4 and IL-10 levels in T helper 2 cells. Transplant 56:978–985

Why HJF, Meany TB, Richardson PJ, Olsen EGJ, Bowles NE, Cunnigham L, Freeke CA, Archard LC (1994) Clinical and prognostic significance of detection of enteroviral RNA in the myocardium of patients with myocarditis or dilated cardiomyopathy. Circulation 89:2582–2589

Wolfgram LJ, Beisel KW, Herskowitz A, Rose NR (1986) Variations in the susceptibility to coxsackievirus B3-induced myocarditis among different strains of mice. J Immunol 136:1846–1852

Woodruff JF, Woodruff JJ (1974) Involvement of T lymphocytes in the pathogenesis of coxsackievirus B3 heart disease. J Immunol 113:1726–1734

Yamamoto M, Fujihashi K, Beagley KW, McGhee JR, Kujono H (1993) Cytokine synthesis by intestinal intraepithelial lymphocytes. Both γ/δ T cell receptor positive and α/β T cell receptor positive T cells in the G1 phase of cell cycle produce IFNγ and IL-5. J Immunol 150:106–114

Zhang M, Lin Y, Iyer DV, Gong J, Abrams JS, Barnes PF (1995) T cell cytokine responses in human infection with Myocbacterium tuberculosis. Infect Immun 63:3231–3234

Subject Index

D

Current Topics in Microbiology and Immunology

Volumes published since 1989 (and still available)

Vol. 184: **Dunon, Dominique; Mackay, Charles R.; Imhof, Beat A. (Eds.):** Adhesion in Leukocyte Homing and Differentiation. 1993. 37 figs. IX, 260 pp. ISBN 3-540-56756-9

Vol. 185: **Ramig, Robert F. (Ed.):** Rotaviruses. 1994. 37 figs. X, 380 pp. ISBN 3-540-56761-5

Vol. 186: **zur Hausen, Harald (Ed.):** Human Pathogenic Papillomaviruses. 1994. 37 figs. XIII, 274 pp. ISBN 3-540-57193-0

Vol. 187: **Rupprecht, Charles E.; Dietzschold, Bernhard; Koprowski, Hilary (Eds.):** Lyssaviruses. 1994. 50 figs. IX, 352 pp. ISBN 3-540-57194-9

Vol. 188: **Letvin, Norman L.; Desrosiers, Ronald C. (Eds.):** Simian Immunodeficiency Virus. 1994. 37 figs. X, 240 pp. ISBN 3-540-57274-0

Vol. 189: **Oldstone, Michael B. A. (Ed.):** Cytotoxic T-Lymphocytes in Human Viral and Malaria Infections. 1994. 37 figs. IX, 210 pp. ISBN 3-540-57259-7

Vol. 190: **Koprowski, Hilary; Lipkin, W. Ian (Eds.):** Borna Disease. 1995. 33 figs. IX, 134 pp. ISBN 3-540-57388-7

Vol. 191: **ter Meulen, Volker; Billeter, Martin A. (Eds.):** Measles Virus. 1995. 23 figs. IX, 196 pp. ISBN 3-540-57389-5

Vol. 192: **Dangl, Jeffrey L. (Ed.):** Bacterial Pathogenesis of Plants and Animals. 1994. 41 figs. IX, 343 pp. ISBN 3-540-57391-7

Vol. 193: **Chen, Irvin S. Y.; Koprowski, Hilary; Srinivasan, Alagarsamy; Vogt, Peter K. (Eds.):** Transacting Functions of Human Retroviruses. 1995. 49 figs. IX, 240 pp. ISBN 3-540-57901-X

Vol. 194: **Potter, Michael; Melchers, Fritz (Eds.):** Mechanisms in B-cell Neoplasia. 1995. 152 figs. XXV, 458 pp. ISBN 3-540-58447-1

Vol. 195: **Montecucco, Cesare (Ed.):** Clostridial Neurotoxins. 1995. 28 figs. XI., 278 pp. ISBN 3-540-58452-8

Vol. 196: **Koprowski, Hilary; Maeda, Hiroshi (Eds.):** The Role of Nitric Oxide in Physiology and Pathophysiology. 1995. 21 figs. IX, 90 pp. ISBN 3-540-58214-2

Vol. 197: **Meyer, Peter (Ed.):** Gene Silencing in Higher Plants and Related Phenomena in Other Eukaryotes. 1995. 17 figs. IX, 232 pp. ISBN 3-540-58236-3

Vol. 198: **Griffiths, Gillian M.; Tschopp, Jürg (Eds.):** Pathways for Cytolysis. 1995. 45 figs. IX, 224 pp. ISBN 3-540-58725-X

Vol. 199/I: **Doerfler, Walter; Böhm, Petra (Eds.):** The Molecular Repertoire of Adenoviruses I. 1995. 51 figs. XIII, 280 pp. ISBN 3-540-58828-0

Vol. 199/II: **Doerfler, Walter; Böhm, Petra (Eds.):** The Molecular Repertoire of Adenoviruses II. 1995. 36 figs. XIII, 278 pp. ISBN 3-540-58829-9

Vol. 199/III: **Doerfler, Walter; Böhm, Petra (Eds.):** The Molecular Repertoire of Adenoviruses III. 1995. 51 figs. XIII, 310 pp. ISBN 3-540-58987-2

Vol. 200: **Kroemer, Guido; Martinez-A., Carlos (Eds.):** Apoptosis in Immunology. 1995. 14 figs. XI, 242 pp. ISBN 3-540-58756-X

Vol. 201: **Kosco-Vilbois, Marie H. (Ed.):** An Antigen Depository of the Immune System: Follicular Dendritic Cells. 1995. 39 figs. IX, 209 pp. ISBN 3-540-59013-7

Vol. 202: **Oldstone, Michael B. A.; Vitković, Ljubiša (Eds.):** HIV and Dementia. 1995. 40 figs. XIII, 279 pp. ISBN 3-540-59117-6

Vol. 203: **Sarnow, Peter (Ed.):** Cap-Independent Translation. 1995. 31 figs. XI, 183 pp. ISBN 3-540-59121-4

Vol. 204: **Saedler, Heinz; Gierl, Alfons (Eds.):** Transposable Elements. 1995. 42 figs. IX, 234 pp. ISBN 3-540-59342-X

Vol. 205: **Littman, Dan R. (Ed.):** The CD4 Molecule. 1995. 29 figs. XIII, 182 pp. ISBN 3-540-59344-6

Vol. 206: **Chisari, Francis V.; Oldstone, Michael B. A. (Eds.):** Transgenic Models of Human Viral and Immunological Disease. 1995. 53 figs. XI, 345 pp. ISBN 3-540-59341-1

Vol. 207: **Prusiner, Stanley B. (Ed.):** Prions Prions Prions. 1995. 42 figs. VII, 163 pp. ISBN 3-540-59343-8

Vol. 208: **Farnham, Peggy J. (Ed.):** Transcriptional Control of Cell Growth. 1995. 17 figs. IX, 141 pp. ISBN 3-540-60113-9

Vol. 209: **Miller, Virginia L. (Ed.):** Bacterial Invasiveness. 1996. 16 figs. IX, 115 pp. ISBN 3-540-60065-5

Vol. 210: **Potter, Michael; Rose, Noel R. (Eds.):** Immunology of Silicones. 1996. 136 figs. XX, 430 pp. ISBN 3-540-60272-0

Vol. 211: **Wolff, Linda; Perkins, Archibald S. (Eds.):** Molecular Aspects of Myeloid Stem Cell Development. 1996. 98 figs. XIV, 298 pp. ISBN 3-540-60414-6

Vol. 212: **Vainio, Olli; Imhof, Beat A. (Eds.):** Immunology and Developmental Biology of the Chicken. 1996. 43 figs. IX, 281 pp. ISBN 3-540-60585-1

Vol. 213/I: **Günthert, Ursula; Birchmeier, Walter (Eds.):** Attempts to Understand Metastasis Formation I. 1996. 35 figs. XV, 293 pp. ISBN 3-540-60680-7

Vol. 213/II: **Günthert, Ursula; Birchmeier, Walter (Eds.):** Attempts to Understand Metastasis Formation II. 1996. 33 figs. XV, 288 pp. ISBN 3-540-60681-5

Vol. 213/III: **Günthert, Ursula; Schlag, Peter M.; Birchmeier, Walter (Eds.):** Attempts to Understand Metastasis Formation III. 1996. 14 figs. XV, 262 pp. ISBN 3-540-60682-3

Vol. 214: **Kräusslich, Hans-Georg (Ed.):** Morphogenesis and Maturation of Retroviruses. 1996. 34 figs. XI, 344 pp. ISBN 3-540-60928-8

Vol. 215: **Shinnick, Thomas M. (Ed.):** Tuberculosis. 1996. 46 figs. XI, 307 pp. ISBN 3-540-60985-7

Vol. 216: **Rietschel, Ernst Th.; Wagner, Hermann (Eds.):** Pathology of Septic Shock. 1996. 34 figs. X, 321 pp. ISBN 3-540-61026-X

Vol. 217: **Jessberger, Rolf; Lieber, Michael R. (Eds.):** Molecular Analysis of DNA Rearrangements in the Immune System. 1996. 43 figs. IX, 224 pp. ISBN 3-540-61037-5

Vol. 218: **Berns, Kenneth I.; Giraud, Catherine (Eds.):** Adeno-Associated Virus (AAV) Vectors in Gene Therapy. 1996. 38 figs. IX,173 pp. ISBN 3-540-61076-6

Vol. 219: **Gross, Uwe (Ed.):** Toxoplasma gondii. 1996. 31 figs. XI, 274 pp. ISBN 3-540-61300-5

Vol. 220: **Rauscher, Frank J. III; Vogt, Peter K. (Eds.):** Chromosomal Translocations and Oncogenic Transcription Factors. 1997. 28 figs. XI, 166 pp. ISBN 3-540-61402-8

Vol. 221: **Kastan, Michael B. (Ed.):** Genetic Instability and Tumorigenesis. 1997. 12 figs.VII, 180 pp. ISBN 3-540-61518-0

Vol. 222: **Olding, Lars B. (Ed.):** Reproductive Immunology. 1997. 17 figs. XII, 219 pp. ISBN 3-540-61888-0

MIX
Papier aus verantwortungsvollen Quellen
Paper from responsible sources
FSC® C105338

If you have any concerns about our products,
you can contact us on
ProductSafety@springernature.com

In case Publisher is established outside the EU,
the EU authorized representative is:
Springer Nature Customer Service Center GmbH
Europaplatz 3, 69115 Heidelberg, Germany

Printed by Libri Plureos GmbH
in Hamburg, Germany